CHEMICAL ANALYSIS

CHEMICAL ANALYSIS

A SERIES OF MONOGRAPHS ON
ANALYTICAL CHEMISTRY AND ITS APPLICATIONS

Editors

P. J. ELVING · I. M. KOLTHOFF

VOLUME 38

WILEY-INTERSCIENCE

A Division of John Wiley & Sons, Inc.
New York / London / Sydney / Toronto

Luminescence Spectrometry in Analytical Chemistry

J. D. WINEFORDNER

Department of Chemistry
University of Florida
Gainesville, Florida

S. G. SCHULMAN

School of Pharmacy
University of Florida
Gainesville, Florida

and

T. C. O'HAVER

Department of Chemistry
University of Maryland
College Park, Maryland

WILEY-INTERSCIENCE

A Division of John Wiley & Sons, Inc.
New York / London / Sydney / Toronto

PREFACE

Luminescence spectrometry is useful for trace analysis of both elements and molecules. Most elements, particularly metals, will luminesce when atomized in a suitable cell and excited by suitable light sources. Similarly, when many metals and inorganic ionic species are present as chelates at room temperature or below, they can be measured by luminescence of the organic part or of the metal part of the chelate. Many organic molecules in their native state or after suitable reaction (at room or lower temperature) luminesce when excited with an external light source. The analytical possibilities of luminescence spectrometry are almost unlimited, and it is not unreasonable to measure atomic or molecular concentrations as low as $0.0001 \ \mu g \ ml^{-1}$.

The topics and approach to luminescence spectrometry in this book are considerably different from those of other books on the subject. For example, no other book on any aspect of luminescence spectrometry or on spectrochemical methods of analysis or instrumental methods of analysis covers the following topics: both atomic fluorescence and molecular luminescence spectrometric methods, and a general approach to radiance expressions for both atoms and molecules. This book also provides an extensive listing of concentrational and absolute limits of detection of atoms that fluoresce in the vapor state and of inorganic species that luminesce in the condensed phase state; a listing of organic species that luminesce in the condensed phase state is given, as well. In addition, we furnish a fairly thorough discussion of the fundamental concepts necessary to an understanding of spectra-structure correlations, of the factors affecting the radiances of atomic fluorescence and molecule luminescence, and of the instrumental components of luminescence spectrometers. No discussion of analytical applications of luminescence spectrometry is given here because these will be covered in a subsequent volume in this series.

This book should be useful as a textbook to teach the principles, instrumentation, and methodology of luminescence spectrometry. It is also hoped that this book will be useful either to the beginner or to the practicing analyst and researcher who wish to understand luminescence spectrometry better, possible to use it for a certain application or to improve the luminescence spectrometric method now in use. Finally, it is believed that this volume would also be useful for applied nonanalytical chemists (e.g., the clinical chemist, the agricultural chemist), who use analytical methods, and for more physically oriented nonanalytical chemists (e.g., the physical chemist or physicist), who uses luminescence spectrometric methods to determine rate constants, collisional cross sections, and other more fundamental parameters.

v

This volume is basically divided into three rather independent chapters (not including the Introduction). The first chapter is on both the qualitative aspects of spectra and on the quantitative relationships between luminescence radiance and analyte concentration. The second chapter deals with the principles and types of instrumental components used in luminescence spectrometry, the use of signal and signal-to-noise ratio theory to obtain optimal quantitative results (best precision and lowest detection limits), and the important measurements performed in luminescence spectrometry. The final chapter concerns methodology (limits of detection, interferences, shapes of analytical curves, and the species that luminesce). It should also be possible to read any given chapter in the volume without having to read either of the other two chapters.

J. D. Winefordner
S. G. Schulman
T. C. O'Haver

Gainesville, Florida
College Park, Maryland
August 1971

CONTENTS

INTRODUCTION

Luminescence spectrometric methods are among the most sensitive analytical methods for trace analysis. In this volume, *only* photoluminescence of atoms and molecules is discussed. (Luminescence is excited in atoms and molecules by an external light source as opposed to bioluminescence, chemiluminescence, and electroluminescence which are excited by a biological process, by a chemical reaction, and by electrical energy, respectively.) Atomic fluorescence spectrometry of atoms in flames and nonflames and molecular fluorescence and phosphorescence spectrometry of molecules (organic molecules and inorganic complexes) at room and lower temperatures are considered. Inorganic phosphors are not covered, however, because they are presently of limited analytical use.

This volume is divided into three basic chapters (not including the Introduction): theory, which deals with the physical basis for fluorescence of atoms in hot gases, molecular luminescence of molecules in the condensed phase, and the expressions relating luminescence radiance to spectral, instrumental, and analytical (analyte concentration) parameters; instrumentation, which covers the types and principles of operation of luminescence spectrometers and of the individual instrumental components, signal-to-noise ratio theory (which is needed to understand the means of improving limits of detection and precision of measurements), and luminescence measurement techniques (which are useful in experimental measurement of spectra, quantum yields, lifetimes, and analyte concentrations); and uses of luminescence spectrometry which involves the types of results expected in each kind of luminescence spectrometry and some of the types of atomic and molecular species previously studied by various luminescence spectrometric methods. This book presents no applications of luminescence spectrometry to real problems (e.g., benz(a) pyrene in cigarette smoke or magnesium in blood), because application will be covered in a future volume of this series.*

In addition to the three basic chapters, a *glossary of symbols*, with their definitions and units used in each main chapter section, and a listing of *pertinent references* for each main chapter section appear at the end of each

---

* P. A. St. John is currently writing a book on the application of luminescence spectrometry which should appear in about 1973.

1

chapter section. The listing of symbols minimizes loss of time on the part of the reader when symbols are used more than once but are only defined the first time they appear in the text. The authors list only the most pertinent references for each section of each chapter. Because this volume is written as a textbook on luminescence spectrometry rather than a review of luminescence spectrometry, most of the references given are to other review articles or books. Only those review articles or books which the authors of this volume felt were comprehensive, clear, concise, and critical to their discussion are listed; a large number of references would be only confusing and of little real use to the analyst wishing to use luminescence spectrometry. Occasionally specific references must be given because of their uniqueness, because of their recent appearance, or because of their omission in the review articles. If the reader wishes an extensive list of references on most topics covered in this volume, he should simply refer to the journal reviews and book references at the end of each chapter.

Also at the end of the volume, a number of appendices are given. These appendices are concerned with concepts or expressions not necessary to the understanding of the text but important in extending the concepts or expressions given in the text to more specific cases.

II

THEORY

A. PHYSICAL BASIS OF ATOMIC FLUORESCENCE IN HOT GASES[1-3,5]

Atomic fluorescence as an analytical method is based on the emission of light by excited atoms. This phenomenon derives from the return of electronically excited atoms to their ground electronic states, with the energy of excitation being discarded in the form of visible or ultraviolet light. Because the frequencies as well as the intensities of the light emitted by electronically excited atoms are determined by the electronic configurations of the atoms in both their ground and excited states, it is worthwhile, at this point, to consider briefly some of the essential features of atomic electronic structure and its effects on the observed optical spectra of atoms. It is assumed here that the reader has had some previous introduction to the fundamental concepts of quantum mechanics.

1. ELECTRONIC STRUCTURE OF ATOMS

a. ATOMIC QUANTUM MECHANICS

An atom consists of a positively charged nucleus and a number of negatively charged electrons. In a neutral atom, the total negative charge of all the electrons is equal to the total positive charge of the nucleus. The forces holding the atom together are predominantly electrostatic, consisting of attractions between each electron and the nucleus and repulsions among all the atomic electrons. The classical theory of electrostatics, however, fails to account for the stability of atoms and for the characteristics of their spectra. According to the classical theory, electrons should describe curved paths about the nucleus, whereby they accelerate, continuously emitting electromagnetic radiation, lose energy and eventually fall into the nucleus, resulting in the annihilation of the atomic electrons and the transmutation of the nucleus to that of a different element. Atoms should, according to the classical theory, be capable of absorbing all frequencies of electromagnetic radiation and should radiate all frequencies of electromagnetic radiation in returning to the ground state. Because the ground states of atoms are obviously stable and because atomic absorption and emission spectra consist of relatively sharp lines rather than continua, a better theoretical approach is required to describe the electronic structure of atoms.

Early attempts at a plausible theory of atomic electronic structure were made by Bohr and de Broglie. Bohr treated the hydrogen atom by presuming that the classical laws of electrostatics were obeyed but also that the electron moved in circular orbits in which the orbital angular momentum of the electron was confined to integral multiples of $h/2\pi$, where h is Planck's constant, 6.625×10^{-27} erg sec. This assumption was quite arbitrary, but oddly enough the energies calculated by Bohr's method were not a continuum of values but were restricted to multiples of $1/n^2$, where n was the integer associated with the orbital angular momentum. Thus the Bohr model predicted discrete stationary states, and the lowest of these corresponding to $n = 1$, was assigned to the ground state. Spectral features were presumed to originate from transitions between these discrete energy levels, which nicely accounted for the line spectrum of hydrogen. de Broglie arrived at an expression for the energy levels of hydrogen that was identical with Bohr's, but he used a different approach. Purely by drawing analogies between the laws of optics and the laws of mechanics, de Broglie was able to associate the momentum of an atomic electron with its "hypothetical" wavelength. By assuming that an atomic electron in a stable state had the properties of a standing wave and would, by analogy with a standing wave, lose no energy as long as it was in this condition, de Broglie was able to state that the condition for an electron to be in a stationary state was that an integral number of wavelengths had to just fit into the atomic orbit. The latter condition in de Broglie's "wave mechanics" is equivalent to Bohr's quantization of angular momentum. Hence the identity of their results.

Not only were these approaches valid for the spectrum of the hydrogen atom, they were also capable of predicting the spectra of atomic ions containing only one electron (e.g., He^+, Li^{2+}). The principal failing of the Bohr and de Broglie models appeared in their application to the spectra of many electron atoms. This was mainly due to the inability of the models to account for repulsions between atomic electrons. Furthermore, it was shown by Heisenberg that it was impossible to specify simultaneously the position and the momentum of an electron. In the Bohr method, given the momentum of the orbital electron, it would be possible to calculate the location of the electron at any instant of time. Thus the notion of well-defined atomic orbits violated the Heisenberg uncertainty principle.

The early quantum theories of Bohr and de Broglie were replaced by the much more sophisticated, probabilistic quantum mechanics of Heisenberg, Schrödinger, and others. These are rather abstract mathematical approaches whose mechanical aspects are concerned with the solutions of some rather complicated partial differential equations. Although we do not want to become involved with the mathematical details of quantum mechanics, it is necessary to use the language and the results of that formalism to describe atomic electronic structure and the features of atomic fluorescence spectra. Consequently, some of the essential features of the Schrödinger method of quantum mechanics are now presented.

In the Schrödinger method, the atom is described by the fundamental differential equation of quantum mechanics, the Schrödinger equation

$$H\psi = E\psi \tag{II-A-1}$$

In equation II-A-1, ψ is a function of the coordinates of the system called the atomic wavefunction (the nucleus is at the center of the coordinate system) and characterizes the state (electronic configuration) of the atom. The H is a differential operator called the Hamiltonian operator of the atom and is composed of the operators for the kinetic energy of each particle in the atom, the electrostatic attractions between each atomic electron and the nucleus, and the electrostatic repulsions between each atomic electron and every other electron in the same atom. If ψ is an exact solution of equation II-A-1, the operation of H upon ψ generates ψ multiplied by a constant E. The latter term corresponds to the energy of the atom when it is in the state characterized by ψ. Although ψ has no real physical significance, the product of ψ and its complex conjugate ψ^* represent a probability distribution function for the atom in the state ψ. In accordance with the probabilistic interpretation of $\psi^*\psi$, $\int_{\tau=-\infty}^{\tau=\infty} \psi^*\psi \, d\tau = 1$, where $d\tau$ is the volume element in configuration space. The average value of the energy of an atom in the state ψ is given, in quantum mechanics, by

$$\bar{E} = \frac{\displaystyle\int_{-\infty}^{\infty} \psi^* H\psi \, d\tau}{\displaystyle\int_{-\infty}^{\infty} \psi^*\psi \, d\tau} \tag{II-A-2}$$

In practice, it is the differences between the average energies of states of different ψ that are determined spectroscopically. Correspondingly, the average value of any dynamical variable $\bar{a}$ ($\bar{a}$ = position, momentum, angular momentum, etc.), in the state ψ, which is represented by the operator α, is given by

$$\bar{a} = \frac{\displaystyle\int_{-\infty}^{\infty} \psi^* \alpha\psi \, d\tau}{\displaystyle\int_{-\infty}^{\infty} \psi^*\psi \, d\tau} \tag{II-A-3}$$

Since it is a rather simple matter to write down the Hamiltonian operator for any atom, no matter how many electrons it contains, we should now be in a position to solve the Schrödinger equation for the wavefunctions of all the elements. Then we can calculate any desired physical variables associated with the wavefunctions and possibly even the electronic structures of the atoms in these states. This would indeed be the case if it were not for the complexity of the many-electron Hamiltonian, which renders the Schrödinger equation insoluble for all but those atoms (or ions) containing one electron—the so-called hydrogenic atoms. Consequently, the present state of knowledge of the structure and dynamics of atoms is largely incomplete, for it is based on approximations to the exact solutions of the Schrödinger equation for hydrogenic atoms. It is within the framework of these approximations that we are constrained to proceed from here.

b. HYDROGENIC ATOMS

The solutions of the Schrödinger equation for hydrogenic atoms correspond almost exactly to the solutions of the Schrödinger equation for a single electron moving in a radially symmetric, positive, electrostatic field. Hence in this case (and this case only), the atomic wavefunctions obtained from the solution correspond to one-electron wavefunctions.

The hydrogenic solutions consist of a family of functions, each expressed in terms of the coordinates of the system (usually spherical polar coordinates) and also in terms of three so-called quantum numbers n, l, and m, which arise in the course of the solution of the differential equation. The physical significance of these quantum numbers is that the average energy and radius of the hydrogenic atom, in a state associated with a particular set of quantum numbers, are determined by n, whereas the average angular momentum is determined by l, and the z component of the angular momentum of the atom, in an external magnetic field, directed along the z axis, is governed by m. The latter property is related to the relative spatial attitude of states having the same n and l but different m.

The solution of the Schrödinger equation also determines the possible values of n, l, and m and their relation to one another. All positive integral values are possible for n. For any wavefunction having a given value of n, the value of l for that wavefunction can be

$$l = 0, 1, 2, \ldots, n-1 \qquad \text{(II-A-4)}$$

Thus associated with any value of n there are n different values of l. A set of wavefunctions all corresponding to the same value of n are said to belong to the *shell* characterized by the *principal* quantum number n. A set of wavefunctions having the same values of n and l are said to belong to the same *subshell* characterized by the *azimuthal* quantum number l. For any ψ_{nlm}, the possible values of m are determined by the value of l for the subshell and can be

$$m = -l, \, -l+1, \, \ldots, \, 0, \, \ldots, \, l-1, \, l \tag{II-A-5}$$

Hence for any value of l there are $2l + 1$ possible values of m, the *magnetic* quantum number. A one-electron wavefunction characterized by a set of the three quantum numbers n, l, and m is said to be an *atomic orbital* (cf. orbital angular momentum). Hydrogenic orbitals belonging to the same shell are degenerate (i.e., they have the same energy). This is not directly observable; however, in a strong magnetic field the degeneracy is removed and this phenomenon can be seen through the splitting of certain spectral lines.

Hydrogenic subshells corresponding to $l = 0, 1, 2, 3, \ldots$, respectively, have been labeled $s, p, d, f, \ldots$, respectively. These letters are a legacy from old spectroscopic terminology and have been assigned to the subshells quite arbitrarily. A subshell having $n = 1$ and $l = 0$ would be called $1s$; one with $n = 2$, $l = 0$ would be called $2s$; one with $n = 2$, $l = 1$, $2p$, and so on. This nomenclature system is used throughout the book. The ground state of the hydrogen atom is then the $1s$ state, or the state produced by the electron occupying the $1s$ orbital.

Through the solution of a wave equation like Schrödinger's, but one which included the effects of special relativity on the motion of the electron, Dirac was able to derive not only the quantum numbers n, l, and m, but also a fourth quantum number s, called the spin quantum number. This quantum number has been interpreted to be related to the angular momentum of the electron spinning on its own axis, and it can only have two values, $+\frac{1}{2}$ and $-\frac{1}{2}$. Thus an electron occupying a hydrogenic orbital of given n, l, and m can have either $s = +\frac{1}{2}$ or $s = -\frac{1}{2}$. As we soon see, in many electron atoms all four quantum numbers play a fundamental role in the electronic configurations of these atoms in their ground and excited states.

C. MANY-ELECTRON ATOMS

Having the one-electron orbitals and energy levels of the hydrogenic atoms, we are now in a position to construct a model of the electronic structure of many-electron atoms. The model for the ground state of a neutral atom of nuclear charge Z is constructed by assigning Z electrons to the hydrogenic orbitals in such a way that the electronic configuration of lowest potential energy is obtained. In order to accomplish this, it is necessary to know how many electrons each orbital can accommodate and in what order the orbitals are filled. Before proceeding to this matter, a word about the relation between one-electron orbitals and many-electron orbitals is in order.

The one-electron orbitals of the hydrogenic atoms are synonymous with the states of these atoms. Furthermore, the energies of these atoms are equivalent to the energies associated with the occupied orbitals. A hydrogenic atom in a $2s$ state has essentially the same energy as a hydrogenic atom in a $2p$ state. This is because of the dependence of energy on n and not on the other quantum numbers for hydrogenic atoms. On the other hand, many-electron atoms have wavefunctions (states) that are, at least approximately, products of one-electron wavefunctions. Hence the state of a many-electron atom is specified by the orbitals occupied by all the electrons and not just one. The energies of many-electron atoms are not simply equal to the sum of the energies of the occupied one-electron orbitals. The energies of one-electron orbitals are derived on the basis of electrostatic attraction between a single electron and the nucleus. In a many-electron atom, there are repulsions between each of the electrons. The hydrogenic orbitals do not account for this. Furthermore, the degeneracies of occupied orbitals of the same n but different l may be removed in many-electron atoms. This is partially because of the effect of repulsions by inner-shell electrons and partially because electrons having low values of l spend more time near and are thus more strongly attracted by the nucleus than those of the same n but higher values of l. The degeneracies of orbitals having the same value of n and l but different values of m are not appreciably removed, in many-electron atoms, in the absence of external magnetic fields.

The orbital capacities and order of filling of atomic orbitals are governed, respectively, by the Pauli exclusion principle and the Hund rule of maximum multiplicity. The Pauli exclusion principle was derived by W. Pauli in 1926 in an attempt to reconcile electron spin with the nonrelativistic Schrödinger equation. In its simplest form, the exclusion principle states that no two electrons in the same atom can have four identical quantum numbers. Hence if an orbital is specified by n, l, and m, it can accommodate a maximum of two electrons, one with $s = +\frac{1}{2}$ and one with $s = -\frac{1}{2}$. A third electron would have n, l, m and s equal to that of one of the electrons already in that orbital and hence would not be allowed to occupy the orbital. For the shell with $n = 1$ there is only one orbital, with $l = 0$. Thus, the $n = 1$ shell is closed to further occupation when it contains two electrons. A closed $n = 1$ shell is denoted by $1s^2$. For the $n = 2$ shell there are four orbitals; one with $l = 0$ and $m = 0$ and three with $l = 1$ ($m = -1, 0, 1$). The $n = 2$ shell is therefore closed when it contains eight electrons and is denoted by $2s^2 2p^6$. In general, the maximum number of electrons any shell can accommodate is $2n^2$. The maximum number any subshell can hold is $2(2l + 1)$.

The Pauli exclusion principle thus determines the electronic configurations of the closed shells as well as the electronic configuration of the closed orbitals. Chemical reactivity is thought to result from the tendency of atoms to complete all partially filled shells, thereby maximizing the atomic entropy by populating all thermally accessible energy levels of the system. Population of shells higher than the highest occupied ones requires energies considerably in excess of thermal energy, hence the low reactivity of the rare gases and atoms whose valences are already satisfied.

The Hund rule of maximum multiplicity enables prediction of the electronic configurations of partially filled shells. This rule has its basis in classical electrostatics (Coulomb's law) and states that in the case of degenerate orbitals, the configuration of minimum potential energy will be achieved by allowing the electrons occupying these orbitals to stay as far apart as possible. This means that, in filling degenerate orbitals, each orbital accepts one electron before spin pairing (double occupancy) occurs, because the separate orbitals occupy different regions of space (whereas two electrons, paired, in one orbital are close together, resulting in greater repulsive potential energy).

The combined application of the Hund rule and the Pauli exclusion principle has been quite successful in predicting the ground state electronic configurations of the lighter elements. Thus the ground electronic states of nitrogen $(Z = 7)$ and sodium $(Z = 11)$ are, respectively, $1s^2 2p^2 2p^3$ (with one unpaired electron in each of the $2p$ orbitals) and $1s^2 2s^2 2p^6 3s^1$. These configurations correlate well with the observed chemical properties of these elements. The electronically excited states of the elements can be constructed by removing an electron from the highest occupied shell (it is this shell which is of interest in optical spectroscopy) and promoting it to an orbital in a higher unoccupied shell. For example, an excited state of sodium could be represented by the configuration $1s^2 2s^2 2p^6 4p^1$.

The foregoing approach has not been, by itself, quite as successful in the assignments of the ground state electronic configurations of the heavier elements, especially those of the transition elements. For example, iron $(Z = 26)$ would be assigned a configuration of $1s^2 2s^2 2p^6 3s^2 3p^6 3d^8$ (with two unpaired electrons and three sets of paired electrons in the five $3d$ orbitals). This configuration is inconsistent with the chemical and magnetic properties of iron. From measurements of paramagnetic susceptibilities, it has been possible to assign to the ground state of iron a configuration of $1s^2 2s^2 2p^6 3s^2$-$3p^6 3d^6 4s^2$ (with four unpaired and two paired $3d$ electrons and two paired $4s$ electrons).

This discrepancy is due to repulsions produced on the $n = 3$ electrons by the inner-shell electrons, raising the energy of the $n = 3$ shell above its relative position in a hydrogenic atom. The $3s$ and $3p$ electrons are more strongly attracted by the nucleus than the $3d$ electrons. The occupied $3d$ orbitals are repelled to energy greater than that of the unoccupied $4s$ orbital, so that the $4s$ orbital becomes occupied by two of the electrons in preference to the $3d$ orbitals. The removal of two electrons from iron, however, drops the energy of the $3d$ orbitals below that of the $4s$ orbital, so that it appears that the $4s$ electrons are ionized first. In general, it is necessary to blend empiricism with the Hund rule and the Pauli exclusion principle to obtain the correct order of filling of atomic orbitals. A scale of the order of filling of atomic orbitals is given in Figure II-A-1. The lowest orbitals in Figure II-A-1 are filled first.

Before leaving the Pauli exclusion principle and Hund's rule, it is interesting to consider the effect of these rules on the electronic configurations of electronically excited states. Let us consider a beryllium atom ($Z = 4$). The ground state configuration of beryllium is $1s^2 2s^2$. Both electrons in the $2s$ orbital of this atom are spin paired in accordance with the exclusion principle. If, however, one of the $2s$ electrons is excited to, say, a $3p$ orbital, the electrons will no longer be in the same spatial orbital and will no longer be required to have opposite spins. Thus two excited states with the $1s^2 2s^1 3p^1$ configuration are possible. One of these will have a $2s$ electron and a $3p$ electron with opposite spins and one of these will have the $2s$ electron and the $3p$ electron with the same spin. Now Hund's second rule tells us that two electrons with the same spin tend to stay farther apart than two electrons with opposite spins. Consequently, the energy of repulsion will be lower in the atom with two unpaired electrons, as will be the energy of the excited state of that atom compared with that of the atom with spin-paired electrons. In general, if two excited states can be formed by the promotion of an electron to a given higher energy orbital, in a particular atom, the excited state with the greatest number of unpaired electrons will be closest in energy to the ground state of that atom. This subject is discussed in greater detail later in the text.

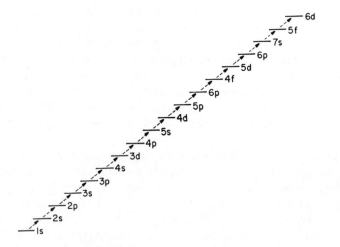

Fig. II-A-1.   Order of filling of Atomic orbitals.

d. CLASSIFICATION OF THE ELECTRONIC STATES OF ATOMS

The simple orbital model of atomic electronic structure, discussed in the previous sections, predicts that the frequencies of the spectral lines observed in atomic spectra should be equal to the differences in energies between the orbital configurations involved in the transitions, divided by Planck's constant, and that all transitions between the same configurations should have the same energy. In practice, this approach serves to locate the approximate positions of spectral lines but does not account for the broadening of the energy level or for the multiplet structure of these lines seen under high resolving power or in the presence of an external magnetic field. Obviously, refinements to the orbital model are needed. These refinements are based on the fact that atomic electrons interact with each other, through their electric and magnetic properties, to perturb the hydrogenic orbitals.

Atomic electrons, being moving charges, act as small magnets. They produce an orbital magnetic moment, by virtue of their orbital motion, which is located at the center of the orbit (i.e., at the nucleus) and is directed at right angles to the orbital plane, collinearly with the orbital angular momentum vector associated with the occupied orbital. The magnitude of the orbital magnetic moment, due to a single electron, is directly proportional to the l of the subshell occupied by that electron

$$\mu_l = \frac{|l|eh}{2\pi m_e c} \tag{II-A-6}$$

where m_e and e are the mass and charge of the electron, h is the Planck's constant, and c is the speed of light. In the classical sense, the electron may also be thought of as spinning on its own axis, thus producing a magnetic moment at the electron whose direction is perpendicular to the plane of rotation of the electron and is either "up" or "down" depending on whether $s = +\frac{1}{2}$ or $s = -\frac{1}{2}$. The magnitude of the spin magnetic moment is directly proportional to $|s|$ and thus always has only one value

$$\mu_s = \frac{|s|\,eh}{2\pi m_e c} = \frac{eh}{4\pi m_e c} \tag{II-A-7}$$

In quantum mechanics, the location of the spin magnetic moment vector would have to be given by a probability distribution, in keeping with the Heisenberg uncertainty principle. Now it is well known, in classical physics, that magnets interact with each other. That is to say, the magnetic moments of several individual magnets will add vectorially (couple) to give a total magnetic moment for the system of magnets. This is also the case for atomic electrons. The individual spin and orbital magnetic moments, may be added vectorially to give a resultant atomic magnetic moment. The addition of the individual electronic spin and orbital magnetic moments is accomplished by defining the line of the direction of the resultant atomic magnetic moment as the " z " axis. The z components of the electronic spin and orbital magnetic moments then add algebraically to give the z component of the resultant atomic magnetic moment. Because the z axis, defined in this way, is identical with the direction in which the resultant atomic magnetic moment would be aligned if an external magnetic field were imposed across the atom in an arbitrary z direction, the " z " components of the individual electronic orbital magnetic moments

$$\mu_{lz} = \frac{l_z e h}{2\pi m_e c} = \frac{m e h}{2\pi m_e c} \qquad \text{(II-A-8)}$$

where m is the magnetic quantum number of the occupied orbital (i.e., $l_z = m$). For a one-electron atom (or one with one electron in a degenerate subshell), l_z is always positive, because the positive z direction is obviously chosen with respect to the single electron. The only possible values of the " z " components of the individual electronic spin magnetic moments are

$$\mu_{sz} = \frac{s_z e h}{4\pi m_e c} \qquad \text{(II-A-9)}$$

where s_z is either $+\frac{1}{2}$ or $-\frac{1}{2}$. Because the magnitudes of the spin and orbital moments of each electron are simple multiples of $eh/2\pi m_e c$ and s_z or l, the practice of adding s and l values vectorially (or s_z and m values algebraically) has been adopted to represent the addition of magnetic moments. This convention, which makes the addition of magnetic moments synonymous with the addition of angular moments, is employed here.

There are two ways in which individual electronic l and s values may be added vectorially to give the resultant atomic angular momentum quantum number J. First, the l and s of each electron may be added vectorially to give a resultant one-electron angular momentum quantum number j. The j values of all atomic electrons may then be added vectorially to give J. This is known as the j-j coupling scheme. Second, the individual l values of each electron may be added vectorially to give a total orbital angular momentum quantum number L. Then the s values of each electron may be added vectorially to give a total spin angular momentum quantum number S. The values of L and S may then be added vectorially to give J for the atom. This is known as the Russell Saunders coupling scheme.

Although the values of J derived by either method of addition may be identical, the latter method implies that L and S are good quantum numbers for the atom and, consequently, that orbital angular momentum and spin angular momentum are well-defined "constants of the motion" of atoms. The former method of addition, on the other hand, does not define a total orbital angular momentum and a total spin angular momentum but only a resultant atomic angular momentum. The two addition methods correspond to different physical situations—one in which L and S as well as J can be used to describe atomic electronic states and one in which L and S have no meaning and J is the only good angular momentum quantum number.

Whether Russell Saunders or j-j coupling is used, the coupling of the angular moments of atomic electrons is determined by the relative degrees of importance of electrostatic and magnetic interactions in the atoms. In the classical sense, the angular momenta correspond to the magnetic moments of the spin and orbital magnetic dipoles. Interaction between them falls off as $1/r^3$ (where r is the distance between them). In the nonclassical sense, because the orbital angular momentum vector of an electron is located at the nucleus whereas the spin angular momentum vector is located, roughly, on the electronic orbit, the coupling of the spin angular momentum vector and the orbital angular momentum vector (spin-orbital coupling) is very weak unless that electron spends a considerable portion of its time near the nucleus (i.e., has a high probability density near the nucleus). Generally only atoms with high nuclear charges (heavy atoms) exhibit appreciable spin-orbital coupling.

Classically, electrostatic interactions fall off only as $1/r$ (where r is now the distance between charge monopoles). In the nonclassical sense, interelectronic repulsion is strong if several electrons have high probability densities in the same region of space. This is generally the case for many electron atoms and results in the large energy separations between states of different L and different S belonging to the same electronic configuration of the atom. Now if a many-electron atom has a nucleus of low nuclear charge, the individual electronic orbital angular momenta, all located at the nucleus, will couple strongly (as will the individual electronic spin angular momenta, located on the electronic distribution), when the overlap of the individual orbitals is appreciable. The coupling of spin and orbital angular momenta will be secondary to the coupling of the l's and s's, and hence Russell-Saunders coupling will adequately describe the situation. On the other hand, if a many-electron atom has a high nuclear charge, spin-orbital coupling will be appreciable, resulting in displacement of the orbital-only or spin-only energy levels from their relative positions in the absence of spin-orbit coupling. In other words, spin-orbital coupling results in the mixing of spin levels with orbital levels so that some electrons change spin. Because the restrictions on the values of L and S are relaxed when there is a change in spin of the individual electrons, and hence in the individual electronic orbital angular momenta (Hund's rule), L and S lose their significance as quantum numbers for the atom. In this case, J is the only good angular momentum quantum number, and j-j coupling best describes the situation. Because for most atoms, even when appreciable spin-orbital coupling occurs, the Russell-Saunders scheme is retained, with an appropriate perturbation treatment to account for the spin-orbital coupling, the remainder of this discussion is confined to the Russell-Saunders scheme. We are now ready to proceed to a formal description of atomic energy levels.

Because the energy of an atom having a particular orbital configuration is determined by the principal quantum numbers of the occupied orbitals, the total orbital angular momentum L, the total spin angular momentum S, and the resultant atomic angular momentum J, these are the parameters used to specify the energy levels of atoms. The electrons of interest to us here are those which occupy the partially filled shells, for these electrons are responsible for the transitions seen in optical spectra. In determining the energy levels of atoms, all electrons in completely filled shells and subshells are neglected because they contribute nothing to the angular momentum of the atom (i.e., $L = \sum_i l_i = 0$, $S = \sum_i s_i = 0$, $J = L + S = 0$) and can be assumed to form a spherically symmetrical charge distribution so that all outer-shell electrons are affected equally by them. The energy levels of the atom are then determined by the electronic configurations of the partially filled subshells.

The definition of the *atomic term* that corresponds to a collection of *energy levels* of the atom all having the same atomic orbital angular momentum quantum number L and the same atomic spin angular momentum quantum number S is $^{2S+1}L_J$. In the same way that the one-electron orbital angular momentum quantum numbers $l = 0, 1, 2, 3, \ldots$, were noted conventionally by $s, p, d, f, \ldots$, the atomic orbital angular momentum quantum numbers $L = 0, 1, 2, 3, \ldots$, are designated by $S, P, D, F, \ldots$, respectively. Thus an atom like boron having an outer-shell electronic configuration of $2s^2 2p^1$ (or an outer-subshell configuration of p^1) has a term corresponding to $L = 1$ an $S = \frac{1}{2}$, which would be designated a ^2P term. Because J can have values of $L + S, L + S - 1, \ldots, L - S$, this term contains two energy levels, corresponding to $J = \frac{3}{2}$ and $J = \frac{1}{2}$, which are designated $2^2P_{3/2}$ and $2^2P_{1/2}$. The quantity of $2S + 1$ is called the *multiplicity* of the term and is equal to the number of energy levels contained in the term when L is greater than S. Since it occurs in energy levels, the multiplicity gives the number of degenerate states contained in each energy level when $L > S$ for the level. These states become apparent only in spectra taken in the presence of an external magnetic field. If L is less than S, the multiplicities of terms and energy levels have no significance other than being the number of unpaired electrons plus one.

The Hund rules can sometimes be used to predict the relative energies of terms derived from the same configuration. For example, an atomic p^2 configuration is split into ^1D, ^3P, and ^1S terms. Because the ^3P term has the highest multiplicity (spin), it has the lowest energy. This term contains the ground state of the atom. Of the singlet terms, the ^1D term has the highest orbital angular momentum and hence lies at lower energy than the ^1S term. The ^3P term is composed of three energy levels; both singlet terms have only one energy level. The energy levels of the ^3P term are $^3P_2$, $^3P_1$, and $^3P_0$. To predict the relative energies of these levels, it is necessary to invoke the third Hund rule. This rule states that "for less than half-filled subshells, the lower the value of J, the lower the energy of the level. For more than half filled subshells, the higher the value of J, the lower the energy of the level." Hence the $^3P_0$ level is lowest in energy and contains the ground state, and the $^3P_2$ level is highest. For further information on atomic terms, levels, and states, the reader is advised to refer to the many excellent references on the subject (see References at the end of this section).

2. ELECTRONIC TRANSITIONS IN ATOMS

a. TYPES OF ATOMIC TRANSITIONS

Atomic electronic spectra are of two types, absorption and emission (or fluorescence). Absorption spectra arise from the absorption of light through the promotion of an atomic electron from an occupied orbital (in the valence shell) to an unoccupied orbital, raising the atom from its ground state to an electronically excited state. Of course excitation from an excited state to a higher excited state may occur, but this is rare compared with the incidence of ground to excited state absorptions. Also, many-electron excitations are very rare compared with the incidence of single-electron excitations. Emission spectra arise from the emission of light accompanying demotion of an electron from a high energy orbital to a lower energy orbital, possibly one in the valence shell of the atom. This drops the fluorescing atom from a high excited state to a lower one or to the ground state. As a result of the much more frequent involvement of intermediate excited states in atomic fluorescence, the emission spectrum of an atom contains many more lines than the absorption spectrum of the same atom.

There are four basic types of atomic fluorescence, and these and several variations are shown in Figure II-A-2. *Resonance fluorescence* results when atoms are excited from the ground electronic state to an excited state (resonance absorption) and then undergo a radiational deactivation process to the ground state, emitting radiation of the same wavelength as that absorbed.

16   THEORY

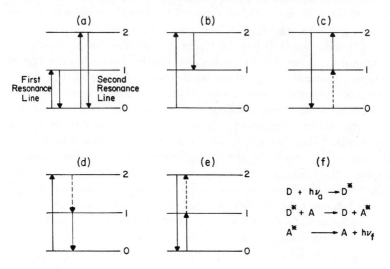

Fig. II-A-2. Types of atomic fluorescence (solid lines represent radiational processes and dashed lines represent radiationless processes): (*a*) resonance fluoresence, (*b*) normal direct-line fluorescence, (*c*) thermally assisted direct-line fluorescence, (*d*) normal stepwise-line fluoresence, (*e*) thermally assisted stepwise-line fluorescence, (*f*) sensitized fluorescence.

The most intense fluorescence for many atoms occurs when an atom undergoes the resonance fluorescence process—first resonance fluorescence line(s) —involving the first excited state (e.g., zinc—213.8 nm; cadmium, 228.8 nm; sodium, 589.0–587.6 nm; copper, 324.7–327.4 nm; (refer to Table IV-A-1). The radiances of the first resonance fluorescence lines are usually greater than for other fluorescence lines because the transition probabilities of the absorption-fluorescence transitions are greater than for other transitions, and so most analytical studies of atomic fluorescence spectrometry have involved these lines.

Normal Direct-line fluorescence results when an atom is raised by radiational means to an excited state and then undergoes a radiational transition to a state above the ground state (see Figure II-A-2*b*). The radiation emitted is of longer wavelength than the radiation absorbed (e.g., absorption of thallium, 377.6 nm, by thallium atoms results in both the 377.6-nm resonance line and the 535.0-nm normal direct-line fluorescence radiation). *Thermally assisted direct-line fluorescence* (see Figure II-A-2*c*) occurs when a metastable state above the ground state is excited by thermal means and radiational excitation of the metastable state results in an upper excited state, which then undergoes a radiational emission transition to the ground state (or state lower than the metastable state involved in the absorption process). For example, thallium atoms in the metastable state can be excited by 535.0-nm radiation and emit 377.6-nm radiation (anti-Stokes emission, because emission radiation has a shorter wavelength than the absorbed radiation). Thermally assisted direct-line fluorescence occurs for atoms (with low-lying excited states) in hot flames (flames with temperatures of order of 2500°K or above).

Normal stepwise-line fluorescence results when an atom is excited from the ground state to an upper excited state and then undergoes a transition to a lower excited state from which a radiational transition to a lower state produces fluorescence (see Figure II-A-2d). The emitted radiation has a longer wavelength than the absorbed; for example, sodium atoms can be excited by the second resonance absorption doublet at 330.2 nm and emission occurs at the 589.9–589.6 doublet. *Thermally assisted stepwise-line fluorescence* (see Figure II-A-2e) takes place for atoms radiationally excited to a metastable state above the ground state and then thermally excited to an upper excited state, which undergoes a radiational transition to the ground state (or to an excited state of lower energy than the radiationally excited state). The emitted radiation has a shorter wavelength than the absorbed radiation.

The final type of atomic fluorescence is *sensitized atomic fluorescence*, which occurs when a radiationally excited species (atom or molecule) transfers energy to an atom that afterward undergoes radiational deactivation (see Figure II-A-2f). This process has been of *no* analytical value for *atomic fluorescence flame spectrometry* because of the difficulty of producing a sufficient donor population to transfer significant amounts of energy and of minimizing collisional deactivation of the excited donors. Sensitized atomic fluorescence, however, has been observed for metal vapors in quartz containers—for example, mercury excited via the 253.7 nm intercombination line transfers energy to other metal atoms (e.g., zinc, cadmium, thallium), as long as some molecula species such as N_2 is present. This process, however, has *not* been used for analytical purposes.*

Spectral lines are characterized by three properties: namely, position in the electromagnetic spectrum, intensity, and shape. These properties are dependent in, varying ways and degrees, on the electronic structure and the environment of the absorbing and emitting atoms and are worthy of some discussion here.

---

* It should be mentioned that an ambiguity in terminology exists between atomic and molecular luminescence. The term fluorescence is used for *all* atomic luminescence processes, whether no spin or a spin change occurs during the radiational process. In molecular luminescence, the emission process is called fluorescence if no spin change occurs and phosphorescence if a spin change occurs.

b. POSITIONS OF SPECTRAL LINES

The frequencies at which atomic spectral lines are observed are, in the first approximation, determined by the differences in energy between atomic states (terms), involved in the transitions. If, in the transition, the atom is initially in state (term) i and finally in state (term) f, the frequency of the line produced by that transition is

$$\nu = \frac{\varepsilon_f - \varepsilon_i}{h} \qquad\qquad \text{(II-A-10)}$$

where ε_f and ε_i are the energies of state (term) f and state (term) i, respectively, and h is the Planck's constant. For absorption spectra, ε_i corresponds to the energy of the lower state (term) and ε_f, to that of the upper state (term). For emission spectra, the opposite is true. Customarily, the frequencies of spectral lines are given by the reciprocals of the wavelengths of the lines, in units of cm^{-1}. When a frequency is expressed in this way, it is denoted by $\tilde{\nu}$ (cm^{-1}) and is related to the true frequency ν (sec^{-1} or Hz), by

$$\tilde{\nu} = \frac{\nu}{c} \qquad\qquad \text{(II-A-11)}$$

where c is the speed of light and ν is expressed in sec^{-1} (or Hz).

Spin-orbital coupling, which splits the terms of $S > 0$ into several energy levels, occurs to some degree in all atoms; therefore, the frequencies of spectral lines are determined by the energy differences between the energy levels involved in the transition to a better approximation than that given previously. For light atoms, the separations between energy levels corresponding to the same term are small because spin-orbital coupling is weak. In this case, the spectral line, which was assumed to have originated from a single transition between two terms, appears to be split into several closely spaced lines, clustered about the frequency of the expected transition between terms. Each line in the so-called *fine structure* actually corresponds to a separate transition between the energy level from which the transition originated and one of the energy levels arising from the splitting of the term to which the transition occurs.

In heavy atoms, where spin-orbital coupling is appreciable, the separations between energy levels corresponding to the same term may be comparable with the separations between different terms. In this case, transitions between different energy levels will have very different frequencies, so that speaking of the fine structure of lines arising from transitions between different terms is not very meaningful.

In the presence of an external magnetic field, the energy levels, each having a particular value of J, are split into $2J + 1$ states. This results in splitting of the lines arising from transitions between two different energy levels as they are now more accurately described as a composite of transitions occurring between the various states. Generally, this type of splitting is very small and is usually seen only under very high resolution.

C. TRANSITIONS PROBABILITIES OF SPECTRAL LINES

The observed intensity* of a spectral line is determined by the incidence of transitions between the initial and final states of an atom, corresponding to the frequency of that line. The incidence of transitions is in turn governed by the intensity of exciting radiation, the path length of exciting radiation through the sample, the concentration of potential absorbers or emitters in the sample, and the probability that a radiative transition will occur between the initial and final states of the absorber or emitter (see Section II-C). In quantitative analysis, it is the concentration of absorbers or emitters in the sample that is of interest. For the present discussion, however, absorber or emitter concentration, illumination intensity, and sample size can all be manipulated externally, thus only the probability of radiative transition, which is a function of atomic electronic structure and the nature of the transition, concerns us here.

The probability of radiative transition between two electronic states is reflected, quantum mechanically, through the transition moment integral M:

$$M = \int_{-\infty}^{\infty} \Psi_i \, er \, \Psi_f \, d\tau \qquad \text{(II-A-12)}$$

In equation II-A-12 Ψ_i is the wavefunction of the atom the initial state (before radiative transition) and Ψ_f is the wavefunction of the atom in the state produced by radiative transition; $d\tau$ is the volume element in configuration space, e is the electronic charge, and r is the operator corresponding to the displacement of electronic charge in the atom as a result of the transition (er is the dipole moment vector). Under given conditions of illumination intensity, sample size, and emitter or absorber concentration, the intensity of the transition is proportional to M^2. Now in general, the calculation of absolute transition probabilities is very difficult because the wavefunctions of many electron atoms are not known exactly. However, one property of the wavefunctions and the transition moment length r enables us to determine whether a transition is allowed or forbidden, that is, whether M is positive or zero, respectively. This property is called *parity* and is derived from the

---

* Intensity is used in the text in a general sense to mean number of transitions (absorption or luminescence) occurring between states per unit time.

behavior of a mathematical function when it is inverted through a center of symmetry (i.e., when x, y, and z are replaced by $-x$, $-y$, and $-z$). Those functions for which

$$f(x, y, z) = f(-x, -y, -z) \qquad \text{(II-A-13)}$$

are said to have *even parity*. Those for which

$$f(-x, -y, -z) = -f(x, y, z) \qquad \text{(II-A-14)}$$

are said to have *odd parity*. The product of two functions having the same parity always results in another function having *even* parity. The product of two functions having different parities always means that another function has *odd* parity. The *integral* over the entire configuration space of a function having *odd parity* is *zero*, and the integral of a function having *even parity* is *nonzero*. Now because the transition moment length r is a vector, it must change sign upon inversion (i.e., it has odd parity). Therefore, for a given transition moment to be nonzero—and for the transition itself to be allowed—the product of the wavefunctions of the two states involved in the transition must also have odd parity. This will be the case only if the two wavefunctions have opposite parities. Consequently, *transitions* occurring between states of the *same parity are forbidden* and should have *zero probability*. Those occurring between states of *different parity* are *allowed* and will have *probabilities* that depend on the *magnitude of charge displacement* and the *states involved* in the transitions.

The parities of atomic electronic states are determined by the quantum numbers that characterize these states. Whether the transitions are allowed or forbidden can thus be expressed through the changes in quantum numbers produced by changes in electronic state. For allowed transitions

$$\Delta n = 0, 1, 2, \ldots$$
$$\Delta L = \Delta l = \pm 1 \text{ (one optical electron)}$$
$$\Delta L = 0, \pm 1 \text{ (for several optical electrons)} \qquad \text{(II-A-15)}$$
$$\Delta J = 0, \pm 1 \text{ (except for } J = 0 \rightarrow J = 0)$$
$$\Delta S = 0$$

All other values of ΔL, ΔJ, and ΔS correspond to forbidden transitions. This set of restrictions is called the *optical selection rules*. These rules are *strictly valid only for the lightest elements in the gas phase at very low pressures. For heavier elements, where spin-orbital coupling is appreciable, and in high pressure gases, where atom-atom interactions occur, the selection rules may be violated.* Thus in the heavier elements transitions with $\Delta S \neq 0$ are often seen, although they are usually less intense than transitions with $\Delta S = 0$. The selection rule for ΔJ, however, is much more rigidly obeyed, there being many fewer exceptions than to the rules for ΔS and ΔL.

As an example of the ways in which atomic terms, energy levels, and selection rules affect the spectra of atoms, let us consider the spectrum of *hydrogen.*

The ground state of the hydrogen atom has the single electron in the $1s$ orbital. The only possible values of L and S are zero and $\frac{1}{2}$, respectively. The only possible value of J is, therefore, $\frac{1}{2}$ and the ground state of hydrogen is 1^2S_1. If the electron is excited to the $n = 2$ shell, the possible values of L^2 are 1 and 0 and the only possible value of S is still $\frac{1}{2}$. For $L = 1$, $J = \frac{3}{2}$ and $J = \frac{1}{2}$; $L = 0$, $J = \frac{1}{2}$. Hence the possible energy levels are $2^2S_{1/2}$, $2^2P_{3/2}$, and $2^2P_{1/2}$. Because there are no electronic repulsions in hydrogen, the $2S$ and $2P$ terms should all have the same energy. Actually, the $^2P_{3/2}$ and the $^2P_{1/2}$ levels are slightly split due to their different resultant angular momenta. If the electron occupies the third shell, then $S = \frac{1}{2}$ and $L = 2$, 1, 0. For $L = 2$, $J = \frac{5}{2}$ and $\frac{3}{2}$ for $L = 1$, $J = \frac{3}{2}$, and $\frac{1}{2}$ and for $L = 0$, $J = \frac{1}{2}$. The possible energy levels are then $3^2S_{1/2}$, $3^2P_{3/2}$, $3^2P_{1/2}$, $3^2D_{5/2}$, $3^2D_{3/2}$. It can be seen that for all terms with $L > 0$, there are two slightly separated energy levels.

Now, remembering the selection rule $\Delta S = 0$, $\Delta L = \pm 1$, and $\Delta J = 0, \pm 1,$— only the absorption transitions $1^2S_{1/2} \rightarrow 2^2P_{3/2}$, $1^2S_{1/2} \rightarrow 2^2P_{1/2}$, $1^2S_{1/2} \rightarrow 3^2P_{3/2}$, $1^2S_{1/2} \rightarrow 3^2P_{1/2}$ are allowed. Because of the close spacings of the $^2P_{3/2}$ and $^2P_{1/2}$ levels, these transitions appear as two doublets; that is, two pairs of slightly separated lines, with the pairs appreciably separated from one another. In the emission spectra, these same pairs of doublets would appear (resonance lines) but also the $3^2D_{5/2} \rightarrow 2^2P_{3/2}$, $3^2D_{3/2} \rightarrow 2^2P_{1/2}$ doublet would be seen.

In hydrogen, resonance transitions involving terms with the same L but different n form series with the spacing between lines being proportional to $(1/n_1{}^2 - 1/n_2{}^2)$. This is the principal difference between hydrogen spectra and alkali metal spectra. Although all have one optical electron and a $^2S_{1/2}$ ground state energy level, the excited states formed by occupation of different subshells of the same shell in the alkali metals do not have the same energy. This is because of the stronger attraction of the nucleus for electrons of the lowest values of l in the same shell, due to more effective screening of electrons of higher l from the nucleus by inner-shell electrons. The effect of this is to destroy the regular spacing between resonance transitions involving energy levels of different n but of the same L because the effect of inner-shell electron screening is not constant but falls off with increasing n. Except for the matter of spacing, the spectra of the alkali metals are similar to those of hydrogen. For example, they both show doublet fine structure in the allowed transitions.

The spectra of the other elements are somewhat more complex owing to the plethora of terms that arise from the existence of several unpaired electrons. In our discussion of hydrogen, excited states having $n > 3$ have been ignored to simplify the treatment. For a discussion of terms and transitions involved in atoms other than hydrogen and the alkali elements, the reader is referred to the fine treatise by Mavrodineanu and Boiteux.[4]

d. QUANTUM YIELDS AND LIFETIMES OF EXCITED STATES OF ATOMS

(1) *The Quantum Yield and Lifetime of Resonance Fluorescence—No Intermediate Excited States—No Quenchers.*   The intensity of an atomic emission line is dependent not only on the symmetries of the states between which the transition occurs but also on the processes competing with radiative depopulation of the upper state. These processes may be *unimolecular* (e.g., radiative decay to an intermediate excited state or nonradiative internal conversion) or they may be *bimolecular*, such as loss of energy by collisions with other species (*quenching*). The relative effectiveness with which excitation from the ground electronic state A to an excited state A^* produces emission from A^* to A is called the relative quantum yield or quantum efficiency Y_F of the transition; it is defined for resonance fluorescence by

$$Y_F{}^o = \frac{\text{number of quanta per unit time emitted from } A^* \text{ to } A}{\text{number of quanta per unit time absorbed by } A \text{ in going to } A^*}$$

(II-A-16)

Because there are no intermediate excited states and no nonradiational means of deactivating A^* (i.e., no quenchers), $Y_F{}^o$ is necessarily unity.

Intimately related to the quantum yield Y_F of atomic fluorescence is the concept of the mean lifetime τ of the excited state from which luminescence arises. Clearly, the longer an atom remains in a given excited state, the greater the number of processes that can compete with luminescence for deactivation of that excited state. Consequently, we might expect some sort of inverse relationship between the mean lifetime of an electronically excited state and the intensity of luminescence arising from that state.

Let us consider an ensemble of noninteracting atoms, in an electronically excited state A^*, which undergo radiative transition to the ground state. If no other processes (i.e., no quenching processes and no radiational transitions to low-lying excited states) compete with the transition $A^* \xrightarrow{h\nu} A$, the rate of transition of A^* to A is given by

$$\frac{-dn_{A^*}}{dt} = k_f\, n_{A^*}$$

(II-A-17)

where n_{A^*} is the concentration of A^* and k_f is the probability of radiative transition of A^* to A. If A^* is being produced and removed under steady state conditions of excitation, then equation II-A-17 also gives the rate of production of A^* from A (i.e., if $A^* \leftrightarrows A$ are resonance transitions). If the exciting light is shut off at time $t = 0$, when the concentration of A^* is the steady state value $n_{A^*}{}^o$ ($n_{A^*}{}^o$ is the denominator of equation II-A-16 if A^* is produced only by direct absorption of A), then after time t the concentration of A^* is found by integration of equation II-A-17

$$n_{A^*} = n_{A^*}{}^o e^{-k_f t} \qquad \text{(II-A-18)}$$

The exponential nature of equation II-A-18 makes it impossible to define a true arithmetic mean lifetime for a single A^* atom. The rate constant k_f, however, has the dimensions of transitions per time (the rate constant k_f is identical with the radiational transition probability A_{ul}, where A_{ul} is the Einstein coefficient of spontaneous emission for the transition between the upper u and lower l states). Because k_f is a constant of the transition, it is convenient to define $k_f{}^{-1}$ as τ^o, the mean inherent radiative lifetime of A^*. The significance of τ^o is that for each time interval of this duration, the number of A^* atoms will be diminished by $1/e = 0.4343$. For allowed transitions, $\tau^o \simeq 10^{-8}$ sec. For forbidden transitions ($\Delta S \neq 0$, $\Delta L \neq 1$; etc.), $\tau^o \geq 10^{-6}$ sec. Alternatively, the time interval for the number of A^* atoms to be halved ($\tau_{1/2}$) may be considered a constant of the emission, since it is independent of $n_{A^*}{}^o$. The relation between the inherent half-life $\tau_{1/2}^o$ and the inherent mean lifetime is

$$\tau_{1/2}^o = \frac{\ln 2}{k_f} = 0.693 \tau^o \qquad \text{(II-A-19)}$$

(2) *The Quantum Yield and Lifetime of Resonance Fluorescence—Available Intermediate Excited States—No Quenchers.*   Let us now turn our attention to a collection of excited atoms that do not interact with one another or with impurities but are capable of dissipating excitation energy by radiative decay to intermediate states and by spin-forbidden intersystem crossing. Such a situation might be found in a low pressure, high purity discharge of a gas consisting of heavy atoms (like mercury) in which spin-orbital coupling is appreciable. If the sum of probabilities per unit time of unimolecular processes deactivating A^* (other than fluorescence to A) is given by $\sum_i k_i$ and that for fluorescence to A by k_f, then the rate of depopulation of A^* is

$$\frac{-dn_{A^*}}{dt} = k_f n_{A^*} + \sum_i k_i n_{A^*} \qquad \text{(II-A-20)}$$

Integration of equation II-A-20 yields

$$n_{A^*} = n_{A^*}{}^o \, e^{-(k_f + \Sigma_i k_i)t} \qquad \text{(II-A-21)}$$

from which the mean lifetime of $A*$ (defined as the time to diminish n_{A*} by $1/e$) is given by

$$\tau_F = \frac{1}{k_f + \sum_i k_i} \tag{II-A-22}$$

Because the denominator of equation II-A-22 is larger than k_f, τ_F must necessarily be smaller than $\tau_F{}^0$. In general, the greater the number of radiative processes competing with the $A* \to A$ fluorescence, the shorter is the mean lifetime of the excited state $A*$.

The quantum yield for $A*$, being the fraction of excited atoms that fluoresce from $A* \to A$, is given by

$$Y_F = \frac{k_f \int_{t=0}^{\infty} n_{A*}\, dt}{n_{A*}{}^0} = \frac{k_f}{k_f + \sum_i k_i} = k_f \tau_F \tag{II-A-23}$$

At a given temperature, the value of Y in the absence of interatomic interactions is a constant of the transition and is the maximum possible value of the ratio of atoms leaving the $A*$ state by $A* \to A$ fluorescence to those leaving by all possible radiative pathways in the same time interval.

(3) *The Quantum Yield and Lifetime of Resonance Fluorescence—Intermediate Excited States—Quenchers Present.* Let us now suppose that the excited atom $A*$ can be deactivated in molecular collisions as well as by atomic fluorescence and monatomic radiationless deactivation processes. For the sake of simplicity, it is assumed here that the only means of production of $A*$ is by direct absorption by A and that the deactivating atoms are greatly in excess of those of $A*$ (i.e., pseudo-first-order kinetics prevail). If the temporal probabilities for molecular deactivation are given by k_j and the concentrations of the collisional deactivating species Q_j are given by n_{Q_j}, then the rate of deactivation of $A*$ is given by

$$-\frac{dn_{A*}}{dt} = k_f n_{A*} + \sum_i k_i n_{A*} + \sum_j k_j (n_{Q_j})(n_{A*}) \tag{II-A-24}$$

which yields upon integration

$$n_{A*} = n_A{}^0{}_* \exp -(k_f + \sum_i k_i + \sum_j k_j n_{Q_j})t \tag{II-A-25}$$

The mean lifetime of $A*$ in this case is

$$\tau_F' = \frac{1}{k_f + \sum_i k_i + \sum_j k_j n_{Q_j}} \tag{II-A-26}$$

which is, of course, smaller than the value of τ_F in equation II-A-22, and the quantum yield Y_F' of fluorescence from A^* to A is given by

$$Y_F' = \frac{k_f}{k_f + \sum_i k_i + \sum_j k_j n_{Q_j}} = k_f k_F' \qquad \text{(II-A-27)}$$

which is obviously smaller than Y_F in equation II-A-23. Consequently, when atomic fluorescence is studied in high pressure situations or in the presence of quenching (impurity) species, the quantum yield of fluorescence is reduced. It can be seen that in the limit where $n_{Q_j} \to 0$; that is, when collisional deactivation becomes negligible, $Y_F' \to Y_F$ and $\tau_F' \to \tau_F$, otherwise $Y_F' < Y_F$ and $\tau_F' < \tau_F$. An expression for the rate constant for a bimolecular process is given in the footnote on page 28.

(4) *The Quantum Yield and Lifetime of Nonresonance Fluorescence—Intermediate Excited States—Quenchers Present.* By a process similar to that used to derive equation II-A-27, a general expression for the quantum yield and the lifetime for any fluorescence process is any atom can be derived. However, because the derivation is no more instructive than the derivations given previously and because the derived expressions are for a specific atom, no derivations are given here; rather, the derived expressions for specific cases are presented in the section on radiances in this chapter.

e. SHAPES OF SPECTRAL LINES

(1) *Natural Half Width of Spectral Lines.* From the classical point of view, it might be expected that spectral lines would be infinitesimally thin, corresponding to one and only one frequency. The Heisenberg uncertainty principle, however, tells us that for an atom in a state with energy E

$$\Delta E \cdot \Delta t \simeq \frac{h}{2\pi} \qquad \text{(II-A-28)}$$

where Δt is the uncertainty in the time the atom exists in the state characterized by E and ΔE is the uncertainty in the energy of the atom in that state. If Δt is taken as being τ^0, the mean lifetime of the state characterized by E, and ΔE the uncertainty in the energy of a transition between that state and the ground state, then

$$\Delta E \cdot \tau^0 = \frac{h}{2\pi} \qquad \text{(II-A-29)}$$

$$\Delta E = h \Delta \nu_u \qquad \text{(II-A-30)}$$

where $\Delta\nu_u$ is the uncertainty in the frequency of the line characteristic of the transition. Then

$$\Delta\nu_u = \frac{1}{2\pi\,\tau^0} \qquad\qquad \text{(II-A-31)}$$

so that in the absence of external effects a spectral line has a finite, *natural width* that is characteristic of a particular transition in a particular atom and is inversely proportional to the mean lifetime of the state from which the transition originates. The natural broadening $\Delta\nu_u$ is the smallest width that a given spectral line can exhibit. If both upper and lower states involved in the transition have an uncertainty in energy, then

$$\Delta\nu_u + \Delta\nu_l = \frac{1}{2\pi\,\tau^0} \qquad\qquad \text{(II-A-32)}$$

where $\Delta\nu_u$ and $\Delta\nu_l$ are the uncertainties in the frequencies of the upper and lower states involved in the transition.

(2) *Doppler Half Width of Spectral Lines.* The Doppler effect, which is manifested as a shift $\Delta\nu'$ in the apparent frequency of radiation emitted or absorbed by an atom (or molecule) in rapid motion relative to the observer, is frequently responsible for the broadening of spectral lines. This is especially so at higher temperatures (e.g., in flames), where atomic velocities are high. If an atom is moving with velocity v at an angle θ between the direction of motion and the line of sight of the observer (the x axis) and if the distribution of absorbing and emitting atoms follows a Maxwellian distribution of velocities in the direction of observation, then

$$\frac{dn}{n\,dv_x} = \left(\frac{m_a}{2\pi k T}\right)^{1/2} \exp\frac{-m_a v_x^{\,2}}{2kT} \qquad\qquad \text{(II-A-33)}$$

where dn/n is the fraction of atoms having velocity v_x, m_a is the atomic weight (g), k is the gas constant (erg $^\circ K^{-1} mol^{-1}$), and T is the absolute temperature $(^\circ K)$. The shift from ν_o corresponding to half the maximum value of dn/ndv_x is $\Delta\nu'$ and is given by

$$\Delta\nu' = \nu_o\left(\frac{v\cos\theta}{c}\right) = \frac{\nu_o v_x}{c} \qquad\qquad \text{(II-A-34)}$$

where ν_o is the frequency when $\nu = 0$ (at the center of the line), v; is the component of velocity along the observer's line of sight, and c is the speed of light.

The Doppler half width is customarily represented by

$$\Delta\nu_D = 2\Delta\nu' = \frac{2\nu_o v_x}{c} \qquad\qquad \text{(II-A-35)}$$

The value of $\Delta \nu_D$ measured in this way is called the half-bandwidth. Now

$$v_x = \sqrt{\ln 2} \sqrt{\frac{2kT}{m_a}} = 0.835 \sqrt{\frac{2kT}{m_a}} \qquad \text{(II-A-36)}$$

and so

$$\Delta \nu_D = 1.67 \left(\frac{v_o}{c} \right) \sqrt{\frac{2kT}{m_a}} = 7.16 \times 10^{-7} \, v_o \sqrt{\frac{T}{M_a}} \qquad \text{(II-A-37)}$$

where M_a is the atomic weight of the species of concern, in atomic mass units (amu). Doppler broadening thus increases with increasing temperature and decreases with increasing atomic weight. It should also be noted that Doppler broadening increases as the energy of the transition increases. Because Doppler broadening is produced only by the component of atomic velocity toward or away from the observer, the observation of an atomic beam at right angles to its path of travel will effectively eliminate Doppler broadening so that fine structure studies can be performed.

(3) *Collisional Half Width of Spectral Lines.* Broadening of spectral lines can also occur from the interactions of absorbing or emitting atoms with foreign species (Lorentz broadening) or other atoms of the same substance (Holsmark broadening). The effects of these interactions on spectral lines fall under the heading of collisional broadening, which is characterized by broadening of the spectral line, shift of the line maximum to the red (lower frequencies), and distortion of the red side of the spectral line (the line becomes asymmetric). The broadening itself can be explained in terms of the changes in velocity, occurring in absorbing or emitting atoms, upon collision with other species. This produces an effect analogous to Doppler broadening. The red shift of the line maximum and the distortion of the line are best explained by quantum mechanics. Thus, the emitting or absorbing atoms form *weak complexes* with ground state species resulting in the perturbation of the atomic energy levels. Perturbation is a process in which the energy levels of the complexed atom are forced closer together, leading to lower energy transitions and thus a red shift in the spectral lines. Because a greater fraction of red-shifted transitions is produced by this mechanism, relative to the fraction of blue-shifted transitions produced solely by elastic collisions, the line maxima are shifted to the red and the broadening extends farther into the red than into the blue.

The half width of a collisionally broadened spectral line is given by

$$\Delta \nu_C = \frac{Z_{QA}}{\pi n_A} \qquad \text{(II-A-38)}$$

where n_A is the concentration of the gaseous analyte A (cm^{-3}) and Z_{QA} is the number of binary collisions between the foreign species Q and the analyte A per sec unit volume (units of $\sec^{-1}\ cm^{-3}$) and is given by

$$Z_{QA} = \sigma_c^2 n_A n_Q \bar{v}_{rel} \tag{II-A-39}$$

where σ_c^2 is the collisional cross section for gaseous A and Q (cm^2) [i.e., for hard spheres of radii r_A and r_Q, $\sigma_c^2 = \pi(r_A + r_Q)^2$], n_A and n_Q are the concentrations of A and Q species (cm^{-3}) and $\bar{v}_{rel}$ is the average relative velocity of the colliding species $(cm\ sec^{-1})$ and is given from the kinetic theory of gases by

$$\bar{v}_{rel} = \sqrt{8\pi k T(1/m_A + 1/m_Q)} \tag{II-A-40}$$

where k is the Boltzmann constant (erg $^{\circ}K^{-1}$ mole^{-1}), T is the gas temperature ($^{\circ}K$), and m_A and m_Q are the masses) (g) of the A and Q species. Combining* the foregoing equations, the collisional half width for Lorentz broadening is given by

$$\Delta \nu_C = \frac{2n_Q\sigma_c^2}{\pi} \sqrt{2\pi k T(1/m_A + 1/m_Q)} \tag{II-A-41}$$

If both A and Q are the same species (i.e., Holzmark broadening), then $m_A = m_Q$ and σ_c^2 is for the process $A^* + A \rightarrow A + A^*$. In cells (flames and furnaces) and sources (hollow cathode and electrodeless discharge lamps) used in atomic fluorescence spectrometry, the concentration of the analyte species should be so low that Holsmark broadening is negligible compared to Lorentz and Doppler broadening.

(4) *Stark Broadening.* In gas systems, where ions are produced, broadening of atomic spectral lines can occur because of the perturbation of atomic energy levels by ionic electric fields (the Stark effect). This type of broadening, however, is *only* appreciable in high energy arcs, sparks, and plasma discharges containing a high concentration of ions. In atomic fluorescence spectrometry, the sources used (e.g., hollow cathode discharge lamps or electrodeless discharge lamps, as well as the cells, flames, or furnaces) do not exhibit appreciable Stark broadening. Similarly, *Zeeman broadening* due to the presence of magnetic fields is negligible because strong magnetic fields are absent.

---

* The second-order rate constant for the bimolecular process $A + Q^* \rightarrow A^* + Q$, where the asterisk indicates an excited species [actually the asterisk could be on the other species $(A^* + Q \rightarrow A + Q^*)$] is given by

$$k_j = \frac{Z_{QA} P \exp(-E_A/kT)}{n_A n_Q} = 2P\sigma_c^2 \sqrt{2\pi k T(1/m_A + 1/m_Q)} \exp - \frac{E_A}{kT}$$

where P is a steric factor to account for deviations from simple classical theory and E_A is the activation energy (erg). The units of k_j are $cm^3\ sec^{-1}$.

(5) *Total Line Half Widths of Spectral Lines.* The total line half width $\Delta\nu_T$ of spectral lines in flame cells used in atomic fluorescence spectrometry is given by the addition of the three main contributors: natural $\Delta\nu_N$, collisional (Lorentz) $\Delta\nu_C$, and Doppler $\Delta\nu_D$.

$$\Delta\nu_T = [(\Delta\nu_C + \Delta\nu_N)^2 + \Delta\nu_D{}^2]^{1/2} \qquad \text{(II-A-42)}$$

where collisional and natural broadening are types of damping* broadening and Doppler broadening is a Gaussian broadening; and so the major types are independent and add quadratically, whereas natural and collisional broadening are dependent and add linearly. Actually for atoms in flames or furnaces, $\Delta\nu_C \gg \Delta\nu_N$ and so

$$\Delta\nu_T = [\Delta\nu_C{}^2 + \Delta\nu_D{}^2]^{1/2} \qquad \text{(II-A-43)}$$

In some furnaces and line light sources with low pressures of foreign species, $\Delta\nu_C$ is often small compared to $\Delta\nu_D$, and so in this special case $\Delta\nu_T \sim \Delta\nu_D$.

The relation between the frequency half width $\Delta\nu_B$ of a spectral line (or band) and the wavelength half width $\Delta\lambda_B$ is given by

$$\Delta\nu_B = -\frac{c}{\lambda_o{}^2}\,\Delta\lambda_B \qquad \text{(II-A-44)}$$

where $\Delta\nu_B$ and $\Delta\lambda_B$ have units of $\sec^{-1}$ and cm, respectively, c is the speed of light ($\mathrm{cm}\ \sec^{-1}$), and λ_o is the wavelength of the line center (cm). (It should be noted that $\Delta\nu_B$ is positive: i.e., the larger frequency minus the smaller frequency; and $\Delta\lambda_B$ is also negative: i.e., the smaller wavelength minus the larger wavelength). The subscript B merely refers to the type of broadening. The approximate half width ranges of atomic lines (absorption or emission) of atoms in analytical flames are about 10^{-6} to 10^{-4} Å for natural broadening, 0.01 to 0.1 Å for Doppler broadening, and 0.002 to 0.2 Å for collisional broadening of most lines of most atoms in analytical flames (H_2/O_2 at about 2650°K, C_2H_2/O_2 at about 2840°K and C_2H_2/air at about 2480°K).

(6) *Profile of Atomic Absorption Lines.* The profile of atomic spectral lines is determined primarily by both collisional and Doppler broadening effects. The most general expression is the Voigt profile and is given by

$$k_v = k_o\,\frac{a}{\pi}\int_{-\infty}^{+\infty}\frac{\exp(-y^2)\,dy}{a^2 + (v-y)^2} \qquad \text{(II-A-45)}$$

---

* According to classical theory, an electron oscillator with frequency ν_o would oscillate forever at ν_o. However, the environment of other electrons results in a damped oscillation, and so the emission of radiation is no longer monochromatic (i.e., there is no longer an infinitely long wavetrain). Collisions of excited atoms with other species results in a weak complex which lowers the excited state and produces a frequency shift toward the red.

where k_o is the peak absorption coefficient for pure Doppler broadening, a is the damping constant,—that is, $a = \sqrt{\ln 2}\,(\Delta\nu_C + \Delta\nu_N)/\Delta\nu_D$, v is the relative frequency compared with $\Delta\nu_D$), that is, $v = [2\sqrt{\ln 2}\,(\nu - \nu_o)]/\Delta\nu_D$ and y is $2\,\delta\sqrt{\ln 2}/\Delta\nu_D$, where δ is an integration variable. (Equation II-A-45 is discussed in greater detail in Section II-C.)

3. GLOSSARY OF SYMBOLS

a	Damping constant, $\sqrt{\ln 2}\,(\Delta\nu_C + \Delta\nu_N)/\Delta\nu_D$.
$\bar{a}$	Average value of any dynamic variable, appropriate units.
A_{ul}	Radiative transition probability (Einstein coefficient of spontaneous emission), $\sec^{-1}$.
c	Speed of light, 3×10^{10} cm $\sec^{-1}$.
dn/n	Fraction of atoms with velocity v_x, no units.
$d\tau$	Volume increment in configurational space, volume units.
e	Charge of electrons, esu.
E	Energy of particle, energy units.
$\bar{E}$	Average energy of particle, energy units.
E_A	Activation energy, erg.
h	Planck constant, 6.6×10^{-27} erg sec.
H	Hamiltonian operator, energy units.
j	One-electron angular momentum quantum number, no units.
J	Resultant (total) angular momentum quantum number, no units.
k	Boltzmann constant, erg $^\circ$K^{-1} mol^{-1}.
k_f	Radiative rate constant, $\sec^{-1}$.
k_j	Nonradiative rate constant (collisional quenching with species j), cm^3 $\sec^{-1}$.
k_o	Peak atomic absorption coefficient for a pure Doppler broadened line, cm^{-1}.
k_ν	Atomic absorption coefficient at any ν, cm^{-1}.
l	Azimuthal quantum number, no units.
l_z	Magnetic quantum number, no units.
L	Total orbital angular momentum number, no units.
m	Magnetic quantum number, no units.
m_A	Mass of species A, g.
M_A	Atomic weight of species A, amu.
m_e	Mass of electron, g.
m_Q	Mass of species Q, g.
M	Transition moment, no units.
M_l	Orbital magnetic moment, erg gauss^{-1}.
M_{lz}	Orbital magnetic moment in z direction, erg gauss^{-1}.
M_s	Spin magnetic moment, erg gauss^{-1}.

M_{sz} Spin magnetic moment in z direction, erg gauss^{-1}.

n Principal quantum number, no units.

n_{A*} Concentration of A^*, cm^{-3}.

n_A Concentration of A, cm^{-3}.

n_Q Concentration of Q, cm^{-3}.

n_{Q_j} Concentration of quencher Q_j, cm^{-3}.

P Steric factor to account for deviations from kinetic theory, no units.

r Distance between magnetic dipoles, cm.

r Operator for displacement of electronic charge in atoms.

r_A Radius of gaseous A, cm.

r_Q Radius of gaseous Q, cm.

R Gas constant, erg °K^{-1} mol^{-1}.

s Spin quantum number, no units.

S Total spin angular momentum quantum number, no units.

t Time after start of decay of fluorescence, sec.

T Temperature of gas, °K.

v Relative frequency of point on spectral line, $2\sqrt{\ln 2}(\nu - \nu_o)/\Delta\nu_D$.

v_a Velocity of atom in any direction, cm sec^{-1}.

v_x Velocity component of atom in direction of observer, cm sec^{-1}.

$\bar{v}_{rel}$ Average relative velocity of colliding species, cm sec^{-1}.

y Variable distance, $2 \ln 2 \, 8/\Delta\lambda_D$, no units.

Y Quantum yield of excited atom with radiative transitions other than fluorescence transition possible, no units.

Y_F^o Quantum yield of excited atom with only fluorescence transition possible, no units.

Y Quantum yield of excited atom with both radiative and nonradiative transitions possible, no units.

z Reference direction for angular momentum, no units.

Z Nuclear charge, esu.

Z_{QA} Number of collisions between Q and A per second per volume, sec^{-1} cm^{-3}.

α Operator used to obtain $\bar{a}$, appropriate units.

ΔE Energy uncertainty of atom in state of energy E, erg.

$\Delta\lambda_B$ Wavelength half width of spectral line or band, cm.

Δt Time uncertainty of atom in state of energy E, sec.

$\Delta\nu'$ Frequency shift in radiation emitted or absorbed by atom because of thermal motion, sec^{-1}.

$\Delta\nu_B$ Frequency half width of spectral line or band, sec^{-1} (or Hz).

$\Delta\nu_C$ Collisional (Lorentz) half width of atomic spectral line, sec^{-1} (or Hz).

$\Delta\nu_D$ Doppler half width of atomic spectral line, sec^{-1} (or Hz).

$\Delta\nu_N$ Natural half width of atomic spectral line, sec^{-1} (or Hz).

$\Delta\nu_T$ Total half width of atomic spectral line, sec^{-1} (or Hz).

$\Delta\nu_u$ Natural half width of atomic level u, sec^{-1} (or Hz).

θ Angle between direction of motion of atom and line of sight of observer, angular degrees.

ν Any frequency of spectral line, sec^{-1} (or Hz).

ν_0 Frequency at line center, sec^{-1} (or Hz).

τ Mean radiative lifetime of excited atom with radiative transitions other than fluorescence transitions possible, sec.

τ^0 Mean radiative lifetime of excited atom with only fluorescence transition possible, sec.

$\tau^0_{1/2}$ Mean radiative half-life of excited atom with only fluorescence transition possible, sec.

τ_f Mean lifetime of excited atom with both radiative and non radiative transitions possible, sec.

$\sigma_c{}^2$ Collisional cross section for gaseous A and Q, cm^2.

ψ Wavefunction, no units.

ψ^* Conjugate wavefunction, no units.

4. REFERENCES

1. J. G. Calvert and J. N. Pitts, *Photochemistry*, Wiley, New York, 1966.
2. E. U. Condon and G. H. Shortley, *The Theory of Atomic Spectra*, Cambridge University Press, London, 1951.
3. R. M. Hochstrasser, *The Behavior of Electrons in Atoms*, Benjamin, New York, 1965.
4. R. Mavrodineanu and H. Boiteux, *Flame Spectroscopy*, Wiley, New York, 1965.
5. A. C. G. Mitchell and M. W. Zemansky, *Resonance Radiation and Excited Atoms*, Cambridge University Press, London, 1961.

B. PHYSICAL BASIS OF MOLECULAR LUMINESCENCE IN SOLUTION[1-6]

Molecular luminescence spectra are similar in origin to atomic fluorescence spectra. Both types of spectra arise from radiative electronic transitions between well-defined energy levels and are observed as discrete transitions rather than as continuous emissions (at all frequencies). Both types of spectra are best explained by quantum mechanical models of electronic structure rather than by classical models. There are, however, many aspects of molecular electronic spectra that are not explicable on the basis of the electronic structure of atoms. For example, atomic spectral lines appear as very narrow bands (i.e., of the order of 0.01–0.1 Å in wavelength units or of order of 0.1–1 cm^{-1}). Molecular electronic spectra, on the other hand, appear as bands, sometimes as much as several thousands of cm^{-1} wide (order of hundreds to thousands of ångstroms wide in wavelength units). The *emission spectrum* of a *single atomic species contains many lines* due to several radiative transitions. In contrast, molecular luminescence spectra almost *never contain more than two bands corresponding* to one *molecular species*. Furthermore, atomic spectra studied in the gas phase do not show pronounced matrix effects* when flames or furnaces are used as sample cells. Molecular electronic spectra, however, are usually studied in condensed phases, where matrix effects bear strongly upon the positions of spectral bands as well as their intensities. Consequently, those aspects of molecular electronic structure and the dynamics of molecular energy levels which account for the electronic spectra, particularly the luminescence spectra, of molecules are considered now.

1. MOLECULAR QUANTUM MECHANICS

If the Schrödinger equation is difficult to solve exactly for a many-electron atom, it is even more so for a molecule. The Hamiltonian operator for a molecule contains nucleus-nucleus repulsions as well as electron-electron repulsions and nucleus-electron attractions. Again, approximations to the true molecular wavefunctions must be used in order to calculate properties derived from molecular, electronic structure. In the case of molecules, however, the central field approximation (hydrogenic orbitals) that served as a reasonable basis for atomic electronic structure is no longer valid. The principal source of this complication is the interaction between nuclei, which results in molecular vibration and rotation. In the absence of molecular vibration, it might be possible to assume as a first approximation that a molecular electron moves in an average electric field produced by all the nuclei at fixed positions; then perhaps electronic repulsion could be included as a second approximation. These assumptions could serve as a basis for constructing molecular wavefunctions from atomic wavefunctions or from the wavefunctions for electrons in potential wells.

a. THE BORN-OPPENHEIMER APPROXIMATION

It is not possible to completely neglect molecular vibration and still obtain reasonable approximate molecular wavefunctions. However, since the nucleus-nucleus terms, the nucleus-electron terms, and the electron-electron terms in the molecular Hamiltonian all appear independently and are added in a linear fashion, it is possible to write the molecular Hamiltonian H_m as the sum of a nuclear Hamiltonian H_m, which is a function of only the nuclear coordinates and an electronic Hamiltonian H_e, which is a function of both the nuclear and electronic coordinates, that is,

$$H_m = H_n + H_e \qquad \text{(II-B-1)}$$

* Matrix effects refer to influence of concomitants on the measured spectral signal.

If it is assumed that the molecular wavefunction ψ_m (rotation is here considered a special case of vibration) is a product of the nuclear wavefunction ψ_n, which is a function of only the nuclear coordinates and an electronic wavefunction ψ_e, which is a function of the nuclear and electronic coordinates (i.e., $\psi_m = \psi_n\psi_e$), then

$$H_m\psi_m = E_m\psi_m = (H_n + H_e)\psi_n\psi_e = E_n\psi_n\psi_e + E_e\psi_n\psi_e \qquad \text{(II-B-2)}$$

where E_m is the total molecular energy, E_n is the vibrational energy, and E_e is the electronic energy. It follows from equation II-B-2 that

$$E_m = E_n + E_e \qquad \text{(II-B-3)}$$

and thus E_n and E_e can be obtained by separate application of H_n and H_e, respectively, to ψ_m (refer to equation II-A-1). To discuss the electronic states of molecules, it is wise to bear in mind that for the lowest vibrational states $E_n \ll E_e$, so that $E_m \approx E_e$ and to a good approximation, the molecular energy level is a *pure electronic level*. This situation occurs at very low temperatures. At higher temperatures, higher vibrational levels, corresponding to a particular electronic state, are populated so that the molecular energy level is more appropriately called a *vibronic level* rather than an *electronic level*. The validity of the simple additivity of E_e and E_n to give E_m is based on the separability of ψ_m into ψ_n and ψ_e (the Born-Oppenheimer approximation). The Born-Oppenheimer approximation works well when there is appreciable separation between the highest vibronic levels belonging to a particular electronic state and the lowest vibronic levels of the next higher electronic state. When there is proximity or overlap between neighboring electronic states through their vibrational levels, coupling of the two electronic states occurs through the vibrational levels, resulting in breakdown of the Born-Oppenheimer approximation. As we show later, the failure of the Born-Oppenheimer approximation is vital to the existence of molecular luminescence involving forbidden transitions. When it does work, the Born-Oppenheimer approximation succeeds in greatly simplifying our understanding of electronic transitions between molecular energy levels.

b. APPROXIMATE MOLECULAR ELECTRONIC WAVEFUNCTIONS AND ENERGY LEVELS

Because it is possible to obtain molecular energy levels, at least approximately, from electronic wavefunctions, it remains to be answered how the approximate molecular electronic wavefunctions will be chosen. Two methods are currently popular: LCAOMO (the *linear combination of atomic orbitals [and] molecular orbital* and FEMO (*free-electron molecular orbital*).

(1) *LCAOMO Method.* In the LCAOMO method, molecular orbitals are constructed from linear combinations of the atomic electronic orbitals be-

longing to the atoms comprising the molecule. The energies of the molecular orbitals are then minimized with respect to the linear combination coefficients. Solutions of the proper linear equations yield the energies of the molecular orbitals and the linear combination coefficients. From the latter, charge densities at the atoms and between the atoms (bond orders) can be calculated, whereas from the former the stability and spectral features can be calculated (approximately). Current LCAOMO calculations require digital computers for solutions, include electronic repulsions, and reiterate the calculation of parameters until self-consistency is achieved.

(2) *FEMO Method.* The FEMO method is based upon the treatment of molecules as if they were "boxes" in which the electrons are free to move (i.e., there is no constraining potential) as long as they remain within the confines of the box (the molecular dimensions). Of course, the uncertainty principle reminds us that the dimensions of the box are defined on the basis of a *most probable extension* rather than in the rigid sense of dimensions. Electrons occupying these *boxes* are presumed to be in *stationary states* when they fulfill certain requirements which give them the properties of standing waves. For *linear boxes* (the model for *linear polyenes*), stationary states correspond to integral numbers of half-wavelengths; and for *circular boxes* (the model for *aromatic hydrocarbons*), the stationary states correspond to integral numbers of whole wavelengths, analogous to the de Broglie model of the hydrogen atom.

The solution of the Schrödinger equation for a particle in a one-dimensional box is of the form

$$\Psi_{\pm k} = \left(\frac{1}{\sqrt{2\pi}}\right) e^{\pm ikx} \tag{II-B-4}$$

where k is an angular momentum quantum number and can have values $k = 0, 1, 2, 3, \ldots$. For a *linear one-dimensional box*, the energy levels are given by

$$E_k = \frac{k^2h^2}{8ma^2} \tag{II-B-5}$$

where a is the length of the box and m is the electronic mass. For a *circular one-dimensional box* (particle on a ring), the energy levels are given by

$$E_k = \frac{k^2h^2}{2ma^2} \tag{II-B-6}$$

where a is now the circumference of the box. Because it is predominantly aromatic hydrocarbons and their derivatives that exhibit molecular luminescence, equation II-B-6 is most relevant to the present discussion.

The interpretation of k, in equation II-B-6, as an angular momentum quantum number implies that there will also be a k_z quantum number corresponding to the clockwise or counterclockwise revolution of the electron and, hence, opposite orientations of the angular momentum vector. For any value of k, k_z can be $\pm k$. For each value of k except $k = 0$, there are thus two values of k_z and hence two orbitals. Each E_k of equation II-B-6 is therefore doubly degenerate, except for $E_k = 0$. As in the case of the hydrogenic orbitals, electron repulsion has been neglected, so that the approximate energy levels and orbitals derived by the FEMO method are, up to this point, applicable only to molecules in which a single valence shell electron moves in a perfectly symmetrical field produced by the atomic nuclei and the inner electrons. Although this physical situation is never realized, the FEMO method, with some refinements, is later shown to be of use in classifying the electronic spectra of aromatic hydrocarbons.

2. ELECTRONIC STRUCTURE OF MOLECULES

The strong interactions between atomic electrons to form chemical bonds are due, almost entirely, to the electrons and orbitals comprising the partially filled outer shells of atoms (the valence electrons). To a very good approximation, the electrons occupying the filled inner shells of atoms belonging to molecules are localized upon the atoms from which they originated and contribute only very weakly, through their repulsive properties, to molecular electronic structure. Because the valence electrons are responsible for the electronic spectra of molecules in the visible and ultraviolet regions of the spectrum, our consideration of molecular electronic structure is confined to those features which arise from the valence electrons and orbitals of molecules.

a. ELECTRONIC STATES—LCAOMO METHOD

In the LCAOMO approximation, a chemical bond (occupied molecular orbital) may be thought to originate from the overlap of atomic orbitals. The geometry of the overlap is used to classify the type of chemical bonding, and the filling of the molecular orbital is governed by the Pauli exclusion principle. When degenerate molecular orbitals are formed, the Hund rules are employed to determine the order of filling of the orbitals.

(1) *σ Orbitals.* The overlap of two atomic orbitals along the line joining the nuclei of the bonded atoms is said to result in a σ orbital. A σ orbital can accommodate two electrons, in accordance with the Pauli exclusion principle. The distribution of charge in a σ bond is strongly localized between the two bonded atoms. Although each atom participating in the σ bond contributes one atomic orbital to the formation of the σ orbital, the two electrons occupy-

ing the σ orbital may originate one from each atom or both from the same atom. In the former case, the σ bond is called a *covalent bond*, and in the latter case, it is called a *coordinate covalent bond*.

If the two atoms joined by the covalent bond exert unequal attractions on the electron pair comprising the σ bond, the electron pair will spend more time near one atom than the other. In this case, the more strongly attracting atom is said to be more *electronegative* than the other, and the bond is said to be a *polar covalent bond*. In some cases, (e.g., the alkali halides), the difference in electronegativity of two covalently bonded atoms is so great that the electron pair is effectively localized on the more electronegative atom. In this extreme case of polar covalent bonding, the bond is said to be *electrovalent*. Because the electronic charge in a σ bond is localized along the line between two atoms, electronic repulsion and the exclusion principle prevent the formation of more than one σ bond between any two atoms in a molecule. Electrons engaged in σ bonding are usually bound very tightly by the molecule. Consequently, a great deal of energy is required to promote these electrons to vacant molecular orbitals. This means that molecular electronic spectra involving σ electron transitions occur well into the vacuum ultraviolet region and are not of interest in conventional luminescence spectroscopy, which is concerned with the region between the near-ultraviolet and the near-infrared spectra (i.e., 2000–10,000 Å or 50,000–10,000 cm^{-1}).

(2) *π Orbitals.* The overlap of two atomic orbitals at right angles to the line joining the nuclei of the bonded atoms is said to result in a π orbital. Pi bonding is a weaker interaction than σ bonding and consequently it usually occurs secondarily to σ bonding. The formation of π bonds always involves atomic p and/or d orbitals but never s orbitals. In a π bond, the distribution of electronic charge is concentrated parallel to and away from the concentration of σ electronic charge, and in accordance with the spatial orientations of the component atomic orbitals. Although σ electrons are strongly localized between the atoms they bind, π electrons, not being concentrated immediately between the parent atoms, are freer to move within the molecule and are frequently distributed over several atoms. If several atoms are σ bonded in series and each has a p or d orbital with the proper spatial orientation to form a π bond with the others (as opposed to the formation of an alternating series of two-atom π bonds), a set of π orbitals is formed whose members are spread over the entire series of atoms. These π orbitals are said to be *delocalized*, and electrons are assigned to them in accordance with the exclusion principle and the Hund rules. In some cyclic organic molecules, π delocalization extends over the entire molecule; *these compounds are said to be aromatic,* and they are the molecules of primary interest in molecular luminescence spectroscopy.*

---

* Aromatic compounds are those containing conjugated double bonds.

Because π electrons are not concentrated between the bonded atoms they are not as tightly bound as σ electrons. Hence their electronic spectra occur at lower frequencies than do σ electron spectra. For molecules containing isolated π bonds, the transitions involving π electrons are still in the vacuum ultraviolet region or at the limit of the near-ultraviolet region. Molecules containing delocalized π electrons usually have π electron spectra in the near-ultraviolet region, whereas the superdelocalized π systems, the aromatic molecules, have π electron spectra ranging from the near ultraviolet for small molecules to the near infrared for large ones.

(3) *Nonbonded Electrons.* In organic molecules, which are our primary interest here, both σ bonding and π bonding occur as the result of spin pairing of two electrons, one arising from each atom, in a vacant low energy molecular orbital. Enough σ and π bonds encompass any atom in a molecule to permit any electrons that were unpaired in the valence shell of the isolated atom to engage in bonding.* In some atoms (e.g., nitrogen), however, there are electrons in the valence shell which are already paired. The orbitals occupied by these electrons are already filled, although the valence shell, in the isolated atom, is not. These electrons are unavailable for conventional covalent bonding and yet have energies comparable to other electrons in the same shell. Consequently, they are called *nonbonding* or n electrons. When the atom is engaged in bonding, the valence shell electrons involved in σ bonding drop in energy, well below the energy of the n electrons. The energy of the electrons involved in π bonding (if any) usually drops below the energy of the n electrons but not quite as much as in the case of the σ electrons. Because the n electrons are higher in energy than either the σ or π electrons, they must be considered as potential contributors to the spectral features of molecules possessing them. The n electrons originate from *pure atomic orbitals*, but the difference in repulsion experienced by these electrons as a result of the difference in electronic environment of a molecular atom compared with that of an isolated atom implies that the n orbitals will have more of the properties of localized molecular orbitals than of atomic orbitals. This is indeed found to be the case. The bonding geometries of atoms in molecules are influenced at least as strongly by the number of nonbonded electron pairs as by the number of bonded electron pairs around them.

(4) *Coordination Complexes.* Although nonbonded electron pairs in molecules do not enter into covalent bonding in the usual sense, they may be transferred into vacant molecular orbitals in suitable acceptor molecules. This results in the formation of an *intermolecular charge-transfer complex* (or coordination complex) in which the bond formed between the nonbonded electron pair donor and the acceptor is said to be a coordinate covalent bond.

---

* We are neglecting free radicals in this statement.

Coordinate covalent bonds are usually of the σ type, but certain molecular species also have the ability to form π type coordination complexes. Bronsted basicity is one well-known type of coordinate covalent bond formation. In this case, the Brønsted base donates a pair of nonbonded electrons to a vacant $1s$ orbital of a hydrogen ion to form the conjugate acid. The σ bond formed between the base and the hydrogen ion results in the loss of identity of the nonbonded pair on the base. The formation of coordination complexes has great significance in the interpretation of spectra of compounds having nonbonded electron pairs. This subject is discussed at length in a later section.

(5) π *Approximation.* In molecules containing π electrons and n electrons, spectral features and chemical reactivity are due almost exclusively to these electrons. Consequently, it has been customary (although not necessarily quantitatively accurate), to neglect σ electron structure in considerations of spectra and reactivity. This approach, called the π approximation, greatly simplifies the description of molecules to which it applies because we now say that the entirety of the molecular electronic structure consists of the occupied π orbitals—at lowest energy, the occupied n orbitals—slightly higher in energy than the π orbitals, and the unoccupied π^* orbitals (to which n or π electrons are promoted for the electronically excited states of the molecule). The molecules in which we are interested here almost always contain extensively delocalized, cyclic π systems and may or may not contain n electrons. Consequently, the electronic structure of these types of molecules is now considered within the framework of the FEMO model and the simplifications introduced by the π approximation.

b. ELECTRONIC STATES—FEMO METHOD

The FEMO model of aromatic π systems has energy levels given by equation II-A-6. The larger is the π system (i.e., the molecule), the smaller are the values of k^2/a^2, and hence the closer is the spacing of the energy levels E_k.

(1) *Platt System—^1A Term.* The FEMO model of π electronic structure, with some modifications, forms the basis of the Platt system of nomenclature of energy levels and spectral transitions for aromatic molecules. According to the FEMO model, the ground state configuration of any stable aromatic molecule has $(4r + 2)$ π electrons, where r is the value of k for the highest occupied orbital. All π electrons are paired in the ground state so that the multiplicity of the ground state configuration is always a singlet state. Because for each occupied energy level E_k there are as many electrons, with orbital angular momentum given by $k_z = k$ as electrons with orbital angular momentum given by $k_z = k$, the net orbital angular momentum of the ground state configuration is always zero; thus there is only one term (or one energy level) for the ground state configuration. Platt calls this term ^1A.

(2) *Platt System—L and B Terms.* The lowest excited states of an aromatic molecule are derived from the configuration in which one electron has been promoted from the highest occupied orbital ($k = r$) in the ground state of the molecule to the lowest unoccupied orbital ($k = r + 1$) in the ground state of the molecule. This leaves the molecule with three electrons having $k = r$ and one with $k = r + 1$. The contributions to the orbital angular momentum from all electrons in orbitals with $k > r$ are still zero. For the orbitals with $k = r$, there are two possible contributions to the total orbital angular momentum quantum number, $k_r = 2(-r) + r = -r$ and $k_r = 2(r) - r = r$. Likewise, for the orbitals with $k = r + 1$ there are two possible contributions to the total orbital angular momentum quantum number $k_{r+1} = -(r + 1)$ and $k_{r+1} = r + 1$. There are thus four possible values of the total orbital angular momentum quantum number K of the excited molecule: $K = (2r + 1)$, 1, -1, and $-(2r + 1)$.

In keeping with the wave mechanical properties of the molecule, each state with $K = \pm (2r + 1)$ has $2r + 1$ modes (lines in the molecular plane where $\Psi'_M = 0$). Each state with $K = \pm 1$ has one node. One state with $K = \pm (2r + 1)$ has the nodes passing through the bonds; Platt calls this state L_a. The nodes of the other state pass through the atoms (i.e., there is zero electron density at the atoms); Platt calls this the L_b state. In a similar fashion, the B states (for $K = \pm 1$) are labeled B_a and B_b. In the absence of electronic repulsions, it might be expected that all these states, which arise from the same configuration, would have the same energy. However, in a manner analogous to term splitting in atoms, electronic repulsions split the r^3, $(r + 1)^1$ configuration into two terms L and B. According to Hund's second rule, the L term, which has the greater angular momentum, lies at lower energy than the B term. If the nuclear field (and the σ electron field) were truly axially symmetrical, as the *particle on a ring model* implies, the L term and the B term would each be doubly degenerate; that is, there would be two states at the same energy with $K = 2r + 1$ and $K = -(2r = 1)$ and two with $K = 1$ and $K = -1$. Actually the π electrons in aromatic molecules move in an essentially periodic potential which splits the L term into two states L_a and L_b. The degeneracy of the B state is also removed in aromatic compounds. The relative energies of L_a and L_b depend on the size of the aromatic system and substituents thereon. This dependence is due to the generally greater sensitivity of the energy of the L_a state to the chemical environment.

To a first approximation, the π energy (energy due to all π electrons) of a molecule in a given state is altered by a substituent on the ith atom of the ring according to

$$\Delta E_\pi = q_i \Delta\alpha \qquad\qquad\qquad \text{(II-B-7)}$$

where ΔE_π is the difference in π energy between the substituted and unsubstituted molecules, q_i is the charge density at the substituted atom i, and $\Delta\alpha$ is the change in the Coulomb integral (energy of attraction of atom i for each π electron) of atom i on the aromatic ring, on substitution. Because the nodes of the wavefunction corresponding to the L_b state pass through the atoms, the charge densities q_i at each atom are zero, and hence ΔE_π is zero. For the L_a state, ΔE_π is nonzero. Actually ΔE_π for substitution of an L_b state is zero only for certain highly symmetrical aromatic hydrocarbons. For aromatic molecules containing odd numbers of carbon atoms, substituents, or heteroatoms, the nodes of the L_b wavefunctions do not pass exactly through the atoms but are slightly displaced; in this case, ΔE_π is nonzero but is still compared with ΔE_π for the L_a state. *Hence the greater sensitivity of the energy of the L_a to substitution in the aromatic ring.* The B_b transitions generally occur at lower energies than the B_a transitions.

(3) *Platt System—Multiplicity of L and B States.* In the B_a, B_b, L_a, and L_b states, the $k = r + 1$ orbitals are each singly occupied. The Pauli exclusion principle no longer requires that these electrons have opposite spins as they now differ in their orbital angular momentum quantum numbers. Hence to each state there are two possible *multiplicities*, singlet and triplet. The appropriate designations are now ^1B_a, ^1B_b, ^3B_a, ^3B_b, ^1L_a, ^3L_a, ^1L_b, and ^3L_b. Because the unpaired electrons of a triplet state are farther apart than those of a singlet state of the same orbital configuration, the energy of each triplet state is lower than the energy of the corresponding singlet state (i.e., $E_{3Bb} < E_{1Bb}$, $E_{3La} < E_{1La}$, $E_{3Lb} < E_{1Lb}$). In larger aromatic systems (those in which the highest occupied orbital in the ^1A state has $k \geq 3$), transitions arising from states formed by the promotion of an electron for the $k = r$ to the $k = r + 2$ orbital (giving rise to C and M bands for $\Delta K = 2$ and $2r + 2$, respectively) move into the near-ultraviolet region. The treatment of these bands follows similar arguments to that for B and L bands and is not discussed further here.

(4) *Platt System—n Electrons.* The FEMO model just discussed has no provision for the occurrence of n electrons in aromatic molecules or for their involvement in molecular spectra, but it is a simple enough matter to include them by introducing extra filled orbitals, slightly higher in energy than the highest occupied π orbitals. Because n orbitals are frequently orthogonal to π orbitals in the same molecule, the same classification system applied to $\pi - \pi^*$ excited states is not applied to $n - \pi^*$ excited states. The $n - \pi^*$ excited states are produced by promoting an electron from an n orbital to vacant π orbitals, with the lowest vacant π orbital denoted by Platt as ^1W or ^3W, depending on whether the $n - \pi^*$ state is a singlet or a triplet state. Higher $n - \pi^*$ transitions are obscured by the $\pi - \pi^*$ spectrum and are not

normally observed, but they may fall into the optical region under consideration here.

(5) *Influence of Vibrational States.* As we proceed to a consideration of electronic transitions, it is worthwhile to keep in mind that each electronic state, ground or excited, is "accompanied" by a manifold of vibrational (and rotational) states. The populations of the vibrational states for any electronic state are governed, to a fair approximation, by the corrected Maxwell-Boltzmann statistics. That is, if n_j is the number of molecules in the jth vibrational state of a given electronic state and n_T is the total number of molecules in that electronic state, then

$$n_j = \frac{n_T g_j \exp\left(-\varepsilon_j/kT\right)}{\Sigma g_i \exp\left(-\varepsilon_i/kT\right)} \tag{II-B-8}$$

where g_i is the degeneracy and ε_i the energy of the ith vibrational state, k is the Boltzmann constant, and T is the absolute temperature. At very low temperatures, only the lowest vibrational states are appreciably populated; but at higher temperatures, the most probable distribution of vibrational energies may favor a higher vibrational state. Also, to a fair approximation, the vibrational wavefunctions for molecules can be obtained (at least for the lower vibrational states) from the solution of the Schrödinger equation for an *anharmonic oscillator.* The importance of this treatment is that it predicts that for the lowest vibrational state, the probability distribution function $\Psi_n^*\Psi_n$ of the system has a maximum at the center of the vibration (the point of maximum vibrational velocity). The higher vibrational states have maximum probabilities ($\Psi_n^*\Psi_n$) near the turning points (where vibrational velocity is smallest). The classical theory of anharmonic vibrations predicts maximum probabilities near the turning points for all vibrational states.

The populations and the probability distributions of vibronic states are of great consequence to the shapes of molecular electronic spectral bands.

3. ELECTRONIC TRANSITIONS IN MOLECULES

Although we are primarily concerned with molecular photoluminescence spectra, all molecular electronic spectra, whether absorption or luminescence, always originate as the result of the absorption of visible or ultraviolet light by molecular species in their ground states.* Consequently, a full understanding of luminescence processes can be obtained only by considering the nature of the absorption of light, and the history of the excited states produced by absorption in addition to the luminescence processes themselves.

* An exception to this statement exists; that is, molecular emission as a result of collisional activation (or chemical reaction) of molecules in flames of analytical use for some elements.

With this in mind, we discuss the principal features of electronic absorption spectra and the interconversion of electronically excited states of molecules before considering the luminescence processes in detail.

a. ELECTRONIC ABSORPTION SPECTRA

The absorption of visible or ultraviolet light by a molecule is accompanied by the promotion of an electron from an occupied orbital in the ground state electronic configuration to an orbital that was previously unoccupied. The absorption process thereby leaves the molecule in an electronically excited state. In the smaller aromatic molecules in solution, having no n electrons, three transitions are normally observed in the visible and near-ultraviolet regions. These absorptions correspond to transition of the molecule from the ^1A ground state to the ^1L_a, ^1L_b, and ^1B_b states respectively. The transitions from the ^1A state to the ^3L_a, ^3L_b, and ^3B_b states are highly forbidden by the spin selection rule ($\Delta S = 0$) and therefore do not appear to a measurable extent in the absorption spectrum. In the Platt nomenclature system, transitions from the ^1A state to the ^1L_a, ^1L_b, and ^1B_b states are called the ^1L_a, ^1L_b, and ^1B_b transitions, respectively (see Figure II-B-1).

If the aromatic ring also contains n electrons, the lower energy $^1(n - \pi^*)$ transitions will appear in the absorption spectrum in addition to the ^1L_a, ^1L_b, and ^1B_b transitions. The transitions to the $^3(n - \pi^*)$ states from the ground state are also highly spin forbidden and consequently are too weak to appear in the absorption spectrum.

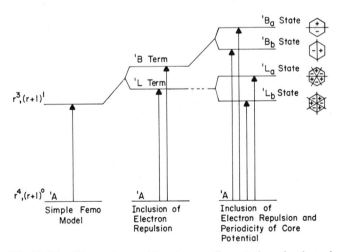

Fig. II-B-1. Electronic transitions in a small aromatic molecule as described by the Platt perimeter model in successive stages of refinement.

Electronic absorption bonds are characterized by their *positions* in the electromagnetic spectrum, their *intensity probabilities*, and their *shapes*. Each of these factors is considered in turn.

(1) *Positions of Molecular Electronic Absorption Bands.* (a) *Selection of Band Type.* The frequencies at which molecular electronic absorption bands occur are determined by the energy separations between the ground state and the various excited states. In an aromatic compound, both the ^1L_a and ^1L_b bands arise from transitions that occur between the ground state and two states of high angular momentum. The ^1B_b transition arises from a transition between the ground state and an excited state of low angular momentum. By Hund's second rule, the higher angular momentum states must be lower in energy than the lower angular momentum states. Hence the ^1L_a and ^1L_b states are *lower* in *energy* than the ^1B_b state and the ^1L_a and ^1L_b transitions thus take place at lower frequencies than does the ^1B_b transition. The relative positions of the ^1L_a and ^1L_b transitions cannot be predicted by the simple FEMO theory but may be determined from the effects of substituents on the spectra of the aromatic ring. The π energy of the ^1L_a state is more sensitive to external perturbations (because of its higher charge densities on the atoms of the ring) than is that of the ^1L_b state. Consequently, the band that shifts most as the result of substitution is generally assigned the ^1L_a designation [see Section II-B-2-b-(2)].

(b) *Position of* $^1(n-\pi^*)$ *and* $^1(\pi-\pi^*)$ *Bands.* Aromatic molecules having n electrons on atoms participating in the aromatic structure may show $^1(n-\pi^*)$ transitions at lower frequencies than any of the $^1(\pi-\pi^*)$ transitions. Such transitional characteristics are due to the n orbitals in the ground state of the molecule, which are higher in energy than the occupied π orbitals (see Section II-B-2-a-3). Actually, the generalization that n orbitals are higher in energy than occupied π orbitals and that $^1(n-\pi^*)$ transitions always occur at lower frequencies than $^1(\pi-\pi^*)$ transitions in the same molecule, needs some qualification. This generalization would be acceptable if molecular absorption spectra were studied only in the gas phase or in nonpolar solvents, where secondary bonding interactions are minimal.

In condensed phases, however, the solvents are very often polar. The positive end of a solvent dipole exerts an attractive force on n electrons in a solute molecule, thereby lowering their potential energy (hydrogen bonding is a common example of this). This has the effect of lowering the energy of the n orbital relative to that of the π orbitals. The π orbitals, being more symmetrically bound by the molecule as a whole, do not experience this electrostatic interaction to a considerable extent. *The excited $n-\pi^*$ and $\pi-\pi^*$ states, however, are frequently considerably more polar than is the ground state of the molecule*; thus interactions of the excited states of the aromatic molecule with a polar solvent tend to lower the energies of the excited states. In molecules having n electrons, the effect of a polar solvent is generally more pronounced on the energy of the ground state n orbitals than it is upon the n orbital excited states. Consequently, the more polar the solvent, the greater is the difference in energy between the highest occupied n orbital in the ground state and the excited states of the molecule.

On the other hand, since the ground state π orbitals are almost unaffected by solvent polarity, the greater the polarity of the solvent, the smaller the energy difference between the highest occupied π orbital of the ground state and the excited states of the molecule. Hence in polar solvents, $^1(n - \pi^*)$ *transitions are shifted to higher frequencies (blue shifted), (whereas* $^1(\pi - \pi^*)$ *transitions are shifted to lower frequencies (red shifted).* If the solvent becomes polar enough, the lowest $^1(\pi - \pi^*)$ transition may actually overtake the $^1(n - \pi^*)$ transition and become the lowest energy band in the spectrum.

In the extreme case of solvent–n electron interaction, a coordinate covalent σ bond is formed, and the energy of the $^1(n - \pi^*)$ transition—which is now really a $^1(\sigma - \pi^*)$ transition—is so great that the absorption does not appear in the visible or near-ultraviolet region but rather in the vacuum ultraviolet.

(c) Effect of Substituents in Aromatic Rings on Band Position. The effects of substituents in aromatic rings on the positions of absorption bands cannot be predicted, to any degree of accuracy, by the FEMO theory alone. For our purposes, it will suffice to say that a perturbation treatment can be applied to the FEMO theory in which substituents alter the π and n orbital energies and therefore alter the energies of the electronic states of the unsubstituted aromatic molecules. The effects of these perturbations are best seen in the shifting of the ^1L_a transition that occurs in the substituted ring relative to the position of the ^1L_a transition in the unsubstituted ring.

(d) Effect of Size of Aromatic Ring on Band Position. The effects of the size of the aromatic system on the positions of the absorption bands can be predicted qualitatively by the FEMO model. As the number of atoms in the π system increases, so does the circumference a of the circular box representing the molecule in the FEMO theory. The energy separation of the lowest excited configuration and the ground configuration of the aromatic molecule are given by

$$\Delta E_{r \to r+1} = \left(\frac{h^2}{2m}\right)\left(\frac{(r+1)^2 - r^2}{a^2}\right) = \left(\frac{h^2}{2m}\right)\left(\frac{(2r+1)}{a^2}\right) \qquad \text{(II-B-9)}$$

Because a^2 increases more rapidly than does $2r + 1$ for increasingly larger aromatic molecules, $\Delta E_{r \to r+1}$ *decreases with increasing molecular size. Consequently, the larger the aromatic molecule the lower the frequency of any given absorption band.* Because molecular electronic spectra are not pure electronic spectra, but rather occur from transitions between various vibrational states as well as electronic states, we must consider the intensity probabilities of the vibronic transitions if we are to understand the observed intensities of the spectral bands. Because the vibrational structure of electronic transitions gives spectral bands their form, the intensity probabilities and shapes of these bands are discussed together.

(2) *Intensity Probabilities and Shapes of Absorption Bands.* (a) *Transition Moment Expression.* The intensity of a vibronic transition from the ground state Ψ'_{mg} to an excited state Ψ'_{me} is determined by the transition moment integral $M_{eg} = \int_{-\infty}^{\infty} \Psi'_{me} er \Psi'_{mg} \, dT$, where e is the electronic charge and r is the operator corresponding to the vector along whose direction charge is displaced in the transition (an electronic transition must be accompanied by a change in electric dipole moment of the molecule). Because the overall state wavefunctions are functions of electronic coordinates, nuclear coordinates, and spin coordinates (in the Born-Oppenheimer approximation), the overall wavefunctions are given by

$$\Psi'_{mg} = \phi_g X_g \sigma_g \tag{II-B-10}$$

and

$$\Psi'_{me} = \phi_e X_e \sigma_e \tag{II-B-11}$$

where ϕ_g and ϕ_e are the electronic wavefunctions, X_g and X_e are the vibrational wavefunctions, and σ_g and σ_e are the spin wavefunctions for ground g and excited e, states, respectively.

The transition moment M_{eg} can be found by integrating separately over the electronic coordinates, the nuclear coordinates, and the spin coordinates, respectively, to obtain

$$M_{eg} = \int \phi_e er \phi_g \, d\tau_e \int X_e X_g \, d\tau_v \int \sigma_e \sigma_g \, d\tau s \tag{II-B-12}$$

where $d\tau_e$, $d\tau_v$, and $d\tau s$ are the volume elements in electronic coordinate space, vibrational coordinate space, and spin coordinate space, respectively. Note that the electronic dipole moment operator er appears only in the electronic term because it is assumed to be a function of only the electronic coordinates. If *any of the three integrals* in equation II-B-12 is *zero*, the entire expression for M will vanish, and the *transition will be forbidden.*

(b) *Symmetry-Forbidden Transitions.* The electronic part of the transition moment will be zero if the parities of ϕ_e and ϕ_g are identical. The transition is then said to be electronically forbidden. For electronically allowed transitions, the *selection rule* $\Delta K = 1$ *must be obeyed.* This means that, in the Platt model for aromatic spectra, the ^1L_a and ^1L_b bands with $\Delta K = 2r + 1$ $(r \geq 1)$

are electronically forbidden, whereas the ^1B_b band, having $\Delta K = 1$, is electronically allowed. The appearance of the ^1L_a and ^1L_b bands at all in the spectra of aromatic compounds is due to the *mixing of vibronic states* with the pure electronic states (i.e., the Born-Oppenheimer approximation is not rigorously obeyed). The ^1B_b band is always observed to be much more intense than either the ^1L_a or the ^1L_b bands.

Another important feature governing the intensities of electronic transitions in aromatic molecules is the *spatial orientations of the orbitals between which transitions occur*. Orbitals that do not have high probability densities (high orbital overlaps) in the same regions of space are likely to have small values of the transition moment integral and may, therefore, be at least partially forbidden. In heterocyclic molecules like pyridine and quinoline, the ground state n orbitals are directed into the molecular plane, whereas the π^* orbitals are directed perpendicular to the molecular plane. Consequently, there is poor overlap between n and π^* orbitals, and thus $^1(n - \pi^*)$ transitions are somewhat forbidden and appear weakly, compared with $\pi - \pi^*$ transitions. Another important example is provided by the relative intensities of the ^1L_a and ^1L_b bands. The ^1A ground state has appreciable probability density at the atoms of an aromatic molecule as does the ^1L_a state. The ^1L_b state, on the other hand, has nodes at or near the atoms. Consequently, there is much better overlap between the ^1A state and the ^1L_a state than between the ^1A and the ^1L_b states. *The ^1L_a band is therefore more intense* than the ^1L_b band. *Transitions that are forbidden either because $\Delta K \neq 1$ or because of poor orbital overlap are said to be symmetry forbidden.*

(c) *Spin-Forbidden Transitions.* The transition moment M may also vanish if the parity of the product of the spin function (σ_g and σ_e) is odd. If $\sigma_g = \sigma_e$, $\sigma_g\sigma_e$ will have even parity; otherwise the parity of $\sigma_g\sigma_e$ will be odd. Hence allowed transitions occur only between the same spin states. This gives rise to the selection rule $\Delta S = 0$, for allowed transitions. The spin selection rule is obeyed much more rigorously in aromatic molecules than is the symmetry selection rule. In fact, for absorption spectra, $\Delta S = 0$ is violated only under the influence of perturbations such as heavy atoms and paramagnetic species, which encourage spin-orbital coupling and spin-spin interactions, respectively. Consequently, for ordinary aromatic absorption spectra, which arise from singlet ground states, only those transitions are observed, which lead to the production of singlet excited states.

(d) *Franck-Condon Principle—Coarse Structure of Electronic Spectra.* Electronic transitions in molecules occur very quickly. The time required for a molecule to pass from the ground state to an electronically excited state upon absorption of a quantum of light is of the order of 10^{-15} sec. In contrast, vibrational transitions that require rearrangement of the positions of the atomic nuclei take about 10^{-12} sec. This is the basis of the *Franck-Condon principle*, which dictates that it is reasonable to assume that *electronic transitions occur without change of position of the nuclei*. If the populations of the various vibrational levels of the ground electronic state —the separations between the vibrational levels in the ground and excited electronic states, respectively—and the variations of potential energy with nuclear configuration of the ground and excited electronic states, respectively, are known for a given molecular species, the Franck-Condon principle is extremely helpful in predicting which vibrational transitions accompany the electronic transition from ground to excited state and the intensity with which each vibronic transition occurs. This can, in turn, be used to account for the shape and structure of absorption bands.

Qualitatively, there are three possible orientations of the variation of potential energy with nuclear configuration of the electronically excited state relative to that of the ground electronic state: that for which the ground state equilibrium nuclear configuration is the same as that of the electronically excited state, that for which the ground state equilibrium nuclear configuration is more compressed than that of the electronically excited state, and that for which the excited state equilibrium nuclear configuration is more compressed than that of the ground state.

Before considering each case separately, we should recall that the vibrational states (for any electronic state) where $v = 0$ (v is the vibrational quantum number from the energy levels of an anharmonic oscillator) have their maximum probability density at the equilibrium nuclear configuration. Vibrational states with $v < 0$ have maximum probability densities near the extrema of the vibrations. The greater the value of v, the greater is the probability of finding the oscillator near the extrema of the vibration. These results are in accordance with classical mechanics, which dictates that a vibrating system moves slowest, and therefore spends more time, near the vibrational extrema. The result that the vibrator spends most of its time near the center of the vibration (where it is moving fastest) for $v = 0$ is purely a quantum mechanical result, since classical theory predicts that there is no vibrational energy at all in this state. Because the intensity of a vibronic transition depends vibrationally on $\int X_e X_g \, d\tau_v$, the most intense vibronic transitions occur when X_e and X_g (or the probability densities $X_e{}^2$ and $X_g{}^2$) are appreciable in the same region of space (i.e., when there is strong overlap between X_e and X_g at the nuclear configuration at which transition occurs).

For the case of identical or nearly identical equilibrium nuclear configurations of ground and excited electronic states, which is true of many aromatic molecules, there is strong overlap between the $v = 0$ wavefunctions of ground and excited states. If the variations of potential energy with nuclear configuration are similar in ground and excited states, there is generally strong overlap between vibronic states having the same value of v. Consequently, in this case, the 0–0, 1–1, 2–2, and so on, vibronic transitions will be most intense. The 0–1, 1–2, 2–3, and so on, transitions will also appear, but less intensely than the transitions between states of identical v (if the equilibrium nuclear configurations are exactly identical). For transitions with $\Delta v = 2$ or more, overlap is progressively weaker as Δv increases, and these are forbidden or partially forbidden transitions. Since the $v = 0$ level of the ground state is most densely populated at very low temperatures, the spectral position of the 0–0 band is located at the maximum of the entire absorption band. At higher temperatures (e.g., room temperature) however, vibronic levels of the ground electronic state with $v < 0$ are more densely populated than that for $v = 0$, and thus the absorption band maximum is shifted away from that of the 0–0 band. It should be noted that the 0–0 band is the lowest energy vibronic transition that can occur in the absorption band. The 0–0 *band* therefore *defines the low frequency* (or *high wavelength*) *limit* of the absorption band. At higher temperatures, the $v = 0$ level of the ground electronic state may be so sparsely populated, relative to levels with $v < 0$ that the 0–0 band may not measurably appear in the absorption spectrum; also, radiation of longer wavelength than the 0–0 band will be absorbed. Figure II-B-2 schematically illustrates the Franck-Condon principle.

For our purposes, it is sufficient to say that in the molecules where the equilibrium nuclear configuration of the ground electronic state is appreciably different from that of the electronically excited state, the most intense vibronic transitions are not those for which $\Delta v = 0$. Very often the most intense transitions correspond to large values of Δv. Because this corresponds to the production of highly excited vibrational states, it is possible for a molecule of this type to dissociate within the lifetime of a single vibration. In this case, the absorption spectrum appears weakly as a discrete spectrum (well-defined bands) at lower energies, while showing strong continuous absorption at higher energies.

Experimentally, the intensity of absorption at a particular frequency is given by the Beer-Lambert law

$$\Phi_t = \Phi_o 10^{-\varepsilon Cl} \tag{II-B-13}$$

where Φ_o is the radiant flux of the incident light source and Φ_t is the radiant flux transmitted through the sample at the frequency of absorption assuming monochromatic radiation is used. The concentration of absorbers is given by C, the thickness of the sample by l, and the molar extinction coefficient, a constant for a particular species absorbing at a given frequency, by ε. Actually the radiant flux of absorbed light is given by $\Phi_o - \Phi_t$, although the logarithm of Φ_o/Φ_t, which is called the absorbance A, is the quantity of practical value in analytical chemistry because it is related linearly to the *absorber concentration*.

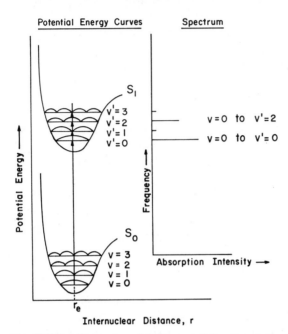

Fig. II-B-2. Illustration of influence of the Franck-Condon principle (for a diatomic molecule) in determining the shape of absorption band produced by electronic excitation from a state represented by potential energy curve S_o to a state represented by potential energy curve S_1 for the case where S_o and S_1 are similar in shape and equilibrium nuclear configurations. The shaded areas represent vibrational probability densities calculated from the integrals of the squares of the vibrational wavefunctions corresponding to $v = 0$, $v = 1$, and so on. It is assumed that all absorptions originate from the $v = 0$ level of the ground electronic state. If the equilibrium nuclear configurations differ, then the most probable Franck-Condon transitions will differ from the ones shown in the figure (maximal overlap of the electronic transition with vibrational probability occurs for levels other than $v' = 0$ and $v' = 2$).

For a given absorption band, ε varies with frequency, having a maximum value ε_{max} at the frequency corresponding to the absorption band maximum. The value of ε_{max} is roughly proportional to the band intensity. A better measure of the overall band intensity, especially if the band is asymmetric, is the oscillator strength f, which is given by

$$f = 4.32 \times 10^{-9} \int \varepsilon_{\bar{\nu}} \, d\bar{\nu} \qquad \text{(II-B-14)}$$

where $\bar{\nu}$ (the wavenumber) is the frequency divided by the velocity of light and ε is expressed as a function of $\bar{\nu}$.

Quantum mechanically, f is given by

$$f = \frac{8\pi^2 c \bar{\nu}_{av} m}{3he^2} \, |M|^2 = 4.70 \times 10^{29} \bar{\nu}_{av} |M|^2 \qquad \text{(II-B-15)}$$

where c is the velocity of light, h is the Planck constant, m and e are the electronic mass and charge, respectively, $\bar{\nu}_{av}$ is the average wavenumber of the absorption band, and $|M|$ is the magnitude of the quantum mechanically calculated transition moment, expressed in electrostatic units (esu). The oscillator strength is a dimensionless quantity and is of the order of unity for strongly allowed transitions. Equations II-B-14 and II-B-15 permit a comparison of experimental band intensities with those calculated by quantum mechanical methods.

b. PHYSICAL PROCESSES IN ELECTRONICALLY EXCITED STATES

The absorption of a quantum of light by a molecule leaves the molecule in one of its electronically excited states. Although the absorption process is extremely rapid (ca. 10^{-15} sec), the sequence of events that returns the excited molecule to its ground state is considerably slower, taking from 10^{-13} to several seconds. Return to the ground state via molecular luminescence is among the slowest of processes in electronically excited states, requiring from 10^{-9} to a few seconds. Consequently, other excited state processes always compete, with varying degrees of success, against molecular luminescence for depopulation of the various excited states. In this section, the sequences of events in the lifetimes of electronically excited states of molecules are discussed and their effects on the nature of luminescence from excited molecules in solution examined.

(1) *Deactivation of Electronically Excited Singlet State—Neglect of Triplet State.* (a) *Vibrational Relaxation.* Excitation of a molecule from its ground electronic state to an excited electronic state of the same multiplicity (an excited singlet state for organic molecules) often results in the production of a vibrationally excited state within the excited electronic state. The natural tendency of the vibrationally excited molecule is to dissipate its excess vibrational energy and thereby arrive at the lowest vibrational level belonging to the excited electronic state that is compatible with the ambient temperature. For simplicity, vibrational relaxation of this type is assumed to proceed to the lowest vibrational level ($v = 0$) of the electronically excited state, regardless of the solution temperature.

The process of *vibrational relaxation* within the excited state is very rapid, requiring less than 10^{-13} sec for completion. Because the vibrational levels of any electronic state are quantized (i.e., discrete), the loss of vibrational energy must proceed in a stepwise fashion and is governed by the selection rule

$\Delta v = 1$. The excess vibrational energy is dissipated in collisions with molecules in the environment (the solvent), each collision removing a quantum of vibrational energy. The less vibrationally excited the electronically excited molecule is, the fewer collisions are required to produce relaxation to the lowest vibronic state.

(b) *Internal Conversion.* Once an excited molecule has relaxed to the lowest vibrational level of the electronically excited state, it can lose excitation energy only by going to a lower electronic energy level. This can be accomplished in one of three ways. If the higher vibrational levels of the lower electronic state *overlap* the lower vibrational levels of the higher electronic state (i.e., if the nuclear configurations and energies of the two electronic states are identical during a low energy vibration of the upper electronic state and a high energy vibration of the lower electronic state), the upper and lower electronic states will be in a transient thermal equilibrium that will permit population of the lower electronic state. Vibrational relaxation of the lower electronic state then follows as before.

If the lower vibrational levels of the upper electronic state do not overlap the higher vibrational levels of the lower electronic state but are separated by a small gap (a few vibrational quanta wide), the molecule in the upper electronic state may still convert to the lower electronic state by a *tunneling* mechanism. *Tunneling, the transmission of particles through regions which are classically forbidden to the particles, is permitted quantum mechanically* because the wavefunction characterizing a particle does not vanish at distances away from the most probable distribution; rather, it decreases asymptotically. Hence there is always some overlap of vibrational levels in different electronic states. The probability of tunneling occurring decreases as the difference in energy between the lower vibrational levels of the upper electronic state and the upper vibrational levels of the lower electronic state increases.

(c) *Fluorescence.* If the energy separation of the upper and lower electronic states is relatively great, so that direct vibrational coupling is impossible and tunneling is improbable, a third process competes with the other two for direct depopulation of the upper electronic state. The third process consists of a *drop in energy of the excited molecule* from that of the lowest vibrational level of the upper electronic state to that of any one of a number of vibrational levels of the lower electronic state. The excess energy in this case is released as a photon of visible or ultraviolet light whose frequency depends on the difference in energy between the lowest vibrational level of the upper electronic state and the vibrational level of the lower electronic state to which the radiative transition occurs. Following radiative transition, the molecule undergoes vibrational relaxation to the lowest vibrational level of the lower electronic state.

The electronic relaxation processes that employ direct vibrational coupling and quantum mechanical tunneling between electronic states are called *internal conversion processes*. Those which employ *radiative transitions* between electronic states are called *luminescence processes*. More specifically, if the *upper and lower electronic states have the same multiplicity, the radiative transition is called fluorescence.*

(d) *Factors Influencing Deactivation of Excited Singlet States.* Whether an excited molecule employs internal conversion or fluorescence to pass from a higher electronic state to a lower one depends on the difference in energy between the upper and lower states and on the number of vibronic states associated with each electronic state. The latter property is important because it determines the probability of vibrational overlap of the electronic states. If a molecule has a large number of modes of vibration in the lower state, then the lower vibrational modes of the higher electronic state are able to excite higher vibrational modes of the lower electronic state with high probability. Hence vibrational dissipation of excitation energy and thus internal conversion will be favored. It is for this reason that fluorescence is rarely exhibited by aliphatic molecules and others that do not have rigid molecular skeletons and thus possess many vibrational degrees of freedom. In addition, the absorption bands of aliphatic molecules appear at high frequencies, and so photodecomposition is often a problem. The aromatic molecules, with their *rigid ring structures*, are the class of compounds that most frequently shows fluorescence.

In most molecules, the excited electronic states are much closer together than are the ground state and the lowest excited state. Consequently, efficient internal conversion almost always precludes the possibility of fluorescence arising as the result of electronic transition between states other than the ground state and the lowest excited state of the same multiplicity as the ground state. Exceptions to this are azulene and some of its derivatives, which fluoresce from the second excited singlet state to the ground state.

The internal conversion process is very rapid, taking about 10^{-12} sec. The average lifetime of the lowest excited singlet state, however, is of the order of 10^{-8} sec. Therefore, even if a molecule cannot pass efficiently from its lowest excited singlet state to the ground state by internal conversion, it may undergo other processes during the lifetime of the lowest excited singlet state and these processes may compete with fluorescence.

(2) *Deactivation of Electronically Excited Singlet State—Consideration of Triplet State.* (a) *Intersystem Crossing.* One of the processes that competes with fluorescence for deactivation of the lowest excited states of aromatic molecules is intersystem crossing or change of multiplicity (spin) in the excited state. The $\Delta S = 0$ selection rule for electronic transitions may break down, in varying

degrees, under the influences of perturbations caused by heavy atoms or paramagnetic ions either in the optical molecule or in the immediate environment (the solvent). Even in the absence of perturbing species some spin-orbital coupling takes place in almost all atoms, and therefore some degree of intersystem crossing should occur in the excited states of essentially all molecules. The extent to which intersystem crossing can occur from higher excited singlet states to form higher triplet states is severely limited by the great rate of the internal conversion process compared with all other excited state processes. Even if some intersystem crossing results in population of higher excited triplet states, internal conversion among the triplet states is so rapid that only the lowest singlet-triplet state can be observed by ordinary spectroscopic means. (Absorption spectra of excited triplet states of molecules have been studied using flash photolysis techniques, however.)

(b) *Deactivation of Triplet State.* The lowest excited triplet state of a given molecular species can return to the ground state (singlet) by either a radiative or a nonradiative transition. Because a change in multiplicity is involved in the transition from excited triplet to ground singlet state, the transition is forbidden and will occur only with low probability. Spin-forbidden transitions of either the radiative or nonradiative type occur with 10^{-3} to 10^{-6} of the rates of their spin-allowed counterparts. Because the probabilities of spin-forbidden transitions are low, the lifetimes of triplet states are high, lasting from 10^{-5} to several seconds.

(c) *Phosphorescence.* The radiative transition from an upper electronic state to a lower electronic state of different multiplicity is called *phosphorescence*. Like fluorescence, phosphorescence is most apt to occur in molecules having *rigid molecular skeletons* and large energy separations between the lowest excited triplet state and the ground singlet state. Because triplet states are so long-lived, chemical and physical processes in solution, as well as internal conversion, compete effectively with phosphorescence for depopulation of the lowest excited triplet state. In fact, except for the shortest-lived phosphorescences, collisional deactivation by solvent molecules, quenching by paramagnetic species (e.g., oxygen), photochemical reactions, and certain other processes preclude the observation of phosphorescence in fluid media at room temperature. Rather, phosphorescence is normally studied in glasses at liquid nitrogen temperature or in solution in very viscous liquids, where collisional processes cannot completely depopulate the triplet state.

(d) *Quenching Processes.* The fluorescence or phosphorescence of molecules may be diminished or even eliminated because of the depopulation of the excited states responsible for luminescence by *weak complex* formation of either the ground or excited state of the luminescing species with certain extraneous species in solution. The mechanism of luminescence quenching is believed to be *reversible electron transfer* between members of the complex, because the quenching abilities of certain substances correlate well with their ionization potentials.

Two types of quenching are distinguished. In *dynamic (diffusional) quenching,* the complex forms between the luminescing species in the excited state and the quencher. Subsequent intermolecular electron transfer or photochemical reaction removes the luminescer from the excited state, thereby diminishing the luminescence yield. In *static quenching,* the complex forms in the ground state between the quencher and the potential luminescer. Excitation of the complex to the luminescing state is only partially effective in producing luminescers, since dissociation of the excited complex may be slow and may, moreover, have efficient competition from intermolecular electron exchange or photoreaction.

(*e*) *Photochemical Processes.* Electronically excited molecules are usually highly polar and therefore very reactive chemical species. Thus we should expect chemical reactions involving these species (photochemical reactions) to compete with luminescence for depopulation of the electronically excited states.

For excited singlet states, the lifetimes (10^{-8} sec) are short compared with the times required for most chemical reactions, especially those in which the *concentrations of reactants are low and steric effects are rate determining.* There are, however, several classes of reactions which occur rapidly compared with the mean lifetime for fluorescence. Among these are isomerization, photoionization, photodissociation, and acid–base equilibrium in the electronically excited singlet state.

For example, the lowest excited singlet state of the β-naphtholate anion and of the undissociated β-naphthol exist in prototropic equilibrium with a pK of 2.8 compared with the pK of ~ 9.5 in the ground state; as a result, fluorescences from both the naphthol and naphtholate ions are seen at pH 2.8, whereas only the absorption of the naphthol occurs at this pH. The rate constant for dissociation of β-naphthol in the lowest excited singlet state is about 10^8 sec^{-1}, and the pseudo-first-order rate constant for protonation of β-naphtholate is of the order of 10^{11} (H$^+$)^1 sec^{-1}. The effect of the prototropic equilibrium in the excited singlet state is to shift the emission maximum to a different frequency upon adding base to a solution of the fluorescing acid. The shift in the fluorescence may or may not be accompanied by quenching. But in any event, it must be reckoned with if analytical data based on the fluorescence are to be properly interpreted.

The long lifetimes of excited triplet states make it more likely that these states will participate in photochemical reaction than return to the ground state by phosphorescence. This is the basis of photochemistry, and this is also why the *triplet state is more important than the singlet state in most photochemical reactions*. Phosphorescence is of analytical importance only in *rigid glasses or extremely viscous media*, where photochemical reactions do not provide a serious interference.

C. CHARACTERISTICS OF ELECTRONIC LUMINESCENCE SPECTRA

(1) *Number of Emission Bands.* Molecular electronic emission spectra are similar to molecular electronic absorption spectra in that they originate from radiative transitions between molecular electronic states. However, they differ in that the absorptive transitions originate from the ground electronic state and terminate in any of a number of excited electronic states, whereas the emissive transitions originate in the lowest excited singlet or triplet state and terminate in the ground state. Consequently, a given molecular species may show a number of bands in its absorption spectrum, b*ut it will never have more than one fluorescence band and one phosphorescence band.* In solution at room temperature, only the fluorescence band can be seen; in rigid glasses, however, both the phosphorescence and the fluorescence may appear.

The appearance of more than one emission band in the solution spectrum or *more than two bands* in the spectrum taken in a rigid matrix implies that *extraneous species are present in the sample*. Most often, extra emission bands are caused by impurities in the sample.

Frequently, however, chemical reactions in the excited singlet or triplet state result in the production of additional luminescing species. *Acid–base reactions, isomerization, and excimer formation* are three common examples of chemical reactions in electronically excited states, which result in the appearance of extra emission bands.

Acid–base reactions in electronically excited states yield conjugate acids and bases, each of which may be luminescent. The variations of the intensities of emission bands with changing pH value usually indicate the occurrence of acid–base equilibria in the state from which luminescence arises.

Isomerization of a molecular species in the electronically excited state is exemplified by salicylic acid. The fluorescence spectrum of salicylic acid displays two bands. It has been shown that one of these bands is due to the neutral salicylic acid molecule and the other is due to a zwitterion formed in the excited state by intramolecular proton transfer. The zwitterion, which has the carboxyl group doubly protonated and the hydroxyl group deprotonated, is not appreciably present in the ground state. In polar solvents (alcohol), the fluorescence of the neutral species is eliminated by the complete conversion of the salicylic acid to the zwitterion in the excited state.

The appearance in the spectrum of a molecular species of an extra emission band that increases in intensity with increasing ground state (absorber) concentration while the other band decreases with increasing absorber concentration indicates the occurrence of *excimer (excited state polymer) formation*. The excimer fluorescence (the one that increases with increasing absorber concentration) generally takes place at lower frequencies than the monomer fluorescence, although there is no detectable absorption spectrum due to the polymer.

Molecular electronic emission spectra, like their absorption counterparts, are characterized by the *positions, intensities, and shapes* of their bands. These features are determined in part by the processes that compete with luminescence for deactivation of the excited states and reflected through the lifetimes of the excited states and the luminescence yields. Furthermore, the features of molecular emission spectra exhibit some definite relationships to the absorption spectra of the same molecular species. Some of the important details of these features are considered now.

(2) *Positions of Molecular Emission Bands.* (*a*) *Relative Band Positions.* The frequencies at which molecular electronic luminescence bands appear are proportional to the differences in energy between the ground state and the lowest excited singlet and triplet states for fluorescence and phosphorescence, respectively. Hund's first rule dictates that, for two electronic states with the same orbital properties but different spin properties, the state with the higher multiplicity is at lower energy. Consequently, the lowest excited triplet state is at lower energy than the lowest excited singlet state, for any molecular species, so that *phosphorescence always occurs at lower frequencies than fluorescence for the same molecule.*

(*b*) *Mirror Image Characteristic of Fluorescence-Absorption Spectra.* Although fluorescence from the lowest excited singlet state can occur subsequent to excitation to the ^1B_b, ^1L_a, or ^1L_b state of an aromatic compound, the absorption band corresponding to absorption to the lowest excited singlet state (^1L_a or ^1L_b in an aromatic compound containing no *n* electrons) has a special relation to the fluorescence band. If it is assumed that the aromatic molecule has a similar equilibrium nuclear configuration in the lowest excited singlet state to that in the ground state and that the spacings between vibronic levels are similar in ground and excited states, then the absorption and fluorescence bands will be *mirror images* of each other, with the 0–0 vibronic band defining the high frequency limit of the fluorescence band and the low frequency limit of the absorption band.

Moreover, the most probable distribution among the vibronic levels at any given temperature will be the same for ground and excited states; thus the absorption and fluorescence band maxima will be equally displaced from the position of the 0–0 band. Therefore, the position of the 0–0 band can be determined at any temperature by averaging the frequencies of the maxima of the fluorescence and lowest singlet absorption bands. The assumption of similar configurations and vibronic spacings in ground and lowest excited singlet states is apparently reasonable for aromatic molecules because most show a mirror image relationship between their fluorescence and lowest excited singlet absorption bands. Although phosphorescence bands do not exhibit a mirror image relationship with singlet-singlet absorption bands, they do show such a relationship with singlet-triplet absorption bands of aromatic compounds which may, on rare occasion, be obtained by using very long cells to increase the length of the path of the exciting radiation. The latter approach increases the absorbance of the sample and may render the singlet-triplet absorption measurable in spite of its low value of molar extinction coefficient.

That the 0–0 band of fluorescence is the high frequency limit of fluorescence and the low frequency limit of the absorption band, corresponding to the same electronic states, implies that the fluorescence maximum always occurs at lower frequency than the corresponding absorption maximum. This is found to be universally true, except, of course, for the case in which the 0–0 band is the fluorescence maximum and the absorption maximum. Even with this exception, the bulk of the fluorescence band occurs to the red (at lower frequencies) with respect to the absorption band. The difference in frequency between the absorption and fluorescence maximum is called the Stokes shift of the fluorescence.

(c) *Sequence of Events in Activation and Deactivation of Molecules.** The reason for the Stokes shift in fluorescence spectra involves the solvent environments and vibrational structures of the ground and lowest excited singlet states and the sequence of events in the absorption and emission processes. Absorption originates from a low vibronic level of the ground electronic state in which the molecule is surrounded by a stable solvent environment, is governed by the polarities of the ground electronic state and the solvent molecules. The absorption process terminates with the molecule arriving in any one of a number of possible vibronic levels of the excited state, still surrounded by the ground state equilibrium solvent cage. Within 10^{-12} sec, the electronically excited molecule relaxes vibrationally to a low vibronic level of the excited state while the solvent cage reorients itself to achieve the state of lowest potential energy (equilibrium) compatible with the new molecular polarity (usually greater than that of the ground state). The vibrational and solvent cage relaxation processes are accompanied by the loss of thermal energy by the excited molecule. Emission then occurs from a low vibronic state of the elec-

---

* Refer to Figure II-B-3 for a schematic representation of processes occurring in activation and deactivation of molecules.

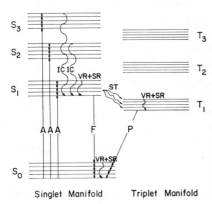

Fig. II-B-3. Electronic transitions in molecules: arrows with straight lines correspond to radiative processes, arrows with wavy lines correspond to nonradiative processes, and A = absorption, F = fluorescence, P = phosphorescence, IC = internal conversion, VR = vibrational relaxation, ST = singlet-triplet intersystem crossing, SR = solvent relation, S_x = singlet state (0 is ground, 1 is first excited, etc.), T_x = triplet state (1 is first excited, 2 is second excited, etc.).

tronically excited molecule, in the excited state equilibrium solvent cage, dropping the molecule into one of a number of possible vibronic levels of the ground electronic state, still surrounded by the excited state equilibrium solvent cage. The molecule finally arrives in its original state after losing energy by vibrational relaxation and relaxation of the solvent cage to the ground state equilibrium configuration.

The energy of the absorption process is thus equal to the energy of the fluorescence process plus the energy associated with vibrational and solvent relaxation in ground and excited states and is therefore greater than the energy of the fluorescence process. Fluorescence, then, occurs at lower frequencies than absorption. For $^1(\pi - \pi^*)$ transitions, the red shifting of the fluorescence maximum increases with increasing solvent polarity. This increase in red shift is due to the greater polarity of the excited singlet state, which drops substantially in energy through electrostatic interactions with polar solvents (*solvent cage relaxation*), reducing the emission energy as the magnitude of the solvent-excited dipole interaction increases.

For $^1(n - \pi^*)$ fluorescence, which is rare and extremely weak when it does occur, the decrease in energy of the ground state accompanies electrostatic perturbation of the nonbonded pair by the solvent and is generally greater than the decrease in energy of the $^1(n - \pi^*)$ state because of solvent relaxation. Consequently, $^1(n - \pi^*)$ fluorescence is blue shifted with increasing solvent polarity. In the limit where the nonbonded pair enters into coordinate covalent bonding with the solvent, the energy of the n orbital, which is transformed into a σ orbital, drops well below the energy of the occupied π orbitals. As a result, $^1(n - \pi^*)$ transitions are eliminated from the visible and near-ultraviolet spectra and fluorescence, if any, will be of a $^1(\pi - \pi^*)$

nature. In nonpolar solvents, however, owing to the greater energy of the n orbital relative to that of the π orbitals, the lowest excited singlet state is $^1(n - \pi^*)$; thus if fluorescence occurs it must originate from this state.

If $^1(n - \pi^*)$ fluorescence is observed for aromatic molecules containing n electrons, it might be anticipated that $^3(n - \pi^*)$ phosphorescence would also be observed for these types of molecules. In fact, $^3(n - \pi^*)$ phosphorescence is extremely rare, even though intersystem crossing from both the lowest $^1(\pi - \pi^*)$ state and the $^1(n - \pi^*)$ state to the $^3(n - \pi^*)$ state are very efficient processes. The $^3(n - \pi^*)$ state fails to phosphoresce, even weakly, because the probability distributions of the n orbital and the π^* orbitals are directed in different planes (in nitrogen heterocyclics the n and π^* orbitals are perpendicular to one another), whereas the probability distributions of the π and π^* orbitals are pointing in the same direction. As a result, the difference in repulsion energy between the $^1(n - \pi^*)$ and the $^3(n - \pi^*)$ states is not as great as that between the $^1(\pi - \pi^*)$ and $^3(\pi - \pi^*)$ states, and therefore the splitting of the $^1(n - \pi^*)$ and $^3(n - \pi^*)$ states is smaller than that of the $^1(\pi - \pi^*)$ and $^3(\pi - \pi^*)$ states. The differences in splittings are such that, although the $^1(n - \pi^*)$ state lies below the $^1(\pi - \pi^*)$ state in energy, the $^3(\pi - \pi^*)$ state almost always lies below the $^3(n - \pi^*)$ state, and intersystem crossing to the $^3(n - \pi^*)$ state usually produces phosphorescence from the $^3(\pi - \pi^*)$ state.

(d) *Effect of Substituents on Luminescence Bands.*   The effect of substituents in aromatic rings on the frequencies of fluorescence and phosphorescence can be treated, only very approximately, by perturbations upon the FEMO model. The changes in π electron energy of ground and excited states may be calculated approximately from equation II-B-7, provided we know the charge density at the substituted atom (this usually requires LCAOMO calculations) and the change produced by substitution in Coulomb integral at the substituted atom (usually determined empirically from absorption spectra).

Because lowest excited singlet states of aromatic molecules usually have much greater dipole moments than the lowest excited triplet states of the same molecules, *substituents*, through their *inductive effects*, generally cause somewhat greater shifting of fluorescence bands than they do phosphorescence bands. Fluorescence bands corresponding to ^1L_a absorption bands should be especially sensitive to substituents in this regard. However, since fluorescence is usually ^1L_b in nature, in most small aromatics, this hypothesis has not been well studied.

Although it is impossible to predict exactly the effect of a given substituent on the frequency of emission of a luminescing molecule, some limited generalizations can be made. Substituents that show appreciable transfer of π electron charge to the aromatic ring in the ground state (e.g., electron releasing groups $-OH$, $-SH$, $-NH_2$) do even more so in the lowest excited state. These substituents produce a red shift of the luminescence relative to that of the unsubstituted molecule. Substituents that withdraw π charge from the aromatic ring in the ground state (e.g., electron withdrawing groups as

$$\overset{\text{O}}{\underset{\|}{}}$$

—C—) do likewise to a greater extent in the lowest excited state. No rigid rules are appropriate because these and other substituents, which may act as either π acceptors or donors (e.g., chlorine, bromine, iodine) depending on their electronegativities and the availability of vacant orbitals capable of interacting with the π system of the ring, may shift the luminescence to the red or blue, depending on the aromatic system.

(e) *Effect of Size of π Electron System on Luminescence Bands.* Increase in size of the aromatic systems increases the size of the FEMO "box" and thus compresses the π orbitals so that, as in the case of absorption spectra, *luminescence spectra tend toward lower frequencies as the size of the π system is increased.* This is illustrated by the polyacene series—benzene, naphthalene, anthracene, tetracene, and pentacene—which consist of one, two, three, four, and five linearly fused rings, respectively. The fluorescence of benzene occurs in the near-ultraviolet region; the fluorescence of naphthalene also appear in the near-ultraviolet region but at lower frequency than that of benzene. Anthracene, tetracene, and pentacene fluoresce in the blue, green, and red, respectively.

(3) *Intensity, Probabilities, and Shapes of Molecular Luminescence Bands.* (a) *Transition Moment Expressions.* Like absorption, the intensity of molecular luminescence is proportional to the square of a transition moment integral M_{ge}. For luminescence, M_{ge} is defined by

$$M_{ge} = \int \phi_g er \phi_e \, d\tau_e \int X_e X_g \, d\tau_v \int \sigma_e \sigma_g \, d\tau_s \qquad \text{(II-B-16)}$$

where all symbols have the same significance as in equation II-B-12. Because the same selection rules that are applicable to electronic absorption also apply to electronic emission, electronically allowed transitions have $\Delta K = 1$ and $\Delta S = 0$.

(b) *Requirements for Fluorescence and Phosphorescence Spectra.* Although molecular fluorescence and phosphorescence do not correspond to transitions for which $\Delta K = 1$ (the lowest excited states give rise to transitions corresponding to the L_a or L_b absorptive transitions and thus have $\Delta K = 2r + 1$, where r is the value of k for the highest occupied orbital in the ground state), the transition moment does not vanish because the Born-Oppenheimer approximation is not rigorously obeyed. Although the $\Delta S = 0$ selection rule is obeyed more rigorously than the orbital selection rules, the admixture of the vibrational wavefunctions $(X_e + X_g)$ with the spin wavefunctions $(\sigma_e + \sigma_g)$, as well as spin-orbital coupling, allow phosphorescence to occur.

In the case of molecular absorption, where the forbiddenness of singlet-triplet absorption (reflected through its low probability per unit time) resulted in few singlet-triplet absorptions compared with the number of allowed singlet-singlet absorptions, it is reasonable to consider the low intensity of the former relative to the latter. This is a consequence of the short time ($\sim 10^{-15}$ sec) in which both types of absorption processes occurred, which means that only a small fraction of all the ground state molecules absorb.

In molecular emission spectra, however, the lowest excited triplet state is not populated to any appreciable extent by singlet-triplet absorption. Rather, it is populated by an efficient intersystem crossing from the lowest excited singlet state, which may be considered to be a spin-forbidden internal conversion mechanism. Furthermore, the fluorescence and phosphorescence transitions, although very different in their probabilities per unit time (phosphorescence is less probable by a factor of about 10^6), are measured on a time scale (sec) that is long by comparison with the time required for all molecules in excited singlet and triplet states to fluoresce or phosphoresce. Consequently, the measured intensities of fluorescence and phosphorescence from a given species may be comparable; or the phosphorescence intensity may even be greater, if more than 50% intersystem crossing occurs. Therefore, the relative intensities of fluorescence and phosphorescence, measured with conventional laboratory equipment, do not reflect the forbiddenness of the phosphorescent transition. A better indication of this is given in the relative lifetimes of fluorescent and phosphorescent decay, the lifetime of phosphorescence being typically of the order of 10^6 times greater than that for fluorescence from the same species.

The much greater lifetime of phosphorescence relative to fluorescence provides a means of identifying the phosphorescent or fluorescent nature of molecular luminescence bands. If the source of excitation is turned off a luminescing sample and the luminescence disappears immediately, it must be fluorescence or very short lived phosphorescence. If the luminescence dies out gradually, it must be phosphorescence. There is, however, an interesting exception to the last statement.

　　(c) *Delayed Fluorescence.*　　Occasionally in rigid and viscous media an extra-long-lived luminescence band is observed in addition to the phosphorescence. The extraneous band is of higher frequency than the phosphorescence and, if the molecule also shows fluorescence, the frequency of the extraneous band coincides with the frequency of fluorescence and its lifetime is similar to that of the phosphorescence. This phenomenon, known as *delayed fluorescence*, is a result of the thermal excitation of a molecule in the lowest excited triplet state back to the lowest excited singlet state (i.e., reverse intersystem crossing) followed by fluorescence. In rigid media, the thermal excitation is accomplished by heating the sample, but in viscous media where there is some mobility, the collision of two triplet molecules results in the thermal excitation of one and the internal conversion of the other to the ground state. The latter process is known as triplet-triplet annihilation.

(d) *Franck-Condon Principle—Coarse Structure of Electronic Luminescence Spectra.* The intensities of vibronic luminescence bands and therefore the shapes of the molecular electronic emission band are governed by the Franck-Condon principle, which was discussed in connection with absorption spectra. In luminescing molecules, transitions originate from the lowest vibrational levels of the lowest electronically excited state. For molecules having identical equilibrium nuclear configurations in ground and excited states, vibronic transitions with $\Delta v = 0$ have the greatest probability, those with $\Delta v = 1$ being weaker and those with $\Delta v = 2$ being weaker still. The observed relative intensities of vibronic transitions are determined by the populations of the vibronic levels involved in each transition at the ambient temperature as well as the probability of transition due to the electronic structure of the molecular species. The populations of the vibronic levels are most responsible for the position of the emission band maximum. At very low temperatures, the levels with $v = 0$ are the heaviest populated in both ground and excited states and the 0–0 band is the emission band maximum. At higher temperatures, vibronic levels with $v > 0$ are the most populated, giving rise to a displacement of the emission band maximum from the 0–0 band to higher frequencies ("hot band"). The 0–0 band is the high frequency limit of the fluorescence or phosphorescence band.

If the equilibrium nuclear configurations of ground and excited states are slightly different, the most probable vibronic transitions will be $\Delta v = 1$, $\Delta v = 2$, or some other low value of Δv, depending on the amount of displacement between the equilibrium nuclear configurations of ground and excited states. At low temperatures, the emission band maximum will then correspond to the position of the 0–1 or 0–2 (etc.) band.

Because at low temperature, almost all transitions originate from the $v = 0$ level of the excited state (or the ground state for absorption), the number of vibronic transitions appearing in low temperature emission (or absorption) spectra is much lower than the number of transitions contributing to a band at room temperature. Low temperature spectral studies, therefore, are frequently used to obtain spectra with much better resolution of the vibrational structure than spectra obtained at room temperature. Also at low temperature and especially in rigid matrices, vibrational, diffusional, and collisional processes are retarded. As a result, internal conversion, diffusion limited quenching, and collisional deactivation compete with fluorescence to a much lower degree than at room temperature. Increased intensity of fluorescence and correspondingly increased sensitivity in fluorimetric analysis is, therefore, frequently achieved in low temperature studies.

Transitions arising between electronic states having very different equilibrium nuclear configurations favor vibronic transitions with large values of Δv. Emissions of this type are unusual because production of the excited state by absorption frequently leads to photodissociation or predissociation. The latter is a dissociative process caused by crossover of the highly excited molecule to a repulsive state of lower energy. Predissociation results in a continuous absorption spectrum over a portion of the wavelength range (see Figure II-B-4).

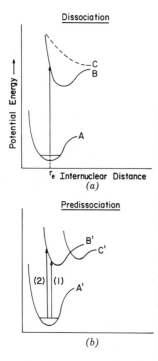

Fig. II-B-4. Representation of dissociation and predissociation (for a diatomic molecule. (*a*) Dissociation. Molecule is excited (state A) to such a high vibrational energy (state B) that during course of vibration, the internuclear distance approaches infinity and the molecule dissociates (a continuum absorption spectrum occurs for frequencies sufficient to produce dissociation). If the excited molecule has no minimum (state C), then every transition produces a continuum. (*b*) Predissociation. The molecule is excited (state A′) to an excited vibrational level in state B′ by process (1), which results in a normal absorption spectrum with vibration and rotational fine structures. However, in transition (2) the vibrational energy of the excited molecule is insufficient to result in dissociation but is sufficient for molecule to cross over to state C′; and during the vibration within state C′, the molecule dissociates. Thus in predissociation the molecular spectrum is normal except for a small segment, which exhibits a continuum nature.

(*e*) *Influence of Heavy Atoms, Paramagnetic Ions, and Heteroatoms on Luminescence Bands.* The intensity of molecular fluorescence may be substantially diminished or even completely quenched by the presence of *heavy atoms, paramagnetic ions, or heteroatoms* in the molecule or in the environment (solvent). Heavy atom substituents enhance quenching as a result of high spin-orbital coupling, which favors intersystem crossing. Heavy atoms in the solvent affect *spin-orbital coupling* in the solute because the solute contains a rather *tightly bound* (and strongly interacting) *solvent sheath*. The quenching of fluorescence by heavy atoms of intermediate size usually produces a substantial enhancement of phosphorescence due to the enrichment of the triplet state. Very heavy elements (e.g., Mercury, thallium, lead, and bismuth in chelates of aromatic ligands), however, quench fluorescence but do not appreciably increase phosphorescence yields and may actually decrease them. This quenching action appears to be caused by the nonplanar nature of heavy metal complexes and the compression of electronic energy levels with increasing nuclear charge, which results in an increase in internal conversion to the ground state from both the lowest excited singlet and triplet states.

Paramagnetic species (e.g., transition metal ions, molecular oxygen) exert a quenching effect on the fluorescence of aromatic compounds by forming *charge transfer complexes* with the potential fluorescers. The charge transfer state mixes the multiplicities of the singlet fluorescer and the high spin quencher in much the same way that spin-orbital coupling mixes singlet and triplet states of the same molecule. Transfer of the energy of excitation of the excited singlet state of the fluorescer to the paramagnetic species is rapid compared with the lifetime for fluorescence and results in efficient radiationless deactivation of the excited singlet state of the potential fluorescer. Whether the energy of excitation is disposed of as phosphorescence depends on the separation between the ground state and the new lowest excited state (either charge transfer or ligand field). If the lowest excited state is the *charge transfer state, phosphorescence may appear.* If the lowest excited state is a *ligand field state* (transition metal ion), *internal conversion or extremely weak phosphorescence is the rule.* This explains why transition metal complexes, even those which are diamagnetic, do not fluoresce and only occasionally phosphoresce.

The *presence* of a heteroatom in an aromatic molecule often induces a lowest excited singlet state that is $^1(n-\pi^*)$ in nature. This is, of course, not the case if the nonbonded electrons are involved in a coordinate covalent bond (e.g., a metal complex, or a protonated heteroatom) or if extremely strong solvation —hydrogen bonding—raises the energy of the $^1(n-\pi^*)$ transition above that for the lowest $(\pi-\pi^*)$ transition. If the lowest excited singlet state is $^1(n-\pi^*)$, fluorescence from this state, which is rare, is extremely weak; indeed, it is about 10^2 to 10^3 times lower in intensity than $^1(\pi-\pi^*)$ fluorescence. This is a consequence of the symmetry forbiddenness of transitions between the $^1(n-\pi^*)$ state and the ground state, which results in a long lifetime and allows intersystem crossing to compete effectively with fluorescence.

The rarity of $^1(n-\pi^*)$ fluorescence is also due to the efficiency of intersystem crossing between the $^1(n-\pi^*)$ state and both the $^3(n-\pi^*)$ and $^3(\pi-\pi^*)$ states. If the $^3(n-\pi^*)$ state is lowest in energy, phosphorescence is possible but very difficult to observe because of the combined symmetry forbiddenness and spin forbiddenness of the $^3(n-\pi^*)$ to ground state transition, which allows triplet-singlet internal conversion to compete effectively with phosphorescence. If the $^3(\pi-\pi^*)$ state is lowest in energy—as it often is, owing to the greater separation between $^1(\pi-\pi^*)$ and $^3(\pi-\pi^*)$ states than between $^1(n-\pi^*)$ and $^3(n-\pi^*)$ states—phosphorescence is likely. Therefore, the presence of a heteroatom in an aromatic molecule frequently enhances the intensity of phosphorescence and precludes fluorescence as well.

(4) *Mean Lifetimes of the Lowest Excited Singlet and Triplet States.* (*a*) *No Quenching Present and No Internal Conversion.* Let us assume that the excitation of a ground state species S_o leads to the production of an excited singlet species S_1 whose concentration under steady state conditions of excitation is $(S_1)_o$ and whose concentration after the exciting light is shut off is (S_1). In the hypothetical situation where S_1 returns to the ground state S_o by fluorescence with no processes competing with fluorescence for deactivation of the excited state, the rate of fluorescence is

$$-\frac{d(S_1)}{dt}=k_f(S_1) \tag{II-B-17}$$

where k_f is the first-order radiative rate constant for fluorescence; that is, k_f is the probability that an S_1 will fluoresce in unit time. Integration of equation II-B-17 yields

$$(S_1)=(S_1)_o\exp\,(-k_f t) \tag{II-B-18}$$

The inherent mean lifetime $\tau_F{}^o$ is defined as the time t required for (S_1) to fall to $1/e$ of its original value; in other words, $\tau_F{}^o=t$, when $(S_1)=(S_1)_o/e$. Consequently

$$\tau_F{}^o=\frac{1}{k_f} \tag{II-B-19}$$

If it is assumed that *all* the S_1 formed by excitation undergoes rapid inert-system crossing to the lowest excited triplet T_1, followed by phosphorescence of all the T_1 molecules, it is possible to define an inherent mean lifetime for phosphorescence $\tau_P{}^o$, as

$$\tau_P{}^o=\frac{1}{k_p} \tag{II-B-20}$$

where k_p is the first-order radiative rate constant for phosphorescence (i.e., k_p is the probability per unit time of phosphorescence.) The inherent mean lifetime for fluorescence can be calculated from the features of the absorption band corresponding to the transition between the ground and lowest excited singlet states, by means of the expression

$$\tau_F{}^0 = \frac{3.5 \times 10^8}{\bar{\nu}_{av}{}^{-2} \int \varepsilon_{\bar{\nu}} \, d\bar{\nu}} \tag{II-B-21}$$

where $\bar{\nu}_{av}$ is the average wavenumber of the band and $\varepsilon_{\bar{\nu}}$ is the molar absorptivity corresponding to the wavenumber $\bar{\nu}$. Equation II-B-21 is derived under the assumption that the intensity of the emission (luminescence) band is identical to that of the absorption band; that is, fluorescence is 100% efficient. If the absorption band corresponding to the lowest singlet-singlet transition is symmetrical, equation II-B-21 may be simplified to

$$\tau_F{}^0 = \frac{3.5 \times 10^8}{\bar{\nu}_{av}{}^2 \varepsilon_m \Delta\bar{\nu}_{1/2}} \tag{II-B-22}$$

where ε_m is the molar absorptivity at the band maximum and $\Delta\bar{\nu}_{1/2}$ is the width of the absorption band (wavenumbers) at half the maximum value of ε_m.

Expressions similar to equations II-B-21 and II-B-22 may be derived for the inherent mean phosphorescence lifetime if the singlet-triplet absorption spectrum can be taken to give the appropriate values of $\bar{\nu}_{av}$, ε_m and , $\bar{\nu}$. Thus $\tau_P{}^0$ is independent of the way in which the triplet state is populated. Because fluorescence and phosphorescence are the most time-consuming processes originating from the lowest excited singlet and triplet states, respectively, the inherent mean lifetimes for fluorescence and phosphorescence represent the maximum theoretical mean lifetimes for the lowest excited singlet and triplet states of a given molecule. Now for a given molecule, $k_f \gg k_p$, so that $\tau_p{}^0$ is always much greater than $\tau_F{}^0$.

(b) *No Quenching Present and Internal Conversion Present.* Actually, the inherent mean radiative lifetime is a purely theoretical concept. In molecules exhibiting fluorescence and phosphorescence, the radiative processes are always in competition with radiationless deactivation processes. In the absence of quenching species, fluorescence is still always in competition with internal conversion and sometimes with intersystem crossing for deactivation of the lowest excited singlet state. Phosphorescence is always in competition with internal conversion and sometimes with delayed fluorescence (reverse intersystem crossing) for depopulation of the lowest excited triplet state.

For the fluorescing species S_1, the rate of loss of S_1 with time is given by

$$-\frac{d(S_1)}{dt} = k_f(S_1) + K_{IC}(S_1) + k_{ST}(S_1) + \cdots = k_f(S_1) + \sum_i k_i(S_1) \tag{II-B-23}$$

where k_{IC} is the probability of internal conversion of S_1 to the ground state per unit time, k_{ST} is the probability of intersystem crossing to the lowest excited triplet per unit time, and $\sum_i k_i$ represents the sum of the probabilities of all unimolecular, radiationless pathways by which S_1 can be deactivated.

Integration of equation II-B-23 yields

$$(S_1) = (S_1)_o \exp -(k_f + \sum_i k_i)t \tag{II-B-24}$$

The mean lifetime of S_1 in this case is

$$\tau_F = \frac{1}{k_f + \sum_i k_i} \tag{II-B-25}$$

where τ_F is the value of t corresponding to $(S_1) = (S_1)_o/e$. Note that if $\sum_i k_i > 0$ (which it always is), then $\tau_F < \tau_F{}^o$. In general, the greater the number of processes (and the greater their rates) that compete with fluorescence for deactivation of S_1, the shorter will be the mean lifetime of S_1.

Similarly, for phosphorescence of the lowest excited triplet state T_1, it is possible to write

$$\tau_P = \frac{1}{k_p + \sum_i k_i} \tag{II-B-26}$$

where τ_P is the mean lifetime of the lowest excited triplet state when only unimolecular radiationless processes compete with phosphorescence for deactivation of T_1 and $\sum_i k_i'$ is the sum of the rate constants (probabilities) for all unimolecular radiationless processes competing with phosphorescence for deactivation of T_1; τ_P is the value of t when $(T_1) = (T_1)_o/e$. Normally $\tau_P > \tau_F$, but in some cases, where the number (and rates) of radiationless deactivating processes competing with phosphorescence is high and the number (and rates) of radiationless processes competing with fluorescence is low, τ_P and τ_F may be of the same order of magnitude. This would correspond to a very long-lived fluorescence and a very short-lived phosphorescence ($\tau_F \approx \tau_P$ 10^{-6} sec).

(c) *Influence of Solvent.* The mean lifetimes defined by equations II-B-25 and II-B-26 are the longest mean lifetimes that can be measured for the lowest excited singlet and triplet states, respectively, of a particular molecule in a given solvent. There is, however, considerable variation in the values of τ_F and τ_P determined in different solvents. Solvents containing heavy atoms (*external heavy atom effect*) tend to enhance singlet-triplet transitions, increasing the probabilities of intersystem crossing from the lowest excited singlet to the lowest excited triplet state, thereby increasing $\sum_i k_i$ and lowering τ_F. The external heavy atom effect also increases the probabilities of phosphorescence and internal conversion from T_1 to S_o. Because k_P and $\sum_i k_i'$ both increase in this case, τ_P is also decreased.

In highly polar solvents if the excited state is more polar than the ground state, the separation between the ground state and the excited $\pi - \pi^*$ state is decreased, thereby enhancing the probability of internal conversion to the ground state from both the lowest $^1(\pi - \pi^*)$ state and the lowest $^3(\pi - \pi^*)$ state. This results in a decrease in both τ_F and τ_P, although the effect on τ_F is more pronounced. Usually the *longest mean lifetimes are thus observed in nonpolar solvents of low molecular weight.*

(d) *Quenching Present and Internal Conversion Present.* If S_1 is involved in bimolecular reactions with a nonabsorbing solute Q which deactivates S_1 and is in much greater concentration than S_1, the rate of deactivation of S_1 is given by

$$-\frac{d(S_1)}{dt} = [k_f + \sum_i k_i + k_Q(Q)](S_1) \qquad \text{(II-B-27)}$$

where $k_Q(Q)$ is the pseudo-first-order rate constant for deactivation of S_1 by Q. For any number of quenchers Q_j with pseudo-first-order rate constants $K_{Q,j}(Q_i)$, equation II-B-27 can be generalized to

$$-\frac{d(S_1)}{dt} = [k_f + \sum_i k_i + \sum_j k_{Q,j}(Q_j)](S_1) \qquad \text{(II-B-28)}$$

which yields upon integration:

$$(S_1) = (S_1)_o \exp\left[-(k_f + \sum_i k_i + \sum_j k_{Q,j}(Q_j))t\right] \qquad \text{(II-B-29)}$$

provided the values of the individual Q_j's are sufficiently great that any loss of Q_j due to reaction with S_1 is negligible compared with the values of the (Q_j)'s—that is, $(S_1) \ll (Q_j)$. The mean lifetime of the lowest excited singlet state S_1 is then given by

$$\tau_F' = \frac{1}{k_f + \sum_i k_i + \sum_j k_{Q,j}(Q_j)} \qquad \text{(II-B-30)}$$

and the mean lifetime of T_1 by

$$\tau_P' = \frac{1}{k_p + \sum_i k_i' + \sum_j k_{Q,j}'(Q_j')} \qquad \text{(II-B-31)}$$

where the $K'_{Q,\,j}(Q_j')$ are the pseudo-first-order rate constants for deactivation of T_1 by the Q'_j species.

Equations II-B-19, II-B-20, II-B-25, II-B-26, II-B-30, and II-B-31 show that in general, the greater the number of processes contributing to the de-activation of the lowest excited singlet and triplet states and the greater the probabilities of these processes, the shorter the mean lifetimes of these elec-tronic states. In the next section, where quantum yields of molecular lumines-cence are considered, we see that the actual lifetimes of the electronically excited states play an important role in the quantum yields of the fluorescence and phosphorescence processes.

(5) *Quantum Yields of Fluorescence and Phosphorescence.* The quantum yield (relative quantum yield) of fluorescence is the fraction of excited molecules that return to the ground state by fluorescent emission. For any given mo-lecular species under given temperature and solution conditions, the quantum yield is a constant of the fluorescence process.

(a) *Fluorescence Quantum Yeild and No Quenching Present—Internal Conversion Present.* If the lowest excited singlet state of the fluorescing molecule is de-activated only by fluorescence and other unimolecular deactivating processes (intersystem crossing and internal conversion), the quantum efficiency of fluorescence is given by

$$Y_F = \frac{k_f \int_{t=0}^{t=\infty} (S_1)\, dt}{(S_1)_o} \tag{II-B-32}$$

Substitution of equation II-B-24 into equation II-B-32 yields

$$Y_F = \frac{k_f \int_{t=0}^{t=\infty} (S_1)_o \exp\left[-(k_f + \sum k_i)t\right]}{(S_1)_o} = \frac{k_f}{k_f + \sum_i k_i} \tag{II-B-33}$$

But from equation II-B-25

$$\tau_F = \frac{1}{k_f + \sum_i k_i} \tag{II-B-25}$$

so that

$$Y_F = k_f \tau_F \tag{II-B-34}$$

Or, in other words, Y_F is affected by the solvent in the same way τ_F is; thus the maximum values of Y_F are normally expected to occur in *solvents of low molecular weight and low polarity.*

(b) *Fluorescence Quantum Yield—Quenching Present and Internal Conversion Present.* Similarly, in the presence of high concentrations of extraneous species which deactivate the lowest singlet state, Y_F depends on the quencher Q concentration according to

$$Y_F' = k_f \tau_F' \qquad \text{(II-B-35)}$$

Because, in general, lowering of temperature tends to weaken vibrational modes of deactivation, the rates of internal conversion and intersystem crossing are diminished at low temperatures, thereby increasing the values of $\tau_F, \tau_F', Y_F,$ and Y_F'.

(c) *The Stern-Volmer Equation.* The validity of equation II-B-35 is based on the assumption that a pseudo-first-order treatment is applicable because the concentrations of quenching species are much greater than the concentration of the fluorescing species. There are many situations, however, in quenching species that seriously affect the quantum yield of a fluorescing species are present in concentrations comparable with that of the fluorescer. In this case, the integration of equation II-B-28 to obtain τ_F' is difficult because the (Q_j)'s are variable, When there is only *one quencher*, this problem can be solved by simple steady rate kinetics. In this instance, Y_F' is given by

$$Y_F' = \frac{1 - \gamma\alpha}{1 + \gamma k_D \tau_F^{\,o}(Q)} \qquad \text{(II-B-36)}$$

where (Q) is the quencher concentration, $\tau_F^{\,o}$ is the mean lifetime of the fluorescing species, k_D is the rate constant for complexation of the fluorescer in the excited state by Q, γ is the probability that the $S_1 Q$ complex will be deactivated rather than dissociate back to S_1 and Q, and α is the fraction of $S_o Q$ complex formed in the ground state that is excited by direct absorption. If complexation in the ground state is unimportant (i.e., $\alpha \to 0$), diffusional quenching is said to prevail and equation II-B-36 reduces to

$$Y_F' = \frac{1}{1 + \gamma k_D \tau_F^{\,o}(Q)} \qquad \text{(II-B-37)}$$

which is known as the *Stern-Volmer equation.* Because the number of molecules that show fluorescence is limited, quantitative fluorimetry may be extended in its range of application by evaluation of the specific quenching properties shown by many molecules which do not themselves fluoresce.

Another important aspect of quenching is relevant to analytical fluorimetry. Systematic errors can arise in analytical fluorimetry if the values of Y_F' are *not* the same in standard and unknown samples: careless sample handling or different workup procedures for standard and unknown samples may produce different quencher concentrations in the various test solutions. This would of course cause serious errors in the interpretation of fluorimetric data.

(*d*) *Phosphorescence Quantum Yields.* The foregoing arguments related to fluorescence quantum yields are also applicable to phosphorescence quantum yields. In the absence of quenchers,

$$Y_P = k_p \tau_P \qquad \text{(II-B-38)}$$

In the presence of a large excess of quencher concentrations

$$Y_P' = k_p \tau_P' \qquad \text{(II-B-39)}$$

and in the presence of a low concentration of quencher

$$Y_P' = \frac{1 - \gamma' \alpha}{1 + \gamma' k_D' \tau_P{}^o(Q)} \qquad \text{(II-B-40)}$$

where γ' is analogous to γ but is defined for the complex $T_1 Q$, and k_D' is the rate constant for complexation of the phosphorescent species by Q.

Quenching of phosphorescence, however, requires some explanation; since phosphorescence is studied in a rigid matrix, the excited species are not free to diffuse during the lifetime of the excited triplet state. Aside from the obvious case of the ground state of the phosphorescer that is complexed and thereby *statically quenched*, quenching of phosphorescence obeying a *Stern-Volmer type (diffusional) relationship* is possible. This is owing to the occurrence of *energy transfer* phenomena, which can take place with the donor (the phosphorescer) as much as 60 Å removed from the acceptor (the quencher).

d. POLARIZATION OF MOLECULAR LUMINESCENCE.

(1) *Polarization of Electronic Transitions.* Light of the type used to excite electronic transitions in molecules consists of traveling waves; each traveling wave has an electric field associated with it whose electric vector oscillates in a plane perpendicular to the direction of travel of the light wave. There is also a magnetic field associated with the light wave whose magnetic vector oscillates in a plane perpendicular to both the direction of travel and the plane of oscillation of the electric vector. It is predominantly the interaction of the electric vector of light with the electronic transition dipole moments associated with the absorbing molecules which changes the electronic dipole moment (i.e., the charge distribution) of the molecule and thereby produces molecular electronic excitation. Because the excitation transition consists of the interaction of two vectors, the orientation of the electric vector of the exciting light relative to the direction of the transition moment vector should be of paramount importance in determining the magnitude of interaction of the two vectors. It is, in fact, the square of the scalar projection of the transi-

tion moment on the electric field vector which is proportional to the number of photons absorbed N_{abs} in the transition

$$N_{abs} \propto (\mathbf{M} \cos \theta)^2 \tag{II-B-41}$$

(i.e., where θ is the angle between the transition moment vector $\mathbf{M}$ of the molecule, and the electric vector of the light; see Figure II-B-5).

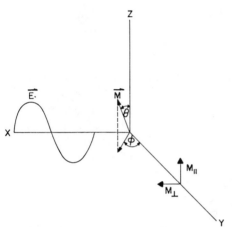

Fig. II-B-5. Relation of the electronic transition moment $\overrightarrow{M}$ of a randomly oriented molecule in a fixed (viscous) medium to the electric vector $\overrightarrow{E}$ of polarized light with which it is excited and the components of $\overrightarrow{M}$ which govern the intensities of fluorescent light measured when the polarizing and analyzing elements have their optical axes aligned parallel to each other ($I_{\parallel}$) and perpendicular to each other ($I_{\perp}$), respectively.

In ordinary light, there is a spherical distribution of electric vectors (of the different component waves) about the direction of travel of the waves in a particular direction. This is a result of the random orientations of the individual oscillators in the light source. Consequently, it is meaningless to speak of directional relationships between the molecular transition moments and the electric vectors of the exciting light waves. If, however, ordinary light is passed through a Nicol prism or other suitable polarizing device, all light waves are filtered out except those whose electric vectors oscillate in (ideally) one plane. In this case, we can show experimentally that the photons of the incident polarized light of a given frequency N_{abs} absorbed per second in producing an electronic transition in a molecule are jointly proportional to the square of the magnitude of the moment M belonging to the transition excited and the square of the cosine of the angle θ between the direction of the transition moment vector and the plane of polarization of the exciting light (i.e., the direction of the electric vector of the exciting light), and

$$N_{abs} \propto M^2 \cos^2 \theta \tag{II-B-42}$$

Now let us consider the emission of fluorescence or phosphorescence from a molecule that has absorbed polarized light. If the luminescence is observed both at right angles to the direction of travel of the exciting light and at right angles to the direction of polarization of the electric vector of the exciting light, the number of photons emitted per second as luminescence N_{lum} (intensity) is proportional to the square of the magnitude of the transition moment according to

$$m_{lum} \propto M^2 \qquad \text{(II-B-43)}$$

where M is the magnitude of a vector which may be resolved into components projected in the direction of observation (the direction of the electric vector of the exciting light) and the direction of travel of the exciting light. The latter two components are of interest to us here.

The component of $\mathbf{M}$ that is pointed in the same direction as (parallel to) the electric vector of the exciting light is hereafter referred to as $M_{\parallel}$, and that pointed in the direction of travel of the exciting light (perpendicular to the electric vector of the exciting light) is referred to as $M_{\perp}$. If θ is the angle between $\mathbf{M}$ and the direction of the electric vector of the exciting light, and if ϕ is the azimuthal angle $\mathbf{M}$ makes with the direction of observation, then

$$M_{\parallel} = M \cos \theta \qquad \text{(II-B-44)}$$

and

$$M_{\perp} = M \sin \theta \sin \phi \qquad \text{(II-B-45)}$$

The photons emitted per second with electric vectors parallel to and perpendicular to the electric vector of the exciting light are $N_{\parallel}$ and $N_{\perp}$, respectively, and are given by

$$N_{\parallel} \propto M_{\parallel}^2 = M^2 \cos^2 \theta \qquad \text{(II-B-46)}$$

and

$$N_{\perp} \propto M_{\perp}^2 = M^2 \sin^2 \theta \sin^2 \phi \qquad \text{(II-B-47)}$$

The polarization p of the emission is defined as

$$p = \frac{N_{\parallel} - N_{\perp}}{N_{\parallel} + N_{\perp}} \qquad \text{(II-B-48)}$$

Because the luminescence signals, measured with the polarizer in front of the source aligned parallel to the analyzer polarizer in front of the detector and with the two polarizers aligned perpendicular to each other, are proportional to $N_{\parallel}$ and $N_{\perp}$, respectively, p can be simply determined. The significance of p is that it is a measure of the extent to which the emitted light has its electric vector pointing in the same direction as that of the exciting light. The widest range of application of the polarization of molecular luminescence is obtained by plotting the polarization p of the emission maximum as a function of the wavelength or frequency of the exciting light. In this way, the polarization of the fluorescence excitation spectrum (fluorescence polarization spectrum) or the polarization of the phosphorescence excitation spectrum (phosphorescence polarization spectrum) is obtained.

Theoretically p can take on values from $+1$ to -1. Values of p approaching these theoretical extrema are sometimes observed in the emission polarization of single crystals and oriented polymers (i.e., samples with the molecules all lined up in the same direction). In these systems, the crystal absorption moments, which are arithmetic sums of the molecular moments, can be lined up with the direction of the electric vector of the exciting light. For absorptions with moments coincident with the emission moment, $N_{lum} = N_{\parallel}$, and so $p = +1$. For absorptions with moments perpendicular to the emission moment, $N_{lum} = N_{\perp}$, and so $p = -1$. For angles between the absorption and emission moments, between 0 and 90°, p is intermediate between $+1$ and -1.

In both fluid and rigid solutions, however, the molecules are randomly oriented, so that the values of $N_{\parallel}$ and $N_{\perp}$ given in equations II-B-46 and II-B-47 must be averaged over all values of θ and ϕ. This treatment, with the approximation that the molecules in solution are rigid and do not interact, yields extreme values of p (designated p_o) of $+\frac{1}{2}$ for coincident absorption and emission transition moments of the molecule and $-\frac{1}{3}$ for perpendicular absorption and emission transition moments, where p_o denotes the polarization in the absence of molecular motion or interaction with other molecules. In general, if the absorption and emission transition moments are situated at an angle, β, with respect to one another, the polarization p_o is denoted by

$$p_o = \frac{3 \cos^2 \beta - 1}{\cos^2 \beta + 3} \qquad \text{(II-B-49)}$$

In the fluorescence polarization spectrum, the most positive values of p_o are observed at or near the excitation maxima of the transitions having their absorption moments pointing in the same (or nearly the same) direction as the fluorescence transition moment. The most negative values of p_o occur at or near the excitation maxima corresponding to transitions having their absorption transition moments perpendicular (or at large angles) to the fluorescence transition moment. Regions near zero polarization correspond to minima in the absorption spectrum. It is to be expected that direct absorption to the state from which fluorescence occurs would show the highest positive polarization, since the transitions are essentially mirror images of each other.

Consequently, fluorescence polarization studies should be invaluable in the assignment of emission bands to the states from which they arise. It was, in fact, the high positive polarization of the second absorption band of azulene relative to the first that confirmed the origin of the fluorescence of azulene from its second excited singlet state. Similar arguments are also applicable to the polarization of phosphorescence. Here, however, the situation is more complicated because the states to which absorption occurs are never the states from which emission occurs. Phosphorescence polarization spectra are not considered further here.

(2) *Depolarization of Molecular Luminescence.* Up to now we have thought that the absorbing and emitting molecules maintained fixed positions in solution and did not interact with one another. This pair of approximations is valid in glassy solutions or very viscous solvents, where molecular rotations are small or very slow compared with the lifetime of the luminescent molecule, and in very dilute solutions, where solute-solute interactions are few and far between. In such solutions, polarization values measured approach the theoretical maxima of p_o. *In solvents of low viscosity or of high solute concentration, the measured values of p are always lower than p_o.*

In the case of low solvent viscosity, the diminution of p is due to appreciable rotation of the emitting molecule during the lifetime of the excited state with resulting loss of molecular orientation and hence loss of polarization of the emission. In the case of high solute concentration, the diminution of p is due to transfer of the energy of excitation to another molecule of the same type but with a different orientation. Emission then occurs from the second molecule. Again the result is an apparent loss of polarization of the emitted radiation.

By observation of the emission of solutions in solutions of low viscosity and low solute concentration, it is possible to study the rotational depolarization apart from the effect due to energy transfer. Similarly, the depolarization due to energy transfer may be studied apart from the rotational effect by observation of the emission of solutions of high concentrations of solute and in viscous or rigid media. Because phosphorescence occurs almost exclusively in rigid media, phosphorescence depolarization studies are almost wholly limited to energy transfer phenomena.

In the absence of energy transfer, the polarization is related to rotational diffusion phenomena according to

$$p = \left[\left(\frac{1}{p_o} - \frac{1}{3} \right) \left(1 + \frac{RT}{\eta} \left(\frac{\tau}{V} \right) \right) + \frac{1}{3} \right] - 1 \qquad \text{(II-B-50)}$$

where R, T, and η are, respectively, the gas constant, the absolute temperature, and the viscosity of the medium, and τ and V are, respectively, the lifetime of the excited state and the molecular volume of the solute. Consequently, if R, T, and η are known and p_o is determined by study in rigid media, measurement of p yields values of τ/V. If the molecular volume (in solution) can be estimated, equation II-B-50 can be used to calculate approximate values of τ. On the other hand, if τ can be measured independently and accurately, as is currently possible with modern instrumentation, accurate values of the molecular volume in solution may be determined.

Depolarization due to energy transfer occurring from noncollisional processes (i.e., Förster energy transfer) and in the absence of rotational depolarization may be calculated from

$$p = \left[\left(\frac{1}{p_o} - \frac{1}{3} \right) \left(1 + \frac{105\ N_A\ R_d{}^6 C}{a_m{}^3} \right) + \frac{1}{3} \right]^{-1} \qquad \text{(II-B-51)}$$

where N_A is Avogadro's number, a_m the effective molecular radius, R_d the distance between parallel dipoles at which the probability of emission is equal to the probability of energy transfer (from the Förster theory), and C is the concentration (moles liter^{-1}). Studies of polarization dependence on energy transfer can thus be useful in determining molecular dimensions a_m or quantities of spectroscopic interest (e.g., R_d).

4. GLOSSARY OF SYMBOLS

a	Length of box (linear or circular), length units.
a_m	Effective molecular radius, cm.
C	Concentration of absorbers, mole liter^{-1}.
$d\tau_e$	Volume increment in electronic coordinate space, cm^3.
$d\tau_v$	Volume increment in vibrational coordinate space, cm^3.
$d\tau_s$	Volume increment in spin coordinate space, cm^3.
e	Electronic charge, esu.
E_e	Electronic energy, erg.
E_k	Energy of level corresponding to angular momentum quantum number k, erg.
E_m	Total molecular energy, erg.
E_n	Vibrational energy, erg.
f	Absorption oscillator strength, no units.
g_i	Statistical weight of vibrational state i.
g_j	Statistical weight of vibrational state j.
h	Planck constant, 6.6×10^{-27} erg sec.
H_e	Electronic Hamiltonian, energy units.
H_m	Total molecular Hamiltonian, energy units.
H_n	Vibrational Hamiltonian, energy units.
k	Angular momentum quantum number, no units.
k	Boltzmann constant, 1.38×10^{-16} erg °K^{-1} mol^{-1}.

k_D — Second-order nonradiative rate constant for complexation of fluorescence species in excited state by quencher Q, liter mole^{-1} sec^{-1}.

k_D' — Second-order nonradiative rate constant for complexation of phosphorescent species in excited state by quencher Q, liter mole^{-1}sec^{-1}.

k_f — First-order radiative rate constant for fluorescence, sec^{-1}.

k_i — First-order nonradiative rate constant for deactivation of first excited singlet state by internal conversion process i, sec^{-1}.

k_i' — First-order nonradiative rate constant for deactivation of lowest triplet state by internal conversion process i, sec^{-1}.

k_{IC} — First-order nonradiative rate constant for deactivation of first excited state by internal conversion, sec^{-1}.

k_p — First-order radiative rate constant for phosphorescence, sec^{-1}.

k_Q — Second-order nonradiative rate constant for collisional deactivation of first excited singlet by Q, liter mole^{-1} sec^{-1}.

$k_{Q,j}$ — Second-order nonradiative rate constant for collisional deactivation of first excited singlet by Q_j, liter mole^{-1} sec^{-1}.

k_{Qj}' — Second-order nonradiative rate constant for collisional deactivation of lowest triplet by Q_j, liter mole^{-1} sec^{-1}.

k_{ST} — First-order nonradiative rate constant for intersystem crossing from first excited singlet to lowest triplet states, sec^{-1}.

k_z — Angular momentum quantum number in z direction, no units.

K — Total angular momentum quantum number, no units.

m — Electronic mass, g.

$\mathbf{M}$ — Transition moment vector, dipole moment units.

M_{eg} — Transition moment, dipole moment units.

n_j — Number of molecules in jth vibrational level, no units.

n_T — Total number of molecules in all vibrational levels of an electronic state, no units.

$\mathcal{N}_{abs}$ — Number of photons absorbed per second, sec^{-1}.

N_{lum} — Number of photons luminesced per second, sec^{-1}.

N_A — Avogadro's number, 6.0×10^{23} mole^{-1}.

$N_{\parallel}$ — Number of photons luminesced per second with electric vector parallel to the exciting light vector, sec^{-1}.

$N_{\perp}$ — Number of photons luminesced per second with electric vector perpendicular to the exciting light vector, sec^{-1}.

p — Degree of polarization, no units.

p_o — Limiting value of degree of polarization for parallel and perpendicular absorption and emission transition moments assuming no interaction with other molecules, no units.

q_i — Charge density of substituted atom i (FEMO method), esu cm^{-3}.

(Q) — Concentration of quencher Q for first excited singlet state, mole liter^{-1}.

(Q_j) Concentration of quencher Q_j for first excited singlet state, mole liter^{-1}.

(Q_j') Concentration of quencher Q_j for lowest triplet state, mole liter^{-1}.

r Operator corresponding to vector along whose direction charge is displaced during transition (vibrational coordinate), length units.

r Value of k for highest occupied orbital.

R Gas constant, 8.3×10^7 erg °K^{-1} mol^{-1}.

R_d Distance between parallel dipoles at which probability of emission equals probability of energy transfer, cm.

(S_1) Concentration of first excited singlet state under decaying conditions, mole liter^{-1}.

$(S_1)_o$ Concentration of first excited singlet state under steady state conditions, mole liter^{-1}.

T Temperature of system, °K.

(T_1) Concentration of lowest triplet state under decaying conditions, mole liter^{-1}.

$(T_1)_o$ Concentration of lowest triplet under steady state conditions, mole liter^{-1}.

v Vibrational quantum number, no units.

x Distance, cm.

Y_F Quantum yield of fluorescence—no quenchers present and internal conversion possible, no units.

Y_F' Quantum yield of fluorescence—quenchers present and internal conversion possible, no units.

Y_p Quantum yield of phosphorescence—no quenchers present and internal conversion possible, no units.

Y_p' Quantum yield of phosphorescence—quenchers present and internal conversion possible, no units.

α Fraction of S_oQ complex formed in the ground state excited by direct absorption, no units.

β Angle between emission and absorption transition moments, angular degrees.

γ Probability that S_1Q complex will be deactivated rather than dissociated to S_1, no units.

γ' Probability that T_1Q complex will be deactivated rather than dissociated to T_1, no units.

ΔE_π Energy difference between substituted and unsubstituted molecule (FEMO method), erg.

ΔS Spin change during transition, no units.

$\Delta\alpha$ Charge in Coulomb integral of atom i of aromatic ring upon substitution, coulomb.

$\Delta \nu_{1/2}$ Halfwidth of absorption band, cm^{-1}.

ε Extinction coefficient of absorber at frequency of absorption, liter mole^{-1} cm^{-1}.

ε_i Energy of ith vibrational state, erg.

ε_j Energy of jth vibrational state, erg.

ε_m Extinction coefficient of absorber at band maximum, liter mole^{-1} cm^{-1}.

θ Angle between transition moments of molecule and electric vector of light, angular degrees.

η Viscosity of medium, g cm^{-1} sec^{-1}.

$\tilde{\nu}$ Wavenumber, cm^{-1}.

$\tilde{\nu}_{av}$ Average wavenumber of absorption band, cm^{-1}.

τ_F Mean lifetime of first excited singlet state—no quenchers present and internal conversion possible, sec.

τ_F' Mean lifetime of first excited singlet state—quenchers present and internal conversion possible, sec.

τ_F^0 Mean lifetime of first excited singlet state—no quenchers present and internal conversion *not* present, sec.

τ_P Mean lifetime of lowest triplet state—no quenchers present and internal conversion present, sec.

τ_P' Mean lifetime of lowest triplet state—quenchers present and internal conversion present, sec.

τ_P^0 Mean lifetime of lowest triplet state—no quenchers present and internal conversion *not* present, sec.

ϕ Azimuthal angle **M** makes with direction of observation in polarization studies, angular degrees.

ϕ_g Electronic wavefunction of ground electronic state, no units.

ϕ_e Electronic wavefunction of excited electronic state, no units.

Φ_o Radiant flux incident upon absorber, erg sec^{-1}.

Φ_t Radiant flux transmitted by absorber, erg sec^{-1}.

X_g Vibrational wavefunction of ground electronic state, no units.

X_e Vibrational wavefunction of excited electronic state, no units.

ψ_e Electronic wavefunction, no units.

ψ_e^* Complex conjugate electronic wavefunction, no units.

ψ_m Total molecular wavefunction, no units.

ψ_m^* Complex conjugate total molecular wavefunction, no units.

ψ_n Nuclear wavefunction, no units.

ψ_n^* Complex conjugate nuclear wavefunction, no units.

ψ_{me} Molecular excited state wavefunction, no units.

ψ_{mg} Molecular ground state wavefunction, no units.

σ_g Spin wavefunction of ground electronic state, no units.

σ_e Spin wavefunction of excited electronic state, no units.

5. REFERENCES

1. J. N. Murrell, *The Theory of the Electronic Spectra of Organic Molecules*, Methuen, London, 1963.
2. H. H. Jaffé and M. Orchin, *Theory and Applications of Ultraviolet Spectroscopy*, Wiley, New York, 1962.
3. G. Herzberg, *Electronic Spectra and Electronic Structure of Polyatomic Molecules*, Van Nostrand, New York, 1966.
4. S. F. Mason, *Quart. Rev. (London)*, **15**, 287 (1961).
5. W. West, Ed., *Chemical Applications of Spectroscopy*, Vol. IX in *Technique of Organic Chemistry*, Wiley-Interscience, New York, 1968.
6. N. J. Turro, *Molecular Photochemistry*, Benjamin, New York, 1965.

C. GENERAL LUMINESCENCE RADIANCE EXPRESSIONS

1. NO PREFILTER, NO POSTFILTER EFFECTS PRESENT[1-9]

a. CONDITIONS

We assume that the absorbing medium is in a suitable cell l cm wide (absorption path length), L cm long (luminescence path length), and l' cm high (see Figure II-C-1). The cell* is assumed to be uniformly illuminated over the face Ll' (cm^2); that is, no postfilter effect exists (see Figure II-C-1b and II-C-1c). In the initial stages of the derivation, only the absorption by and the luminescence from a volume increment dx cm wide, dy cm long and l' cm high will be considered (also see Figure II-C-1a). We further assume that the analyte absorbers (atomic or molecular in gas phase, liquid phase, or solid phase) are equally and uniformly distributed within the cell and that no interferences are present (i.e., only the analyte absorbs and only the analyte luminesces).† Finally, reemission of self-absorbed luminescence radiation is disregarded.

Of course, luminescence is isotropic, and so it is emitted in all directions and emanates from the six sides of the parallelepiped cell. The following approach is valid whether the luminescence is observed perpendicular to the exciting radiation (i.e., the *right angle case*, Figure II-C-1b) or the luminescence is observed from the same face that is being illuminated with exciting light

---

* The cell geometry in this chapter is assumed to be a parallelepiped as shown in Figure II-C-1. A cylindrical cell is much more difficult to treat exactly and is not handled here. However, a cylindrical cell of radius r results in about the same luminescence radiance as a parallelepiped with a square cross section of area l^2, where l is related to r by $l = r/\sqrt{\pi}$.

† In the following discussion, absorber, absorbing analyte, and analyte appear interchangeably.

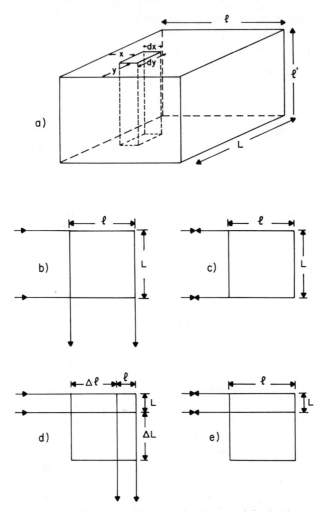

Fig. II-C-1. Schematic diagrams of cell assumed for luminescence radiance expressions: (*a*) three-dimensional diagram of cell to show dimensions of unit cell; (*b–e*) top view of cell to show method of illumination measurement; the cell height in all cases is l' cm; (*b*) Right angle—complete illumination of cell, complete measurement of right angle luminescence; (*c*) front surface—complete illumination of cell, complete measurement of front luminescence; (*d*) right angle—incomplete illumination of cell, incomplete measurement of luminescence; (*e*) front surface—incomplete illumination of cell, complete measurement of luminescence.

(i.e., the *front surface case*,* Figure II-C-1*c*). No intermediate cases are considered here because the two limiting cases just named are analytically useful, because most intermediate cases are approximately represented by one of the two simpler limiting cases, and because the intermediate cases are extremely difficult to treat explicitly. However, in this part, we assume for the right angle case that there is no prefilter effect—that is, all the luminiscence from the face of area *ll'* cm² is measured (Figure II-C-1*b*)—and that there is no post filter effect—that is, the entire front surface of area *Ll'* is fully illuminated. For the front surface case, there are *no* prefilter or postfilter effects. In the next part of this section, the prefilter and postfilter effects for the right angle case are considered.

b. ABSORBED RADIANT FLUX

The spectral radiance† of the excitation source is $B_{S\lambda}{}^\circ$ (erg sec^{-1} cm^{-2} sr^{-1} nm^{-1}) at the excitation wavelength λ. If the source is imaged on the absorption cell with suitable entrance optics, the source spectral radiance flux $\Phi_{S\lambda}{}^\circ$ (erg sec^{-1} nm^{-1}) incident upon the cross-sectional area of $dy\, l'$ cm² of the volume element $dy\, dx\, l'$ cm³ is

$$\Phi_{S\lambda}{}^\circ = B_{S\lambda}{}^\circ\, \Omega_E\, [dy\, l']\, \left(\frac{1}{m_L m_H}\right) [T_E] \qquad \text{(II-C-1)}$$

where Ω_E is the solid angle of exciting radiation collected by the entrance optics and imaged upon the volume element $dy\, dx\, l'$, m_L and m_H are the source image magnification at the cell surface in the longitudinal and the horizontal directions, respectively, and T_E is a factor to account for absorption and reflection losses due to optical components between the source and the absorption cell.‡

According to the Lambert absorption law, the spectral radiant flux at any distance x from the front surface of the cell $\Phi_{S\lambda}(x)$ (erg sec^{-1} nm^{-1}) is given by

$$\Phi_{S\lambda}(x) = \Phi_{S\lambda}{}^\circ\, [\exp -k_\lambda x] \qquad \text{(II-C-2)}$$

---

* The front surface case of 0° illumination measurement is *not* of great analytical importance but is simpler to handle here. However, the expressions to be derived for this case should be approximately valid even if the angle between the incident and measured beams is greater than 0°.

† The symbols used are those which are most commonly used and generally those tentatively agreed upon by the International Union of Pure and Applied Chemistry (IUPAC).

‡ It is assumed for simplicity that Ω_E, m_H, m_L, and T_E are independent of the excitation wavelength, this is a reasonable approximation even for lenses, as long as the optical components are not used near their absorption cutoff wavelengths and as long as the wavelength range is not too large (e.g., greater than about 1000 Å).

where k_λ is the absorption coefficient of the analyte at the excitation wavelength λ (i.e., the fraction absorbed cm^{-1}). The radiant flux absorbed at wavelength λ per unit wavelength interval by the small volume element of path length dx is then given by

$$d\Phi_{S\lambda} = \Phi_{S\lambda}{}^\circ k_\lambda \left[\exp -k_\lambda x\right] dx \qquad \text{(II-C-3)}$$

C. LUMINESCENCE RADIANT FLUX

To convert radiant flux absorbed by the volume element $dx\, dy\, l'$ to radiant flux luminesced, it is necessary to use the luminescence spectral power yield $Y_\lambda(\lambda')$, which is given by

$$Y_\lambda(\lambda') = Y_p f(\lambda') \qquad \text{(II-C-4)}$$

where Y_p is the power yield of luminescence* (i.e. the luminescence radiant flux divided by the absorbed radiant flux) and is dependent on the excitation wavelength λ for gaseous atoms and diatomic molecules with widely separated vibrational levels. (The value of Y_p is generally independent of excitation wavelength λ for molecules in the liquid or solid phases.) Also in equation II-C-4, $f(\lambda')$ is the spectral distribution of luminescence (no units) and is dependent *only* on the luminescence wavelength λ'. The product $f(\lambda')$ $d\lambda'$ represents the probability that an emitted photon has a wavelength between λ' and $\lambda' + d\lambda'$; that is, $\int_0^\infty f(\lambda')\, d\lambda' = 1$. Therefore, $f(\lambda')$ represents the normalized shape† of the luminescence line or band and is given by

$$f(\lambda') = \frac{k_\lambda{}'}{\int_0^\infty k_\lambda{}'\, d\lambda'} \qquad \text{(II-C-5)}$$

where k_λ, is the absorption coefficient (cm^{-1}) for the luminescence wavelength λ'.

The luminescence spectral radiant flux $d\Phi_{L\lambda'}$ from the small volume element $dx\, dy\, l'$ is the product of the flux absorbed $d\Phi_{S\lambda}$, the luminescence spectral power yield $Y_\lambda(\lambda')$, and a factor to account for loss of luminescence due to self-absorption, and so

$$d\Phi_{L\lambda'} = d\Phi_{S\lambda} Y_\lambda(\lambda')\left[\exp -k_{\lambda'} y\right] dy \qquad \text{(II-C-6)}$$

The spectral radiant flux of luminescence can be converted to the spectral radiance of luminescence by dividing $d\Phi_{L\lambda'}$ by 4π; that is, there are 4π sr in a sphere of isotropic luminescence, by the area dA_s of the six surfaces of the

---

* Except in this section, the luminescence power yield is designated by Y (no λ subscript) with the proper subscript designating the process (e.g., Y_F, with the λ subscript omitted).

† The emission coefficient for the luminescence process (given symbol $e_{\lambda'}$) is identical to the absorption coefficient $k_{\lambda'}$.

volume element, and by n^2, where n is the refractive index of the solvent,*
and so

$$dB_{L\lambda'} = d\Phi_{L\lambda'}\left(\frac{1}{4\pi n^2}\right)\left(\frac{1}{dA_s}\right) \tag{II-C-7}$$

The luminescence radiance dB_L (erg sec^{-1} cm^{-2} sr^{-1}) from the volume
element $dx\,dy\,l'$ is therefore given by

$$dB_L = \left[\frac{\Omega_E}{4\pi}\right]\left[\frac{l'\,dx\,dy}{dA_s}\right]\left[\frac{T_E}{n^2 m_L m_H}\right]\left[\frac{1}{\int_0^\infty k_\lambda\cdot d\lambda'}\right]\left(\int_0^\infty \Phi_{S\lambda}{}^0 k_\lambda\,[\exp\,-k_\lambda x]\,d\lambda\right)$$
$$\cdot\left[\int_0^\infty Y_p k_{\lambda'}\,[\exp\,-k_{\lambda'}y]\,d\lambda'\right] \tag{II-C-8}$$

If this equation is integrated with respect to x and y between the limits of 0
and l for x and 0 for L and y, then the luminescence radiance B_L is given by

$$B_L = \left[\frac{\Omega_E}{4\pi}\right]\left[\frac{Ll'}{2Ll' + 2ll' + 2Ll}\right]\left[\frac{T_E}{n^2 m_L m_H}\right]Y_p$$
$$\cdot\left[\int_0^\infty B_{S\lambda}{}^0(1 - [\exp\,-k_\lambda l]\,d\lambda\right]\left[\frac{\int_0^\infty (1 - \exp\,-k_{\lambda'} L)\,d\lambda'}{\int_0^\infty Lk_{\lambda'}\,d\lambda'}\right] \tag{II-C-9}$$

Equation II-C-9 is specifically the luminescence radiance B_L for the *right
angle case* of illumination measurement. For the *front surface case* of illumination
measurement, it is only necessary to change the emission path length from
L to l, and therefore

$$B_L = \left[\frac{\Omega_E}{4\pi}\right]\left[\frac{ll'}{2Ll' + 2ll' + 2Ll}\right]\left[\frac{T_E}{n^2 m_L m_H}\right]Y_p$$
$$\cdot\left[\int_0^\infty B_{S\lambda}{}^0(1 - \exp\,-k_\lambda l)\,d\lambda\right]\left[\frac{\int_0^\infty (1 - \exp\,-k_{\lambda'} l)\,d\lambda'}{\int_0^\infty lk_{\lambda'}\,d\lambda'}\right]$$
$$\tag{II-C-10}\dagger$$

The final ratio of integrals in equations II-C-9 and II-C-10 is the self-
absorption factor, sometimes designated f_s, and it accounts for the reabsorp-

---

* The refractive index term accounts for the increased solid angle of luminescence from
media with refractive indices greater than unity. For a gas, such as a flame, $n \approx 1$; but for
molecules in liquid solutions, n is greater than unity.

† Equation II-C-9 is multiplied and divided by L, and equation II-C-10 is multiplied and
divided by l. This is done so that the final ratio of integrals, called f_s, the self-absorption
factor, is dimensionless.

tion of luminescence by analyte species. The first integral in equations II-C-9 and II-C-10 is the absorbed radiance.

d. EVALUATION OF INTEGRALS IN EQUATIONS II-C-9 AND II-C-10

(1) *Final Ratio.* The final ratio of integrals in equations II-C-9 and II-C-10 (i.e., the self-absorption factor f_s) is unity or near unity for the following cases:

1. Atomic fluorescence involving a dilute atomic gas and transitions terminating in the ground or in any excited state.

2. Atomic fluorescence involving any concentration of atoms except for transitions terminating in states above the ground state (the hotter the gas the higher the level* must be above the ground state for f_s to be unity).

3. Molecular luminescence (fluorescence or phosphorescence) involving any concentration of analyte molecules in liquid or solid solution as long as there is minimal overlap of the luminescence and absorption spectra.†

Therefore, for the cases just described

$$f_s = \frac{\int_0^\infty (1 - \exp -k_{\lambda'} z)\, d\lambda'}{\int_0^\infty zk_{\lambda'}\, d\lambda'} = 1 \qquad \text{(II-C-11)}$$

where z is either L (in equation II-C-9) or l (in equation II-C-10).

However, for atomic fluorescence of concentrated gases and for transitions terminating in the ground or near ground states, the self-absorption factor f_s is less than unity and is given‡ by

$$f_s = \frac{\int_0^\infty (1 - \exp -k_\lambda'z)\, d\lambda'}{\int_0^\infty zk_\lambda'\, d\lambda'} = \frac{2\sqrt{a'}}{\sqrt{(\sqrt{\pi}k_o'z)}} \qquad \text{(II-C-12)}$$

---

* For example, the level should be about 0.5 eV (or higher) above ground state for analytical flames (temperatures between 2000 and 3000°K).

† This is approximately true for most cases of analytical interest. If molecular luminescence of gaseous molecules were to become of analytical interest, then this approximation would not be valid under many conditions. So far molecular luminescence of gases is of no analytical use.

‡ The integrals in f_s are evaluated as follows:

$$\int_0^\infty zk_{\lambda'}d\lambda' = \frac{\sqrt{\pi}k_o' \Delta\lambda_D'}{2\sqrt{\ln 2}}$$

and

$$\int_0^\infty (1 - \exp -k_{\lambda'} z)\, d\lambda' = \Delta\lambda_D' \left(\frac{\sqrt{\pi}k_o'a}{\ln 2}\right)^{1/2}$$

The latter integral is the result of integration for a spectral line exhibiting both Lorentzian (collisional) and Gaussian (Doppler) broadening effects.

where z is either l or L as described previously, k_o is the peak atomic absorption coefficient for pure Doppler broadening for the fluorescence line, and a' is the so-called damping constant for the fluorescence line and is given by

$$a' = \sqrt{\ln 2}\ \frac{\Delta\lambda_C'}{\Delta\lambda_D'} \qquad\qquad \text{(II-C-13a)}$$

$$a = \sqrt{\ln 2}\ \frac{\Delta\lambda_C}{\Delta\lambda_D} \qquad\qquad \text{(II-C-13b)}$$

where $\Delta\lambda_C'$ is the collisional half width of the fluorescence line and $\Delta\lambda_D'$ is the Doppler half width of the fluorescence line. The unprimed terms are identical to the primed terms except they are defined for the absorption (excitation) process and the primed terms are for the fluorescence process. For resonance absorption-fluorescence, $a = a'$, because the two lines should be nearly identical. For a nonresonance process, the absorption and fluorescence lines will not be the same, and so $a \neq a'$. In equation II-C-12, the value of the damping constant for the absorption of the fluorescence line (i.e., a') must be used.

(2) *First Integral.* The first integral in equations II-C-9 and II-C-10 is more difficult to evaluate exactly because of the need to convolute two wavelength-dependent terms $B_{S\lambda}{}^o$ and $1 - \exp -k_\lambda l$. However, again it is possible to simplify this process by several limiting but nevertheless analytically useful cases.

(a) *Continuum-Source Case.* Suppose that the source of excitation has a constant spectral radiance over the entire absorption line or band (e.g., atomic fluorescence being excited with a continuum source with no excitation monochromator or with an excitation monochromator or filter having spectral bandwidths considerably greater than the absorption line half width); for this case, called *continuum source case* (see Figure II-C-2a)

$$\int_0^\infty B_{S\lambda}{}^o\ (1 - \exp -k_\lambda l)\ d\lambda = B_{C\lambda_o}^o \int_0^\infty (1 - \exp -k_\lambda l)\ d\lambda \quad \text{(II-C-14)}$$

where $B_{C\lambda_o}^o$ is the spectral radiance of the continuum source at the peak absorption wavelength. The final integral is readily solved for absorption lines or bands having a Voigt profile and resulting from either very low or very high concentrations of absorbers. A spectral line or band with a Voigt profile (i.e., a line or band broadened by both Lorentzian and Gaussian effects) has an absorption coefficient of

$$k_\lambda = \frac{a}{\pi}\, k_o \int_{-\infty}^{+\infty} \frac{\exp(-y^2)\ dy}{a^2 + (v - y)^2} \qquad\qquad \text{(II-C-15)}$$

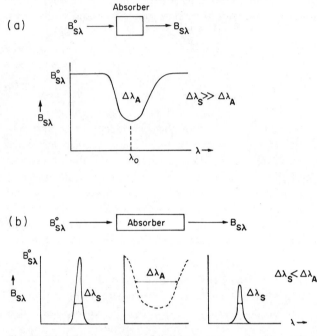

Fig. II-C-2. Representation of limiting source cases: (*a*) continuum-source case, (*b*) line source case ($\Delta\lambda_S$ is source line half width and $\Delta\lambda_A$ is absorption line half width).

where a is the classical damping constant and is given by $\sqrt{\ln 2}$ times the ratio of the Lorentzian half width to the Gaussian half width of the line or band (see equation II-C-13), k_o is the peak absorption coefficient for pure Gaussian broadening (for atoms, Doppler broadening) for the excitation process, v is the relative wavelength of any point in the absorption line in terms of Doppler half widths, and y is a variable relative wavelength taken with respect to v. The variables a, v, and y are given by

$$a = \sqrt{\ln 2}\,\frac{\Delta\lambda_C}{\Delta\lambda_G} \tag{II-C-16}$$

$$v = 2\sqrt{\ln 2}\,\frac{(\lambda - \lambda_o)}{\Delta\lambda_G} \tag{II-C-17}$$

$$y = 2\sqrt{\ln 2}\,\frac{\delta}{\Delta\lambda_G} \tag{II-C-18}$$

where $\Delta\lambda_L$ is the Lorentzian half width (for atoms, collisional half width), $\Delta\lambda_G$ is the Gaussian half width (for atoms, Doppler half width), λ is any wavelength, λ_o is the peak wavelength, and δ is a variable distance from $\lambda - \lambda_o$.

The integral in equation II-C-14 (called the total absorption and given the symbol A_t) can be evaluated for two limiting cases. For a low concentration of atomic or molecular absorbers (the absorption coefficient k_λ is directly proportional to the concentration of absorbers),

$$\int_0^\infty (1 - \exp -k_\lambda l) \, d\lambda = \int_0^\infty k_\lambda l \, d\lambda = \frac{\sqrt{\pi} k_o \, \Delta\lambda_G l}{2\sqrt{\ln 2}} \qquad \text{(II-C-19)}$$

and for a high concentration of atomic absorbers*

$$\int_0^\infty (1 - \exp -k_\lambda l) \, d\lambda = \Delta\lambda_G \sqrt{\left(\frac{\sqrt{\pi} k_o \, la}{\ln 2}\right)} \qquad \text{(II-C-20)}$$

Figure II-C-3 presents theoretical total absorption curves (sometimes called growth curves) versus a modified absorption coefficient curve for a single isolated spectral line (refer to Section II-C-2-f for a discussion of the factors influencing the peak absorption coefficient k_o for pure Gaussian

---

* Equations II-C-19 and II-C-20 are valid for all atomic-molecular systems. However, for atoms with atomic lines having hyperfine structure (i.e., separate simple lines of different intensities), the overall absorption coefficient k_o is composed of the summation of all individual line absorption coefficients comprising the complex atomic "line system," that is,

$$k_o = k_{o1} + k_{o2} + k_{o3} + \cdots$$

where k_{oi} is the absorption coefficient of hyperfine component i. For a spectral line with hyperfine components

$$\int_0^\infty k_\lambda \, d\lambda = \left(\frac{\pi e^2}{mc}\right) n_l f_{lu} l = \left(\frac{\lambda_o^2 g_u}{8\pi g_l}\right) n_l A_{ul} l = \frac{1}{2} \sqrt{\pi/\ln 2} \, \Delta\lambda_D l [k_{o1} + k_{o2} + \cdots]$$

Therefore the total absorption integral is given by equation II-C-19. If self-absorption for *any given* component of the hyperfine structure is great, then the limiting equation II-C-20 must be used for the given component.

The equations for k_o and $\int_0^\infty k_\lambda \, d\lambda$ are *also* valid for molecular bands that consist of many lines resulting from transitions between rotational and vibrational levels of two different electronic states. The individual lines are broadened by collisional type effects—primarily due to interactions between the solvent, oxygen gas, or other major components present in the solution and the analyte molecules. Because of the great width of molecular band (i.e., the large number of broadened lines composing the band), it is rare for the concentration of the analyte to be sufficiently great for k_{oi}, the absorption coefficient for any given line i, to be large enough for the self-absorption equation to be valid—in other words, for equation II-C-20 to be useful. Therefore, for molecular bands of molecules in the condensed phase, the total absorption expression to a good approximation is given by equation II-C-19.

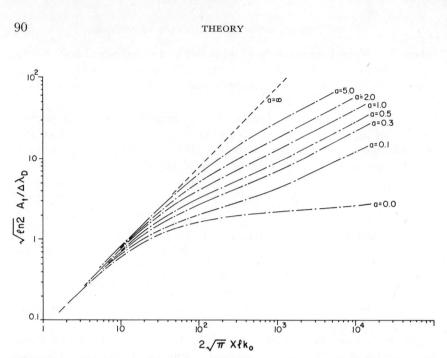

Fig. II-C-3. Theoretical total absorption (growth) curve plots for a single spectral line with various values of the a parameter (plot of $\sqrt{\ln 2}\; A_t/\Delta\lambda_D$, where A_t is the integral in equations II-C-19 and II-C-20 and $\Delta\lambda_D$ is the Doppler half width, versus $2\sqrt{\pi}\; Xlk_o$, where $X = \pi e^2/mc$ (m and e are mass and charge of electron, c is speed of light), $l = $ absorption path length, and k_o is the peak absorption coefficient for pure Gaussian (Doppler) broadening.

(Doppler for atoms) broadening). The growth curves in Figure II-C-3 are particularly useful in atomic spectroscopy for predicting the shape of analytical curves (refer to Section II-C-2-g) and for predicting the characteristics of spectral lines in astrophysics. Although the growth curves in Figure II-C-3 are applicable in principle for both atomic and molecular electronic transitions, such curves are in reality not very useful for molecules because of the need to apply the expressions to each vibrational-rotational line of the electronic transition and the complex solvent broadening processes.

(b) *Line-Source Case.* The source of excitation is a narrow line source (i.e., the half width of the source line or band is less than the half width of the absorption line or band) this is the most common excitation source in atomic fluorescence spectrometry (electrodeless discharge tube sources are used) and *also* the most common excitation source in condensed-phase molecular luminescence spectrometry (continuum source plus an excitation monochromator or narrow band filter having a spectral bandwidth *less* than the absorption band half width). For the *line-source case* (see Figure II-C-2*b*)

$$\int_0^\infty B_{S\lambda}{}^o(1 - \exp -k_\lambda l)\, d\lambda = (1 - \exp -\bar{k}l)\int_0^\infty B_{S\lambda}{}^o\, d\lambda = (1 - \exp -\bar{k}l)B_N{}^o$$

$$\text{(II-C-21)}$$

where $\bar{k}$ is the effective (average) absorption coefficient (cm) over the source line half width and

$$\int_0^\infty B_{S\lambda}{}^o\, d\lambda = B_N{}^o$$

where $B_N{}^o$ is the source integrated spectral radiance (integrated over the exciting line or band) or simply the source radiance (erg sec^{-1} cm^{-2} sr^{-1}).

The term in parentheses in equation II-C-21 can be evaluated for two limiting cases. For a low concentration of absorbers (the effective absorption coefficient $\bar{k}$ is directly proportional to the concentration of absorbers),

$$1 - \exp -\bar{k}l = \bar{k}l \qquad \text{(II-C-22)}$$

and for a high concentration of absorbers

$$1 - \exp -\bar{k}l = 1 \qquad \text{(II-C-23)}$$

e. LUMINESCENCE RADIANCE EXPRESSIONS*

(1) *Right Angle Case—Continuum Source—Low Concentration of Analyte.* Equations II-C-11, II-C-14, and II-C-19 are used with equation II-C-9 to give

$$B_L = \left[\frac{\Omega_E}{4\pi}\right]\left[\frac{Lll'}{2Ll' + 2ll' + 2Ll}\right]\left[\frac{T_E}{n^2 m_L m_H}\right] Y_L\, B^o_{C\lambda_o}\left[\frac{\sqrt{\pi}k_o\,\Delta\lambda_G}{2\sqrt{\ln 2}}\right] \quad \text{(II-C-24)}$$

(2) *Front Surface Case—Continuum Source—Low Concentration of Analyte.* Equations II-C-11, II-C-14, and II-C-19 are used with equation II-C-10 to give

$$B_L = \left[\frac{\Omega_E}{4\pi}\right]\left[\frac{l^2l'}{2Ll' + 2ll' + 2Ll}\right]\left[\frac{T_E}{n^2 m_L m_H}\right] Y_L\, B^o_{C\lambda_o}\left[\frac{\sqrt{\pi}k_o\,\Delta\lambda_G}{2\sqrt{\ln 2}}\right] \quad \text{(II-C-25)}$$

(3) *Right Angle Case—Continuum Source—Atomic Fluorescence Terminating In Ground or Near Ground States—High Concentration of Analyte.* Equations II-C-12, II-C-14, and II-C-20 are used with equation II-C-9 to give

---

* From here on in this section, Y_p is replaced with Y_L, where L represents the luminescence process and so Y_L is the luminescence power yield for the desired process. Also Y_L could be a function of the excitation wavelength, λ.

$$B_L = \left[\frac{\Omega_E}{4\pi}\right]\left[\frac{Ll'}{2Ll' + 2ll' + 2Ll}\right]\left[\frac{T_E}{n^2 m_L m_H}\right] Y_L B^o_{C\lambda_o} \left(\frac{k_o \, la \, a'}{k'_o \, L \ln 2}\right)^{1/2} \Delta\lambda_G$$

$$(\text{II-C-26})$$

(4) *Right Angle Case—Continuum Source—Atomic Fluorescence Terminating in Excited States or Molecular Luminescence*—High Concentration of Analyte.* Equations II-C-11, II-C-14, and II-C-20 are used with equation II-C-9 to give

$$B_L = \left[\frac{\Omega_E}{4\pi}\right]\left[\frac{Ll'}{2Ll' + 2ll' + 2Ll}\right]\left[\frac{T_E}{n^2 m_L m_H}\right] Y_L B^o_{C\lambda_o} \left(\frac{\sqrt{\pi k_o} \, la}{\ln 2}\right)^{1/2} \Delta\lambda_G \quad (\text{II-C-27})$$

(5) *Front Surface Case—Continuum Source—Atomic Fluorescence Terminating in Ground or Near Ground States—High Concentration of Analyte.* Equations II-C-12, II-C-14, and II-C-20 are used with equation II-C-10 to give

$$B_L = \left[\frac{\Omega_E}{4\pi}\right]\left[\frac{ll'}{2Ll' + 2ll' + 2Ll}\right]\left[\frac{T_E}{n^2 m_L m_H}\right] Y_L B^o_{C\lambda_o} \left[2\left(\frac{k_o \, aa'}{k'_o \ln 2}\right)^{1/2}\right] \Delta\lambda_G$$

$$(\text{II-C-28})$$

(6) *Front Surface Case—Continuum Source—Atomic Fluorescence Terminating in Excited States or Molecular Luminescence*—High Concentration of Analyte.* Equations II-C-11, II-C-14, and II-C-20 are used with equation II-C-10 to give

$$B_L = \left[\frac{\Omega_E}{4\pi}\right]\left[\frac{ll'}{2Ll' + 2ll' + 2Ll}\right]\left[\frac{T_E}{n^2 m_L m_H}\right] Y_L B^o_{C\lambda_o} \left(\frac{\sqrt{\pi k_o} \, la}{\ln 2}\right)^{1/2} \Delta\lambda_G \quad (\text{II-C-29})$$

(7) *Right Angle Case—Line Source—Low Concentration of Analyte.* Equations II-C-11, II-C-21, and II-C-22 are used with equation II-C-9 to give

$$B_L = \left[\frac{\Omega_E}{4\pi}\right]\left[\frac{Lll'}{2Ll' + 2Ll' + 2Ll}\right]\left[\frac{T_E}{n^2 m_L m_H}\right] Y_L B_{N^o} \, \bar{k} \quad (\text{II-C-30})$$

(8) *Front Surface Case—Line Source—Low Concentration of Analyte.* Equations II-C-11, II-C-21, and II-C-22 are used with equation II-C-10 to give

$$B_L = \left[\frac{\Omega_E}{4\pi}\right]\left[\frac{l^2 l'}{2Ll' + 2ll' + 2Ll}\right]\left[\frac{T_E}{n^2 m_L m_H}\right] Y_L B_{N^o} \, \bar{k} \quad (\text{II-C-31})$$

(9) *Right Angle Case—Line Source—Atomic Fluorescence Terminating in Ground or Near Ground State—High Concentration of Analyte.* Equations II-C-12, II-C-21, and II-C-23 are used with equation II-C-9 to give

---

* Molecular luminescence here means fluorimetry and phosphorimetry of molecules in the condensed phase. These equations (II-C-27 and II-C-29) are of little analytical value because the "continuum-source" condition is *not normally* used in analytical fluorimetry or phosphorimetry.

$$B_L = \left[\frac{\Omega_E}{4\pi}\right]\left[\frac{Ll'}{2Ll' + 2ll' + 2Ll}\right]\left[\frac{T_E}{n^2 m_L\, m_H}\right]Y_L\, B_N{}^o\left[2\left(\frac{a'}{\sqrt{\pi k_o'}\, L}\right)^{1/2}\right]$$

$$(\text{II-C-32})$$

(10) *Right Angle Case—Line Source—Atomic Fluorescence Terminating in Excited State or Molecular Luminescence*—High Concentration of Analyte.* Equations II-C-11, II-C-21, and II-C-23 are used with equation II-C-9 to give

$$B_L = \left[\frac{\Omega_E}{4\pi}\right]\left[\frac{Ll'}{2Ll' + 2ll' + 2Ll}\right]\left[\frac{T_E}{n^2 m_L\, m_H}\right]Y_L\, B_N{}^o \qquad (\text{II-C-33})$$

(11) *Front Surface Case—Line Source—Atomic Fluorescence Terminating in Ground or Near Ground State—High Concentration of Analyte.* Equations II-C-12, II-C-22, and II-C-23 are used with equation II-C-10 to give

$$B_L = \left[\frac{\Omega_E}{4\pi}\right]\left[\frac{ll'}{2Ll' + 2ll' + 2Ll}\right]\left[\frac{T_E}{n^2 m_L\, m_H}\right]Y_L\, B_N{}^o\left[2\left(\frac{a'}{\sqrt{\pi k_o'}\, l}\right)^{1/2}\right] \quad (\text{II-C-34})$$

(12) *Front Surface Case—Line Source—Atomic Fluorescence Terminating in Excited State or Molecular Luminescence—High Concentration of Analyte.* Equations II-C-11, II-C-21, and II-C-23 are used with equation II-C-10 to give

$$B_L = \left[\frac{\Omega_E}{4\pi}\right]\left[\frac{ll'}{2Ll' + 2ll' + 2Ll}\right]\left[\frac{T_E}{n^2 m_L\, m_H}\right]Y_L\, B_N{}^o \qquad (\text{II-C-35})$$

(13) *Discussion of Equations* II-C-24 *through* II-C-35 *for* B_L. From the foregoing expressions for B_L, several generalizations† can be made.

1. The value of B_L is directly proportional to the concentration‡ of analyte for either type of illumination measurement and for *any type* of atomic or molecular luminescence *as long as the concentration of analyte is low.*

2. The B_L is dependent *only* on the *cell volume (not the cell shape)* as long as the concentration of absorbers is low for the *right angle case,* and as long as the entrance optics are such that the entire cell cross section of Ll' cm² is fully illuminated and the entire luminescence from the right angle face of ll' cm²

---

* Equations II-C-30, II-C-31, II-C-33, and II-C-35 should be useful for analytical molecular luminescence studies. We should stress at this point that in most luminescence spectrometry studies, the absorption area of the sample is not determined by Ll' (or ll') but by WH, the width times the height of the excitation monochromator slit. Also it should be emphasized that $B_N{}^o$ can be replaced by $B^o_{C\lambda_o}s$, where s is the spectral bandpass of the excitation monochromator, when a continuum source-excitation monochromator is employed.

† The generalizations are all first-order approximations—assuming that the previously stated conditions are valid.

‡ The values of k_o, k_o', and $\bar{k}$ are directly proportional to the concentration of analyte (see Sections II-C-3-e and II-C-3-f).

is measured. This dependence on cell volume holds for *any* type of atomic or molecular luminescence.

3. The B_L depends directly on the *spectral radiance* of the *continuum source* $B^o_{C\lambda_o}$ or the *integrated spectral radiance* of the *line source* $B_N{}^o$ depending on the source type utilized for *any* type of atomic or molecular luminescence, for either type of illumination measurement, and for *all* concentrations of absorbers.

4. The B_L depends directly on the *luminescence power yield* Y_L for either type of source, for either type of illumination measurement, and for *all* concentrations of absorbers.

5. The B_L depends directly on the *solid angle of radiation* collected from the excitation source Ω_E (*as long as* the product $m_L m_H$ does *not* change simultaneously) for either type of source, for either type of illumination measurement, and for *all* concentrations of absorbers.

6. The B_L is independent of *concentration of absorbers for high concentration* of absorbers in atomic fluorescence transitions terminating in the ground or near ground state and for a continuum source of excitation; *and B_L is inde*pendent of concentration of absorbers for high concentration of absorbers in atomic fluorescence transitions terminating in excited states or in molecular luminescence* and for a line source of excitation.

7. The B_L is independent of the *shapes of the absorption and luminescence bands* as long as the absorber concentration is low, whatever the illumination measurement method, as long as a continuum source of excitation is utilized. The same statement is approximately true for the case of a line source of excitation, especially if the line source is very narrow compared with the absorption band; if the line source is of the same width as the absorption band, then there is a complex dependence of B_L on the shapes of the absorption and luminescence bands.

8. The B_L is independent of the *shapes of the absorption and luminescence bands* if the absorber concentration is high, if a line source is used for atomic fluorescence transitions terminating in an excited state or for molecular luminescence.*

9. The B_L depends on the square root of the absorber concentration for atomic fluorescence transitions terminating in an excited state if a continuum source is used and if the absorber concentration is high.

10. The B_L depends on the reciprocal of the square root of the absorber concentration for atomic fluorescence transitions terminating in the ground or near ground state if a line source is used and if the absorber concentration is high.

---

* Molecular luminescence here means fluorimetry or phosphorimetry of molecules in the condensed phase.

2. PREFILTER AND POSTFILTER EFFECTS PRESENT[3-5,8]

a. CONDITIONS

All assumptions given in Section II-C-1-a are retained here except for the method of illumination and measurement of luminescence. If the sample cell surface of area Ll' is not completely illuminated as in Figure II-C-1d, then the luminescence radiant flux emerging from the surface (toward the detector) of area ll' is reduced by the post-filter effect. If the luminescence emerging from the surface (toward the detector) of area ll' is not completely measured, then the incident radiant flux from the source impinging on the measured portion of the cell is reduced and so is the luminescence radiance. Both prefilter and postfilter effects are only significant for the right angle case and for high concentrations of analyte.

b. POST-FILTER EFFECT

The reabsorption in the cell volume of luminescence of an analyte species that is *not* being illuminated by excitation radiation decreases the luminescence radiance as estimated by the equations in the previous section by the factor f_I; that is,

$$B_L' = f_I B_L \qquad \text{(II-C-36)}$$

where B_L is the luminescence radiance calculated from the equations in Section II-C-1 and B_L' is the luminescence radiance accounting for the post filter effect (no prefilter effect is present). Both for atomic fluorescence and molecular luminescence of a *low* concentration of analyte *and* for atomic fluorescence transitions terminating in an excited state and for molecular luminescence of a *high* concentration of analyte, f_I is approximately unity:

$$f_I = 1 \qquad \text{(II-C-37)}$$

In other words, for the systems and conditions named previously, *no appreciable* reabsorption of luminescence occurs.

However, for atomic fluorescence transitions terminating in ground or near ground states and for the right angle illumination measurement case and for a *high concentration of analyte absorbers*, f_I is given* by

$$f_I = \sqrt{1 + \Delta L/L} - \sqrt{\Delta L/L} \qquad \text{(II-C-38)}$$

* The value of f_I is given by

$$f_I = \frac{\int_0^\infty [1 - \exp - k_\lambda L(1 + \Delta L/L)] \, d\lambda - \int_0^\infty [1 - \exp - k_\lambda L(\Delta L/L)] \, d\lambda}{\int_0^\infty (1 - \exp - k_\lambda L) \, d\lambda}$$

where all terms have been previously defined.

where ΔL is the sample cell thickness (actually thickness of sample) nearest the measurement device and not being excited and L is the sample path length being excited (see Figure II-C-1d). As $\Delta L/L$ becomes very large (i.e., as less and less sample is being excited), f_I approaches zero; and as $\Delta L/L$ approaches zero—more and more sample being excited—f_I approaches unity and $B_L{}'$ is equal to B_L, which is given by the equations in Section II-C-1).

For atomic and molecular luminescence for the front surface illumination-measurement case and for all concentrations of analyte,

$$f_I = 1 \tag{II-C-37}$$

C. PREFILTER EFFECTS

The luminescence radiance is decreased by a factor f_E for the right angle case if the luminescence in the portion of the cell nearest the source is *not* measured. The corrected luminescence radiance B_L'' is then

$$B_L'' = f_E B_L \tag{II-C-39}$$

where B_L is the luminescence radiance given by an appropriate equation in Section II-C-1.

The f_E factor for atomic fluorescence excited by a *continuum* source and for *high concentration of analyte* is given by

$$f_E = \sqrt{1 + \Delta l/l} - \sqrt{\Delta l/l} \tag{II-C-40}$$

where Δl is the cell length over which atomic fluorescence *is not* measured and l is the cell length over which atomic fluorescence *is* measured. As $\Delta l/l$ becomes very large, f_E approaches zero; as $\Delta l/l$ becomes very small, f_E approaches unity.

The f_E factor for atomic fluorescence excited by a line source or molecular luminescence excited by a line or narrow band source,* assuming a high concentration of analyte, is given by

$$f_E \simeq \exp -\bar{k}\Delta l = 1 - \frac{\bar{k}\Delta l}{1!} + \frac{(\bar{k}\Delta l)^2}{2!} - \frac{(\bar{k}\Delta l)^3}{3!} + \cdots \tag{II-C-41}$$

where Δl is defined as previously and $\bar{k}$ is the effective (average) absorption coefficient of the analyte atoms or molecules averaged over the half width of the line source.* If $\bar{k}\Delta l \gtrsim 0.05$, then

---

* A line source is either a source that emits spectral lines (e.g., an electrodeless discharge lamp or a hollow cathode lamp) or a continuum source with excitation monochromator having a spectral bandwidth (bandpass) s smaller than the half width of the absorption line or band.

$$f_E \cong 1 - \bar{k}\Delta l \qquad \text{(II-C-42)}$$

and if $\bar{k}\Delta l \gtrsim 0.01$, then

$$f_E \cong 1 \qquad \text{(II-C-43)}$$

For the front surface case, there can be *no prefilter effect*, and so $f_E = 1$ for either atomic fluorescence or molecular luminescence spectrometry for any concentration of analyte.

d. SPECTRAL INTERFERENCES

If foreign species are present which absorb over the wavelength range corresponding to the excitation process, the luminescence radiance is reduced by a factor f_C. Similarly, in the presence of foreign species that absorb over the wavelength range corresponding to the emission (luminescence) process, the luminescence radiance is reduced by a factor f_D. For *atoms*, the factor $f_C f_D$ is generally close to unity; that is,

$$f_C f_D \simeq 1$$

However, for *molecules* in the condensed phase, absorption and luminescence occur over wide spectral ranges, and so spectral interferences are quite common:

$$f_C f_D \ll 1$$

For the line source case† and for right angle illumination measurement of molecular luminescence, f_C is given by

$$f_C = \exp\left[-\sum_i \bar{k}_i l\left(1 + \frac{\Delta l}{l}\right)\right] \qquad \text{(II-C-44)}$$

where $\bar{k}_i$ is the effective absorption coefficient (averaged over the source line width and evaluated at the excitation wavelength) for interferent i and the

---

* The half width of a line source is either the half width of the exciting line or the spectral bandwidth s of the excitation optical system when a continuum source with monochromator is utilized.

† Only the line source case is considered here because it is the most common means of excitation of molecules. The source width in the line source case is simply the spectral bandwidth (bandpass) s of the optical dispersive device used to isolate a narrow wavelength region from the continuum excitation source; or, it is the half width $\Delta\lambda_s$ of the line source (e.g., a mercury arc lamp), assuming that only one line is isolated. If more than one line from a "line" source is isolated and allowed to excite the sample, if the source also has a spectral continuum, or if both circumstances obtain, summation techniques must be used. This can be handled with the previous equations. However, no such special case is given here.

summation is over all interferents. The exponential factor in equation II-C-44 can be expanded to give

$$f_C = 1 - \sum_i \bar{k}_i l \left(1 + \frac{\Delta l}{l}\right) + \frac{\left[\sum_i \bar{k}_i l(1 + \Delta l/l)\right]^2}{2!} - \cdots \quad \text{(II-C-45)}$$

and for small $\sum_i \bar{k}_i$ values

$$f_C = 1 \qquad \text{(II-C-46)}$$

On the other hand, as $\sum_i \bar{k}_i l(1 + \Delta l/l)$ increases, f_C decreases below unity. For the case of front surface illumination measurement, the factor f_C is given by

$$f_C = \exp -\sum_i \bar{k}_i l \qquad \text{(II-C-47)}$$

where l is the total cell path length (see Figure II-C-1e).

Similarly, the factor f_D for the case of right angle illumination measurement is given by

$$f_D = \exp -\sum_j \bar{k}_j L \left(1 + \frac{\Delta L}{L}\right) \qquad \text{(II-C-48)}$$

where $\bar{k}_j$ is the effective absorption coefficient (averaged over the measurement wavelength interval and evaluated at the measured luminescence wavelength) of species j, the summation is over all absorbers of luminescence, and all other terms are as previously defined. The exponential terms in equation II-C-48 can be expanded, and so

$$f_D = 1 - \sum_j \bar{k}_j L \left(1 + \frac{\Delta L}{L}\right) + \frac{\left[\sum_j \bar{k}_j L(1 + \Delta L/L)\right]^2}{2!} - \cdots \quad \text{(II-C-49)}$$

and for small $\sum_j \bar{k}_j$ values.

$$f_D = 1 \qquad \text{(II-C-50)}$$

On the other hand, as $\sum_j \bar{k}_j L(1 + \Delta L/L)$ increases, f_D decreases below unity. For the case of front surface illumination measurement, f_D is given by

$$f_D = \exp -\sum_j \bar{k}_j l \qquad \text{(II-C-51)}$$

which is identical to equation II-C-47 for f_C.

e. COMBINATION OF PREFILTER AND POSTFILTER EFFECTS AND SPECTRAL
   INTERFERENCES

The luminescence radiance $B_L{}^c$ corrected for all pre-filter and postfilter effects and for spectral interferences is given by

$$B_L{}^c = B_L f_E f_I f_C f_D \tag{II-C-52}$$

where B_L is the luminescence radiance (see Section II-C-1) for a system with no prefilter, postfilter, or spectral interference effects.

For atoms in atomic fluorescence transitions, $f_C f_D$ is close to unity (see Section II-C-2-d), and so for the right angle case

$$B_{L\text{ atom}}^c = B_L f_E f_I \tag{II-C-53a}$$

and for the front surface case

$$B_{L\text{ atom}}^c = B_L \tag{II-C-53b}$$

For molecules in the condensed phase in molecular luminescence, f_I is nearly unity, as noted earlier, and so for the right angle case

$$B_{L\text{ mol}}^c = B_L f_E f_C f_D \tag{II-C-54a}$$

and for the front surface case

$$B_{L\text{ mol}}^c = B_L f_C f_D \tag{II-C-54b}$$

f. COMPARISON OF RIGHT ANGLE AND FRONT SURFACE
   ILLUMINATION MEASUREMENT

It is apparent from the foregoing discussion and expressions that the front surface case should be of more use at high concentrations of analyte than the right angle case because $f_E f_I \cong 1$ in the former instance and can be less than unity in the latter for high concentrations of analyte. In the front surface case, incomplete illumination of the front surface merely leads to a reduction in the size of the luminescing surface toward the measurement system. At low concentrations, front surface and right angle illumination measurement should give essentially the same response and same slope of B_L versus analyte concentration. However, the right angle case is considerably less prone to interference from scattering of excitation radiation than the front surface case—especially for molecular luminescence of molecules in the condensed phase. Because of this scattering problem and because of the limited use of luminescence at high analyte concentrations, *the right angle illumination measurement system is recommended for analytical use.*

g. SHAPES OF GROWTH CURVES

Growth curves* consist of plots of the logarithm of the luminescence radi-ance B_L or a function of the luminescence radiance versus the logarithm of the concentration of the absorbers n_o. Since the absorption coefficients k_o and $\bar{k}$ for the analyte are directly proportional to n_o (also refer to Sections II-C-2-e and II-C-2-f), the shape of log B_L versus log n_o curves can be predicted from the preceding relationships. It should be emphasized that the

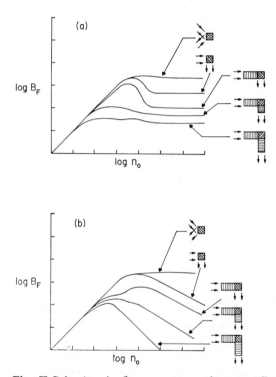

Fig. II-C-4. Atomic fluorescence growth curves [log (luminescence radiance) versus log (concentration of absorbers)] for atoms (a) excited by a continuum and (b) excited by a line source. The influence of the prefilter and postfilter effects is shown.

* Growth curves generally, and in astrophysics especially, refer to plots of a function of total absorption versus a function of absorber concentration, such as in Figure II-C-3. However, in the remainder of this book, growth curves are used to represent the fundamental plots of luminescence radiance B_L versus absorber concentration n_o. This designation of "growth curves" has been used in the past by Alkemade[2,3] and by Winefordner.[5-7] Note that the plots in Figure II-C-3 are useful in predicting atomic fluorescence growth curves for the continuum-source case.

log B_L versus log n_o curves are very important curves in analytical luminescence spectrometry because the shape of the growth curves is the major
factor in determining the shape of the analytical curves (plots of the logarithm of the instrumental signal S_L due to luminescence versus the logarithm
of the concentration C_o of the analyte solution placed in the measurement
cell).

Figure II-C-4 presents typical growth curve shapes for atomic fluorescence
spectrometry of atoms excited with a continuum source (Figure II-C-4a) and
with a line source (Figure II-C-4b) and for a variety of illumination measurement conditions—that is, both prefilter and postfilter effects are shown. In
Figure II-C-5, typical growth curve shapes are given for atomic fluorescence

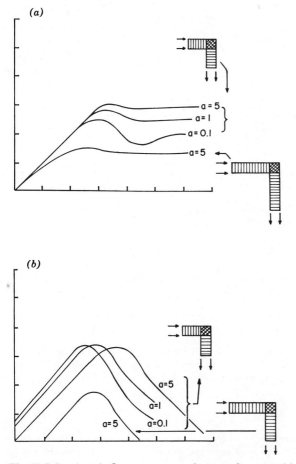

Fig. II-C-5. Atomic fluorescence growth curves for atoms (a) excited by a continuum and
(b) excited by a line source. The influence of the magnitude of the a parameter is shown.

spectrometry of atoms with different a parameter values excited with a continuum source (Figure III-C-5a) and with a line source (Figure III-C-5b). Figure II-C-6 illustrates typical growth curve shapes for molecular luminescence spectrometry of molecules in the condensed phase, assuming that a line source* of excitation is used; the influence of a variety of illumination measurement conditions and the influence of spectral interferences are shown in Figure II-C-6. It should be noted that the prefilter and postfilter effects are *only* present for the right angle case (not the front surface case).

It should be noted that for atomic and molecular luminescence spectrometry, incomplete illumination measurement in the right angle case assuming a constant source flux only leads to a reduction of the luminescence radiance at high concentrations of absorbers. Also in molecular luminescence spectro-

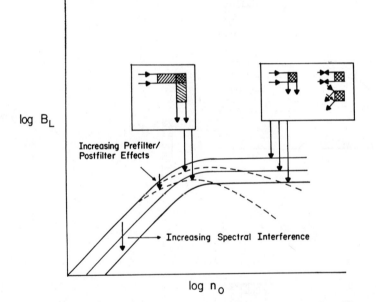

Fig. II-C-6. Molecular luminescence growth curves for molecules in the condensed phase excited by a "line" source. The influence of the prefilter and postfilter effects for the right angle illumination measurement method is shown, as well as the influence of spectral interferents.

* No growth curves are given for the continuum-source case for molecular luminescence; these are of *no analytical use*. Most cases of molecular luminescence can be approximated by the line-source case.

metry, absorbing interferents (if constant in concentration) simply shift the growth curves to the right and down. Finally, the front surface case of illumination measurement does have definite advantages for high concentration of absorbers.

As stated previously, analytical curve plots of $\log S_L$ versus $\log C_o$ should have the same shape as the growth curve plots of $\log B_L$ versus $\log n_o$ as long as S_L is linear with B_L and n_o is linear with C_o. The instrumental signal S_L should be linear with B_L over a large range of radiances (see Chapter III). The concentration of absorbers n_o should also be linear with respect to C_o over a wide concentration range. Deviations from this linearity are discussed in Section II-D. Therefore, growth curves are extremely useful to the analytical spectroscopist both for explaining anomalous behavior of analytical curves and for explaining interferences.

3. CRITERIA FOR APPLYING RADIANCE EXPRESSIONS FOR LUMINESCENCE TO REAL SYSTEMS

a. CONDITIONS

The previous radiance expressions can be used for estimating the radiance of luminescence if the absorption cell is in thermodynamic equilibrium. For thermodynamic equilibrium to exist in a system, the temperature must be the same at all points and all forms of energy must be equilibrated among each other; that is, *the principle of detailed balance* must be valid. Thermodynamic equilibrium is usually rapidly achieved; it is therefore present for molecules in the condensed phase. Thermodynamic equilibrium, however, is more difficult to achieve in high temperature atomizers used in atomic fluorescence flame spectrometry.

b. THERMODYNAMIC EQUILIBRIUM IN FLAME ATOMIZERS[1,2,10-12]

Because of the importance of flames as atomizers in atomic fluorescence spectrometry, a discussion of thermodynamic equilibrium with respect to flames is warranted. In a flame or other high temperature plasma, strict thermodynamic equilibrium is *never* possible because the flame radiates heat to the environment, undergoes secondary combustion at the flame boundaries, and undergoes convection (turbulence); these processes lead to a net transfer of heat, mass, and radiation within the flame.

However, even though net transfer of heat, mass, and radiation occurs within the flame gases, it is still possible to talk about a *local thermodynamic equilibrium* as long as the *transfer rates are small* compared with the rate of equipartitioning of energy over the different energy forms (i.e., over internal degrees of freedom) and over dissociation and ionization products and re-actants. If the flame is produced by a *laminar flow* of gases and if there is an outer flame sheath surrounding the analytical flame, the region of a local thermodynamic equilibrium can be extended over a considerable portion of the flame gases. In turbulent flames produced using total-consumption neb-ulizer burners, the local region can be very small, although we may consider the entire flame to be in approximate thermodynamic equilibrium.

The equilibration of flame energy over the various translational and internal degrees of freedom and over molecular dissociation and atomic ionization products requires an exhange of energy with the surrounding flame gas mole-cules. A species within the flame gases undergoes about 10^6 collisions within 1 msec (the time for the gases to travel above the reaction zone to a height corresponding to the lower edge of the measurement system). Therefore, equilibration of energy within the various degrees of freedom and flame gas products approximately occurs, and so the population of the internal degrees of freedom, the distribution of particle velocities, the distribution of dissocia-tion and ionization products, and the spectral distribution of radiation are governed by a single parameter T, the temperature of the system.

c. THERMODYNAMIC EQUILIBRIUM IN NONFLAME ATOMIZERS[1,2,10–12]

Nonflame atomizers are usually a furnace containing hot gases; thus they should be nearly ideal systems and approximately in thermodynamic equi-librium.

d. THERMODYNAMIC EQUILIBRIUM IN THE CONDENSED PHASE[11–13]

Molecules in fluidly liquid solutions should reach equilibrium rapidly at room temperature. Molecules in rigid solid solutions (e.g., a frozen organic solvent) should be in an equilibrium indicative of the molecules at room temperature prior to freezing.

e. STATISTICAL DISTRIBUTION LAWS FOR A SYSTEM IN THERMODYNAMIC EQUILIBRIUM[1,2]

The five statistical distribution laws for the various forms of energy are the Maxwell law for the velocity distribution of particles, the Boltzmann law for the population of various discrete levels of internal energy of a particle, the mass action law relating con-centration of reaction products to those of reactants at equilibrium, the Saha law relating the concentrations of neutral species to those of the corresponding ionic species and free electrons, and the Planck Law of radiation for a black body.

(1) *Maxwell Law.* The fraction of particles with mass m that have a kinetic energy E_j with a spread of dE_j is given by the Maxwell law

$$f(E_j)\, dE_j = \frac{2}{\sqrt{\pi}}\, \sqrt{E_j/(kT)^3}\, (\exp\, -E_j/kT)\, dE_j \qquad \text{(II-C-55)}$$

where k is the Boltzmann constant.

(2) *Boltzmann Law.* The population of the u level n_u as compared with the total population of all levels n_t is given[4] by the Boltzmann law as

$$\frac{n_u}{n_t} = \frac{g_u \exp\, -E_u/kT}{\sum_i g_i \exp\, -E_i/kT} \qquad \text{(II-C-56)}$$

where the g's are the statistical weights of the various levels designated by subscripts, E_i is the excitation energy of the i level, and the summation is the electronic partition function. The Boltzmann law describes the relative population of the electronic, vibrational, and rotational states.

(3) *Mass Action Law.* The mass action law for the equilibrium

$$AB \rightleftharpoons A + B \qquad \text{(II-C-57)}$$

is given by

$$K_d(T) = \frac{n_A n_B}{n_{AB}} \qquad \text{(II-C-58)}$$

where the n's are concentrations of the respective species (cm^{-3}) and $K_d(T)$ is the equilibrium constant for dissociation which depends on the temperature T.

(4) *Saha Law.* The Saha law for the ionization equilibrium

$$A = A^+ + e \qquad \text{(II-C-59)}$$

is given by

$$K_i(T) = \frac{(n_{A^+})(n_e)}{n_A} \qquad \text{(II-C-60)}$$

where $K_i(T)$ is the temperature-dependent ionization equilibrium constant. The ionization constant $K_i(T)$ and the dissociation constant $K_d(T)$ can be calculated from well-known expressions in thermodynamics (see Appendix 1).

(5) *Planck Law of Radiation.* The spectral radiance $B_{B\lambda}$ of a black body at temperature T in thermodynamic equilibrium is given by

$$B_{B\lambda} = \left(\frac{C_1}{\lambda_o{}^5}\right)\left(\exp\frac{hc}{\lambda_o kT} - 1\right)^{-1} \qquad \text{(II-C-61)}$$

where C_1 is a constant for any spectral line ($C_1 = 8\pi hc^2$, erg sec^{-1} cm^{-3}, where c is the speed of light, h is the Planck constant, and λ_o is the wavelength of the line peak), k is the Boltzmann constant (erg $°K^{-1}$), and T is the temperature of the black body ($°K$). A black body has by definition an absorption factor α_λ of unity at all wavelengths; that is, it completely absorbs any radiation incident upon its surface. The absorption factor α_λ is defined by Kirchoff's law

$$\alpha_\lambda = \frac{B_\lambda}{B_{B\lambda}} \qquad \text{(II-C-62)}$$

where B_λ is the spectral radiance of any radiating body with $\alpha_\lambda < 1$. For a black body, the emissivity ε_λ is the same as the absorption factor α_λ if the body is in thermodynamic equilibrium. The Kirchhoff law holds exactly if the body is enclosed within a furnace at the same temperature as the body and even holds approximately for open flames, where radiation equilibrium rarely exists. The Kirchhoff law also holds for the emission coefficient e_λ, the emissivity per centimeter of thickness of the radiating layer, and the absorption coefficient k_λ, the absorption factor per centimeter of absorbing layer (see equation II-C-15 and subsequent equations and discussion).

f. THE ATOMIC ABSORPTION COEFFICIENT[14]

The equations derived in Sections II-C-1 and II-C-2, feature the maximum absorption coefficient k_o (in equations for B_L, assuming that a continuum source is used) and the average absorption coefficient $\bar{k}$ (in equations for B_L, assuming that a line source is used). It was previously stated that k_o and $\bar{k}$ were directly proportional to the concentration n_o of analyte in the lower state* involved in the absorption transition.

The relation between k_o and n_o is given by

$$k_o = \frac{2\sqrt{\ln 2}\,\lambda_o{}^4 n_o A_{ul}}{8\pi\sqrt{\pi c}\,\Delta\lambda_D} \qquad \text{(II-C-63)}$$

---

* It is assumed that the lower state l is the ground state in the discussion to follow and so $n_o = n_l$. If l is actually an excited state,

$$n_l = n_o\left(\frac{g_l}{g_o}\right)e^{-E_l/kT}$$

or

$$k_o = \frac{2\sqrt{\ln 2}\ \pi e^2 \lambda_o{}^2 n_o f_{lu}}{\sqrt{\pi}\ mc^2 \Delta\lambda_D} \qquad (\text{II-C-64})$$

where the first expression is a consequence of the quantum theory of radiation and the second is a consequence of the classical theory of radiation.* Refer to Appendix 2 for a discussion of relations between A_{ul} and other spectral parameters. The terms in the foregoing equations are: λ_o is the peak absorption wavelength (cm), n_o is the concentration of analyte in the lower state involved in the absorption transition (cm^{-3}), A_{ul} is the Einstein coefficient of spontaneous emission (sec^{-1}), c is the speed of light (cm sec^{-1}), $\Delta\lambda_D$ is the Doppler half width of the atomic line (cm), f_{lu} is the absorption oscillator strength* (no units), m and e are the mass (g) and charge (esu) of the electron, respectively, and π is 3.1418 .... The strict linearity between k_o and n_o should be noted.

The relation between $\bar{k}$ and n_o is given by

$$\bar{k} = k_o \delta_o \qquad (\text{II-C-65})$$

where k_o is related to n_o by the previous two equations and δ_o is a factor to correct for the finite width of the source-line and the relative source line and absorption-line profiles. It has been shown that δ_o can be approximated by the Voigt profile integral evaluated under very special conditions; that is,

$$\delta_o = \frac{a}{\pi} \int_{-\infty}^{+\infty} \frac{\exp{(-y^2)}\ dy}{a^2 + (\bar{v} - y)^2} \qquad (\text{II-C-66})$$

where a is the damping constant for the absorption line and is given (also see equation II-C-13b) by

$$a = \sqrt{\ln 2}\ \left(\frac{\Delta\lambda_C}{\Delta\lambda_D}\right) \qquad (\text{II-C-67})$$

where $\Delta\lambda_C$ and $\Delta\lambda_D$ are the collisional and Doppler half widths of the absorption line, respectively (cm), and $\bar{v}$ is the effective v value determined by the ratio of the source-line half width $\Delta\lambda_S$ and the absorption-line half width $\Delta\lambda_A$ and is

$$\bar{v} = \frac{\Delta\lambda_S}{\Delta\lambda_A} \qquad (\text{II-C-68})$$

---

* From the classical electron theory of dispersion, the optical behavior of n_o atoms per cubic centimeter was represented by η_o quasi-elastically bound electrons per cubic centimeter (i.e., the so-called dispersion electrons). The ratio of η_o/n_o was called f, the oscillator strength. The f value for a spectral line can be regarded as a measure of the degree to which the ability of the atom to absorb or emit this line resembles such an ability on the part of a classical oscillating atom.

The value of $\Delta\lambda_A$ is approximately the quadratic sum of the Doppler and collisional half widths, that is,

$$\Delta\lambda_A = \sqrt{\Delta\lambda_D{}^2 + \Delta\lambda_C{}^2} \qquad \text{(II-C-69)}$$

It is apparent that $\bar{k}$ will never be greater than k_o and will usually be smaller (see Table II-C-1).

Table II-C-1. Values of $\delta_o{}^a$

			a parameter			
v	0	0.5	1.0	2.0	5.0	10.0
0	1.00	0.61	0.42	0.26	0.11	0.057
0.2	0.96	0.60	0.42	0.25	0.11	0.057
0.4	0.85	0.56	0.40	0.25	0.11	0.057
0.6	0.70	0.50	0.38	0.24	0.11	0.057
0.8	0.52	0.43	0.34	0.23	0.11	0.056
1.0	0.37	0.35	0.30	0.22	0.11	0.056

[a] Taken from L. de Galan, W. W. McGee, and J. D. Winefordner, *Anal. Chim. Acta*, **37**, 436 (1967); δ_o has no units.

g. THE MOLECULAR ABSORPTION COEFFICIENT[8,11]

In the equations derived in Sections II-C-1 and II-C-2, the maximum absorption coefficient k_o and the average absorption coefficient $\bar{k}$ appear just as for atomic fluorescence. The values of k_o (and $\bar{k}$) for molecules, however, are much less than for atoms because of the wide wavelength spread of molecular bands compared with atomic lines. This phenomenon can be rationalized as follows; both atoms and molecules absorb light in the ultraviolet-visible wavelength range to about the same extent; but the energy absorbed by atoms is in a range of ~ 0.1 Å and the energy absorbed by molecules is within a wide wavelength range of the order of 100 Å or more. This can be made evident in the following manner. The integrated absorption coefficient for an electronic transition is given by

$$\int_0^\infty k_\lambda \, d\lambda = \frac{\sqrt{\pi} k_o \, \Delta\lambda_G}{2\sqrt{\ln 2}} = \frac{\pi e^2}{mc^2} \lambda_o{}^2 n_o f_{lu} \qquad \text{(II-C-70)}$$

where all terms have been previously defined. For a molecular band, $\Delta\lambda_G$ is of the order of 100 Å or more; but for an atomic line, $\Delta\lambda_G$ is of the order of

0.1 Å. If we assume that the concentration of absorbers n_o and the oscillator strength for the electronic transition f_{lu} are identical for both the atoms and the molecules, then k_o for atoms must exceed k_o for molecules by the ratio of the half width of the molecular band to the half width of the atomic line, or by about 10^3 or greater.

The average molecular absorption coefficient $\bar{k}$ can be given by an equation similar to equation II-C-65 where δ_o is evaluated in the same manner as for atoms, except that $\bar{v}$ is now the ratio of the monochromator spectral bandwidth s to the absorption band half width.

It is customary to separate the concentration of analyte C_o (mole liter^{-1}) from the absorption coefficient $\bar{k}$ and k_o and to define the resulting concentration independent absorption coefficient, called an average or peak molar extinction coefficient $\bar{\varepsilon}$ or ε_o, with respect to common rather than natural logarithms. Therefore

$$\bar{k} = 2.3 C_o \bar{\varepsilon} \tag{II-C-71}$$

$$k_o = 2.3 C_o \varepsilon_o \tag{II-C-72}$$

The first expression is useful for the line-source case and the second is useful for any situation in which the peak molar extinction coefficient is measured.

h. CONCENTRATION OF SPECIES IN THE LOWER STATE INVOLVED IN THE ABSORPTION PROCESS[10]

The concentration n_o of atoms (or molecules) in the lower state involved in the absorption process is determined by the Boltzmann distribution equation and is

$$n_o = n_t \frac{g_o}{Z_T} \exp \frac{-E_i}{kT} = n_t \left(\frac{g_o}{Z_T}\right) 10^{-5040 V_i/T} \tag{II-C-73}$$

where the g_o is the statistical weight of the lower level (usually ground or very near ground level), n_t is the total concentration of atoms (or molecules) in all states, E_i is the excitation energy (erg), V_i is the excitation energy (eV), T is the temperature of the system (°K), and Z_T, the electronic partition function, is given by

$$Z_T = \sum_i g_i \exp \frac{-E_i}{kT} \tag{II-C-74}$$

where g_i is the statistical weight of state i, E_i is the excitation energy of state i, and all other terms are as defined for equation II-C-73.

If $E_i \lesssim 0.5$ volt for the lowest lying excited state and if $T \gtrsim 3000°K$, then $n_o \sim n_t \sim n_g$ within 5% or better. If E_i for the lowest lying excited state is greater, then T can be greater and still $n_o \sim n_t \sim n_g$ within 5% or better. For atoms with many levels (e.g., transition metals, rare earths, and actinides), $n_o < n_t$. For molecules, $n_o \simeq n_t$, since excited electronic states are generally much greater than 0.5 eV above the ground state, and the temperature of condensed phase systems is generally below 300°K. A molecule in the condensed phase at room temperature will be in low lying vibrational levels (mostly in the lowest level) of the ground electronic state. A molecule in the condensed phase at liquid nitrogen temperature (77°K), will of course be predominately in the lowest vibrational level of the ground electronic state. For the purpose of this chapter, we assume that all molecules are in the same lowest vibrational electronic state. If this is not true, then the absorption equations must contain a summation over all possible lower levels. This will not produce any significant departure from the conclusion for a single lower level* but rather will result only in a slightly different spectral profile for the absorption process.

i. GENERAL EXPRESSIONS FOR ATOMIC FLUORESCENCE POWER[3] YIELDS

There are five basic types† of atomic fluorescence: resonance, R; normal direct line, NDL; thermally assisted direct line, TDL; normal stepwise line, NSL; and thermally assisted stepwise line, TSL. These five basic types are discussed with respect to a three-state energy diagram with the uppermost state designated 2, the next lower state designated 1, and the ground state designated o (refer to Figure II-C-7). Luminescence power yield

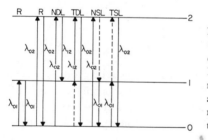

Fig. II-C-7. Basic types of atomic fluorescence (solid lines represent radiational processes; dashed lines represent radiationless processes): R = resonance absorption-fluorescence, NDL = normal direct-line fluorescence, TDL = thermally assisted direct-line fluorescence, NSL = normal stepwise-line fluorescence, TSL = thermally assisted stepwise-line fluorescence.

* Actually k_o and $\Delta\lambda_G$ will vary slightly bacause of the population distribution of lower levels.

† Resonance luminescence generally means the same wavelength of radiation that is absorbed will be emitted. Direct-line luminescence generally means that the excited state is involved in both the absorption and luminescence processes but the energy absorbed is *not* equal to the energy emitted. Stepwise-line luminescence generally means that the excited state of the absorption and luminescence processes are different and, again, the energy absorbed is *not* equal to the energy emitted.

expressions for these specific cases are furnished for completeness, even though general expressions for the quantum yield of atomic fluorescence were given and discussed in Section II-A. For the first resonance $R1$ process involving state 1, (i.e., excitation of a species to state 1 from o and radiational deactivation of 1 to produce species in state o), the power (quantum) yield is

$$Y_{1o} = \frac{A_{1o}}{A_{1o} + K_1} \qquad \text{(II-C-75)}$$

where A_{1o} is the Einstein coefficient of spontaneous emission* from state 1 and K_1 is the total pseudo-first-order rate constant for radiationless deactivation (quenching) for the first level.† The power (quantum) yield for the second resonance $R2$ process involving state 2 (i.e., excitation of a species to state 2 from o and radiational deactivation of 2 back to state o) is given by

$$Y_{2o} = \frac{A_{2o}}{\sum_i A_{2i} + K_2} \qquad \text{(II-C-76)}$$

where A_{2o} is the Einstein coefficient of spontaneous emission‡ between state 2 and state o, A_{2i} is the total Einstein coefficient for spontaneous emission§ from state 2 (i.e., $A_{2i} = A_{2o} + A_{21}$), and K_2 is the total pseudo-first-order rate constant for radiationless deactivation (quenching) from state 2 and is similar to K_1. The resonance processes are of the greatest *analytical importance* in atomic fluorescence.

In normal direct line fluorescence (NDL) the species is excited from state o to state 2, and then the fluorescence between states 2 and 1 is of interest.‖ For the case

$$Y_{\text{NDL}} = Y_{21} \left(\frac{\lambda_{o2}}{\lambda_{12}}\right) \qquad \text{(II-C-77)†}$$

where λ_{o2} is the absorption coefficient, λ_{12} is the fluorescence wavelength,

---

* The A_{1o} is sometimes given symbol k_{1o} or k_F, representing the fluorescence first-order rate constant (see Section II-A).

† The most common means of quenching is via collisional deactivation of the analyte with a foreign species of concentration n_Q and with a second-order rate constant of k_Q. Thus $K_1 = \sum_i n_{Qi} k_{Qi}$, where the summation is over all quenchers of this type (see Section II-A).

‡ The A_{2o} is sometimes designated k_{2o} or k_F, representing the fluorescence first-order radiational rate constant for the process $2 \to o$.

§ The A_{21} is also sometimes designated k_{21}, representing the fluorescence first-order rate radiational rate constant for the process $2 \to 1$.

‖ For simplicity, Y_X is used to represent the luminescence power yield for process X (this designation is used throughout the remainder of the book). Also Y_{ij} is used to represent the luminescence quantum yield for the radiational transition $i \to j$.

and Y_{21} is the quantum yield for the fluorescence process from state 2 to state 1. We then have

$$Y_{21} = \frac{A_{21}}{A_{2i} + K_2} \qquad \text{(II-C-78)}$$

In thermally assisted direct line fluorescence (TDL) the species is thermally activated into state 1, from which radiational activation by wavelength λ_{12} results in the species in state 2. The fluorescence of interest is then between states 2 and o resulting in radiation of wavelength λ_{o2}. Therefore

$$Y_{\text{TDL}} = Y_{2o}\left(\frac{\lambda_{12}}{\lambda_{o2}}\right) \qquad \text{(II-C-79)}$$

where Y_{2o} is given by equation II-C-74. It can be seen that Y_{TDL} will always greatly exceed Y_{NDL}, since $Y_{2o} > Y_{21}$ and $\lambda_{12} > \lambda_{o2}$. However, TDL requires that the absorbers be in state 1, whereas NDL requires that the absorbers be in state o, the ground state. Thus the amount of energy absorbed in NDL will greatly exceed the amount of energy absorbed in TDL, and in most cases, the radiance of NDL will exceed the radiance of TDL as long as the energy of level 1 is 0.5 eV or greater above the ground state. The NDL and TDL processes are of *some analytical importance* in atomic fluorescence spectrometry.

In normal stepwise-line fluorescence (NSL) the species is radiationally excited to some high level, say level 2, and then it undergoes radiational and/or nonradiational deactivation to state 1, and the atomic fluorescence process of interest proceeds from there. Thus in this case, radiation of wavelength λ_{o2} is absorbed and radiation of λ_{o1} is emitted as fluorescence. The fluorescence power yield Y_{NSL} is given by

$$Y_{\text{NSL}} = Y_{1o}\left(\frac{\lambda_{o2}}{\lambda_{o1}}\right)\left(\frac{k_{21}}{k_{21} + \sum_i A_{2i} + K_2}\right) \qquad \text{(II-C-80)}$$

where all terms have been previously defined except k_{21}, which represents the first-order rate constant for intersystem crossing between levels 2 and 1.

In thermally assisted stepwise-line fluorescence, the species is radiationally activated from the ground to the first excited state, whereupon the atom is radiationlessly activated from state 1 to state 2 and fluorescence then occurs, which results in the species in state o. Thus in this case, the atom is excited via radiation of wavelength λ_{o1}, and fluorescence results in radiation of wavelength λ_{o2}. The fluorescence power yield Y_{TSL} is given by

$$Y_{\text{TSL}} = Y_{20}\left(\frac{\lambda_{o1}}{\lambda_{o2}}\right)\left(\frac{k_{12}}{k_{12} + A_{10} + K_1}\right) \qquad \text{(II-C-81)}$$

where all terms have been previously defined except k_{12}, which is the first-order rate constant* for intersystem crossing from state 1 to state 2.

It should be noted that stepwise-line fluorescence requires the *coupling* of states via collisions with gas molecules in the cell. For stepwise-line fluorescence to be analytically useful, it is necessary for the analyte atoms to have large coupling constants with the cell gases; but they must also have small quenching cross sections with the same molecules. These are mutually conflicting requirements, and so stepwise-line fluorescence has so far been of little analytical use.†

j. GENERAL EXPRESSIONS FOR MOLECULAR LUMINESCENCE POWER[8] YIELDS

There are two major types of molecular luminescence resulting from photoactivation: fluorescence and phosphorescence (refer to Figure II-C-8). The expressions for the molecular luminescence power yields are analogous

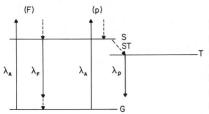

Fig. II-C-8. Basic types of molecular luminescence (solid lines represent radiational processes; dashed lines represent radiationless processes): F = fluorescence, P = phosphorescence, ST = singlet-triplet intersystem crossing, S = first excited singlet state, T = first triplet state, and G = ground singlet state.

to those for NDL for atoms. General expressions for fluorescence and phosphorescence quantum yields were given in Section II-B; however specific luminescence power yields for the specific cases of fluorescence and phosporescence are repeated here for completeness of this section. Assuming that the triplet energy level is appreciably below the first excited singlet energy level so that there is negligible intersystem crossing from the triplet T to the first excited singlet state S, then the fluorescence power yield is

$$Y_F = \left(\frac{\lambda_A}{\lambda_F}\right)\left(\frac{A_F}{A_F + K_S}\right) \tag{II-C-82}$$

---

* The coupling rate constant k_{12} depends on $\exp -\Delta E/kT$ when ΔE is the energy difference between states 1 and 2. The coupling rate constants k_{12} and k_{21} also depend on the probability for the processes to occur.

† In contrast to the lack of analytical use of NSL and TSL, phosphorescence of organic molecules that involves intersystem crossing and therefore coupling of levels is important analytically.

where λ_A and λ_F are the wavelengths for maximal absorption and fluorescence, A_F is the radiational rate constant* for deactivation of the first excited singlet state, and K_S is the total pseudo-first-order rate constant for radiationless deactivation of the first excited singlet state. Also, the phosphorescence power yield is given by

$$Y_P = \left(\frac{\lambda_A}{\lambda_P}\right)\left(\frac{k_{ST}}{A_F + K_S}\right)\left(\frac{A_P}{A_P + K_P}\right) \qquad \text{(II-C-83)}$$

where λ_A and λ_P are the wavelengths of maximal absorption and phosphorescence, k_{ST} is the first-order rate constant for intersystem crossing between T and S (i.e., it is a coupling constant that is dependent on spin-orbital interaction to lessen the restriction of spin change forbiddenness and, A_P and K_P are analogous to A_F and K_F, respectively, but are for the first excited triplet state.†

Therefore, to increase phosphorescence at the expense of fluorescence, the coupling constant k_{ST} must be increased. The denominators in equations II-C-82 and II-C-83, which effectively represent the total first-order rate constant for deactivation of the excited molecule, are decreased by lowering the temperature of the system and avoiding high concentrations of quenchers, such as dissolved oxygen in the solution. Therefore, in the limit

$$A_S + K_S \cong k_{ST} + A_S \qquad \text{and} \qquad A_P + K_P = A_P \qquad \text{(II-C-84)}$$

and then

$$Y_F \cong \left(\frac{\lambda_A}{\lambda_F}\right)\left(\frac{A_S}{A_S + k_{ST}}\right) \qquad \text{(II-C-85)}$$

$$Y_P \cong \left(\frac{\lambda_A}{\lambda_P}\right)\left(\frac{k_{ST}}{A_S + k_{ST}}\right) \qquad \text{(II-C-86)}$$

and so in this highly desirable analytical situation, every excited molecule produces *either* fluorescence or phosphorescence upon deactivation; that is, $Y_F + Y_P = 1$.

k. COMPARISON OF LUMINESCENCE RADIANCE WITH ABSORBED RADIANCE

The ratio of luminescence radiance to absorbed radiance for the same species (atomic or molecular) is the same as the ratio of instrumental signals in luminescence and absorption spectrometry, assuming that the same species

---

* Sometimes A_S is replaced by the symbol k_F, the first-order fluorescence rate constant (see Section II-B).

† Sometimes A_P is replaced by k_P, the first order phosphorescence rate constant (see Section II-B).

is measured by the same instrumental system. If it is also assumed that the noise level in all methods is the same (not valid for these two methods), then the ratio of radiances (or signals) is identical to the inverse ratio of limits of detection.* By using the previously derived expressions for B_L and for B_A† (B_A is the radiance absorbed producing B_L), it is possible to derive an expression for the ratio B_L/B_A. The resulting expression can be further simplified by assuming that the cell is square (i.e., $L = l$, that the solid angle subtended by the measurement system is the same for absorption and luminescence, that the analyte absorber is low in concentration, and that no spectral interferences are present. Finally $\sqrt{\pi/2}\sqrt{\ln 2}$ is assumed to be unity. Therefore, the ratio of luminescence to absorption radiance is given to a first-order approximation (for both atoms in the gas phase or molecules in the condensed phase, assuming that the same instrument and *source type* are used for both the absorption and luminescence measurements of the same species) by the gain factor $G_{L/A}$:

$$G_{L/A} = \frac{B_L}{B_A} = Y\frac{\Omega_E}{4\pi} \qquad \text{(II-C-87)}$$

where Y_L is the power luminescence yield and Ω_E is the solid angle over which radiation from the exciting source is collected and used to illuminate the sample cell containing the analyte.

For a good luminescer (atomic or molecular), $Y_L \sim 0.5$ and for a good experimental system, $\Omega_E/4\pi$ is of the order of 0.01, and so $G_{L/A} \sim 0.005$ at best. Therefore, the *absorption signal* should be of the order of 200 times greater than the *luminescence signal* for the same species measured by the same instrument. *If this same sort of ratio resulted for limits of detection, then luminescence spectrometry would never be used for analytical studies* because absorption spectrometry would be 200 times more sensitive in all cases. Obviously this is not true, since luminescence spectrometry for many atomic and molecular species is more sensitive than the corresponding absorption spectrometric method. The discrepancy is caused by the neglect of the noise levels in each method. For many species, the noise in the luminescence method is more than 200 times lower than the noise in the absorption method and so more than balances the greater signal strength in the absorption method. Also by substituting for Y_L in terms of rate constants, the variation of $G_{L/A}$ with experimental conditions can be predicted.

---

* One would expect the limit of detection to decrease as the signal increases if the noise is the same at all signals.

† $B_A = \int_0^\infty B_{S\lambda}(1 - \exp -k_\lambda l)\, d\lambda$

4. GLOSSARY OF SYMBOLS

a	Damping constant for absorption line, $\sqrt{\ln 2}\ \Delta\lambda_C/\Delta\lambda_D$, no units.
a	Damping constant for absorption line or band, $\sqrt{\ln 2}\ \Delta\lambda_L/\Delta\lambda_G$, no units.
a'	Damping constant for absorption of fluorescent line, $\sqrt{\ln 2}\ \Delta\lambda_C'/\Delta\lambda_D'$, no units.
A_F	Radiational rate constant for molecular fluorescence (radiational transition between first excited singlet and ground singlet), $\sec^{-1}$.
A_P	Radiational rate constant for molecular phosphorescence (radiational transition between first excited triplet and ground singlet), $\sec^{-1}$.
A_{ui}	Einstein coefficient of spontaneous emission (radiational first-order rate constant) for transition $u \rightarrow i$, $\sec^{-1}$.
A_{ul}	(same as A_{ui} except for transition $u \rightarrow l$), $\sec^{-1}$.
A_S	Surface area of cell (see Figure II-C-1), cm^2.
B_A	Radiance* absorbed by analyte in cell, $\mathrm{erg\ sec^{-1}\ cm^{-2}\ sr^{-1}}$.
$B_{B\lambda}$	Spectral radiance* of a black body at wavelength λ, $\mathrm{erg\ sec^{-1}\ cm^{-2}\ sr^{-1}\ nm^{-1}}$.
$B_{B\lambda_0}$	Spectral radiance* of a black body at wavelength λ_o, $\mathrm{erg\ sec^{-1}\ cm^{-2}\ sr^{-1}\ nm^{-1}}$.
$B_{C\lambda_0}^o$	Spectral radiance* of a continuum source at wavelength λ_o, $\mathrm{erg\ sec^{-1}\ cm^{-2}\ sr^{-1}\ nm^{-1}}$.
B_L	Luminescence radiance* of analyte, $\mathrm{erg\ sec^{-1}\ cm^{-2}\ sr^{-1}}$.
B_L'	Luminescence radiance* of analyte corrected for postfilter effect, $\mathrm{erg\ sec^{-1}\ cm^{-2}\ sr^{-1}}$.
B_L''	Luminescence radiance* of analyte corrected for prefilter effect, $\mathrm{erg\ sec^{-1}\ cm^{-2}\ sr^{-1}}$.
$B_L{}^c$	Luminescence radiance* of analyte corrected for postfilter and prefilter effects and for spectral interferences, $\mathrm{erg\ sec^{-1}\ cm^{-2}\ sr^{-1}}$.
$B_N{}^o$	Source line radiance,* $\mathrm{erg\ sec^{-1}\ cm^{-2}\ sr^{-1}}$.
$B_{S\lambda}{}^o$	Source spectral radiance,* $\mathrm{erg\ sec^{-1}\ cm^{-2}\ sr^{-1}\ nm^{-1}}$.
c	Speed of light, 3×10^{10}, $\mathrm{cm\ sec^{-1}}$.
C_1	$8\pi hc^2\lambda_o{}^{-5}$, $\mathrm{erg\ sec^{-1}\ cm^{-3}}$.
C_o	Concentration of analyte molecules in solution, $\mathrm{mole\ liter^{-1}}$.
dA_S	Surface area of volume element $dx\ dy\ l'$, cm^2.

---

* 10^7 erg $\sec^{-1} = 1$ watt.

$dB'_{L\lambda}$	Spectral luminescence radiance emitted by analyte within volume element $dx\, dy\, l'$, erg sec^{-1} cm^{-2} sr^{-1} nm^{-1}.
dE_j	Spread of kinetic energy for Maxwellian distribution of particle energies, erg.
dx	Cell dimension (see Figure II-C-1), cm.
dy	Cell dimension (see Figure II-C-1), cm.
$d\Phi'_{L\lambda}$	Spectral radiant flux luminesced by analyte in volume element $dx\, dy\, l'$, erg sec^{-1} nm^{-1}.
$d\Phi_{S\lambda}$	Spectral radiant flux absorbed by analyte in volume element $dx\, dy\, l'$, erg sec^{-1} nm^{-1}.
$d\Phi_{S\lambda}(x)$	Spectral radiant flux absorbed by analyte in volume element $dx\, dx\, l'$ at point x, erg sec^{-1} nm^{-1}.
$d\Phi_{S\lambda}{}^{o}$	Source spectral radiant flux (evaluated at source surface), erg sec^{-1} nm^{-1}.
e	Electron charge, coulomb or esu.
e_λ	Emission coefficient (emissivity per centimeter path length), cm^{-1}.
E_i	Energy of electronic level i, erg.
E_j	Energy of particle j, erg.
E_u	Energy of electronic level u, erg.
f_c	Correction factor for absorption of exciting light by interferents (for molecular luminescence), no units.
f_D	Correction factor for absorption of luminescence light by interferents (for molecular luminescence), no limits.
f_E	Correction factor for prefilter effect, no units.
f_I	Correction factor for postfilter effect, no units.
f_{lu}	Oscillator strength (absorption) for transition $l \to u$, no units.
f_s	Self-absorption correction factor (fraction of luminescence reabsorbed by absorbing media), no units.
$f(E_j)dE_j$	Fraction of particles with mass m having a kinetic energy of E_j and a spread of dE_j, erg.
$f(\lambda')$	Spectral distribution of luminescence, no units.
g_g	Statistical weight of ground level, no units.
g_i	Statistical weight of level i, no units.
g_o	Statistical weight of lower state o (normally ground), no units.
$G_{L/A}$	Ratio of luminescence to absorption radiances, no units.
h	Planck constant, 6.6×10^{23} erg sec.
k	Boltzmann constant 1.38×10^{-16} erg $^{\circ}$K^{-1} mol^{-1}.
$\bar{k}$	Average (effective) absorption coefficient over source line width, cm^{-1}.

k_{ij}	First-order rate constant for intersystem crossing from state i to state j, sec^{-1}.
$\bar{k}_i$	Average (effective) absorption coefficient at absorption wavelength by interferent i in molecular luminescence, cm^{-1}.
k_o	Absorption coefficient at absorption line center for pure Gaussian broadening, cm^{-1}.
k_o'	Absorption coefficient at fluorescence line center for pure Gaussian broadening, cm^{-1}.
k_{oi}	Absorption coefficient at absorption line center for pure Gaussian broadening for component i, cm^{-1}.
k_λ	Absorption coefficient at wavelength λ, cm^{-1}.
$k_{\lambda'}$	Absorption coefficient at wavelength λ', cm^{-1}.
K_u	Total pseudo-first-order rate constant for radiationless deactivation of level u, sec^{-1}.
$K_d(T)$	Equilibrium constant for dissociation at temperature T, particles cm^{-3}.
$K_i(T)$	Equilibrium constant for ionization at temperature T, particles cm^{-3}.
$\bar{k}_i$	Average (effective) absorption coefficient (averaged over the source line width) at the excitation wavelength for interferent i in molecular luminescence, cm.
$\bar{k}_j$	Average (effective) absorption coefficient at emission wavelength by interferent j in molecular luminescence, cm.
K_P	Total psuedo-first-order rate constant for nonradiational deactivation of first excited triplet state, sec^{-1}.
k_{ST}	Intersystem (first excited singlet to triplet) crossing rate constant for molecules, sec^{-1}.
K_S	Total pseudo-first-order rate constant for nonradiational deactivation of first excited singlet state, sec^{-1}.
l	Absorption path length (see Figure II-C-1), cm.
l'	Cell height (see Figure II-C-1), cm.
L	Emission path length (see Figure II-C-1), cm.
m	Mass of particle, g.
m_L	Magnification of source image in length direction, no units.
m_H	Magnification of source image in width direction, no units.
m_a	Atomic weight of emitting or absorbing atom, g.
n	Refractive index of solvent, no units.
n_x	Concentration of species x, cm^{-3}.
n_i	Concentration of analyte in state i, cm^{-3}.
n_o	Concentration of analyte in state o (ground), cm^{-3}.

n_t	Total concentration of analyte atoms in all electronic states, cm^{-3}.
s	Spectral bandwidth of monochromator, nm.
T_E	Absorption and reflection loss factor for entrance optics, no units.
T	Temperature of system, $^\circ K$.
v	Variable wavelength interval, $2\sqrt{\ln 2}\,(\lambda - \lambda_o)/\Delta\lambda_G$, no units.
I	Relative width of source line to absorption line, $\Delta\lambda_S/\Delta\lambda_A$, no units.
V_i	Excitation potential of state i, eV.
x	Cell dimension (see Figure II-C-1), cm.
y	Cell dimension (see Figure II-C-1), cm.
y	Integration variable, no units.
z	Cell dimension (l or L in Figure II-C-1) cm.
Y_λ (Y_L, or Y_X)	Luminescence power yield, no units.
$Y_\lambda(\lambda')$	Luminescence spectral power yield, no units.
Y_{ul}	Luminescence power yield for transition $u \rightarrow l$, no units.
Z_T	Electronic partition function, no units.
α_λ	Absorption factor (fraction of radiation absorbed), no units.
δ	Variable wavelength distance from $\lambda - \lambda_o$, nm.
δ_o	Factor to correct for finite source line-width and source line and absorption line profiles, no units.
Δl	Prefilter region (see Figure II-C-1), cm.
ΔL	Postfilter region (see Figure II-C-1), cm.
$\Delta\lambda_A$	Absorption line half width, nm.
$\Delta\lambda_C$	Collisional half width of absorption line, nm.
$\Delta\lambda_C{}'$	Collisional half width of fluorescent line, nm.
$\Delta\lambda_D$	Doppler half width of absorption line, nm.
$\Delta\lambda_D{}'$	Doppler half width of fluorescent line, nm.
$\Delta\lambda_G$	Gaussian half width of absorption line or band, nm.
$\Delta\lambda_L$	Lorentzian half width of absorption line or band, nm.
$\Delta\lambda_S$	Source line half width, nm.
$\bar\varepsilon$	Average extinction coefficient for analyte at absorption wavelength λ, liter mole^{-1} cm^{-1}.
ε_o	Extinction coefficient of analyte at absorption line center λ_o for pure Gaussian broadened band, liter mole^{-1} cm^{-1}.
η_0	Number of dispersion electrons per cubic centimeter, cm^{-3}.
λ	Absorption wavelength, cm(or nm).
λ'	Emission wavelength, cm(or nm).
λ_o	Peak absorption wavelength, cm (or nm).

λ_o' Peak emission wavelength, cm (or nm).

Ω_E Solid angle of radaion collected from excitation source and impinging upon sample cell, sr.

5. REFERENCES

1. C. T. J. Alkemade, "Excitation and Related Phenomena In Flames," in *Proceedings of the Xth Colloquium Spectroconicum Internationale*, E. R. Lippincott and M. Margoshes, Eds., Spartan Books, Washington, D. C., 1963.
2. C. T. J. Alkemade and P. T. J. Zeegers, in *Spectrochemical Methods of Analysis*, J. D. Winefordner, Ed., Wiley, 1971.
3. C. T. J. Alkemade, "A Theoretical Discussion of Some Aspects of Atomic Fluorescence Spectroscopy in Flames," in *Proceedings of the International Atomic Absorption Spectroscopy Conference*, Sheffield, England, July, 1969, Butterworth, London, 1970.
4. H. P. Hooymayers, *Spectrochim. Acta*, **23B**, 567 (1968).
5. P. J. T. Zeegers, R. Smith, and J. D. Winefordner, *Anal. Chem.*, **40**, (13), 26A (1970).
6. J. D. Winefordner, V. Svoboda, and L. J. Cline, *Crit. Rev. Anal. Chem.*, **1**, 233 (1970).
7. P. J. T. Zeegers and J. D. Winefordner, *Spectrochim. Acta*, **26B**, 161 (1971).
8. T. Förster, *Fluoreszenz Organisher Verbindungen*, Vandenhoeck and Ruprecht, Göttingen, 1951.
9. S. S. Penner, *Quantitative Molecular Spectroscopy and Gas Emissivities*, Addison-Wesley, Reading, Mass., 1959.
10. R. Mavrodineanu and H. Boiteux, *Flame Spectroscopy*, Wiley, New York, 1965.
11. C. Sandorfy, *Electronic Spectra and Quantum Chemistry*, Prentice-Hall. Englewood Cliffs, N.J., 1964.
12. K. Denbigh, *The Principles of Chemical Equilibrium*, Cambridge Press, Cambridge, 1961.
13. W. T. Simpson, *Theories of Electrons In Molecules*, Prentice-Hall, Englewood Cliffs, N.J., 1967.
14. A. C. G. Mitchell and M. W. Zemansky, *Resonance Radiation and Excited Atoms*, Cambridge, University Press, 1961.

D. PRODUCTION OF SPECIES TO BE EXCITED IN LUMINES-CENCE SPECTROMETRY

1. PRODUCTION OF ATOMS IN FLUORESCENCE SPECTROMETRY[1-4]

a. USE OF FLAMES

In all atomic flame spectrometric methods (atomic emission, atomic absorption, and atomic fluorescence flame spectrometry), a sample solution is aspirated into a flame to produce an atomic vapor. The processes occurring when a sample solution is sprayed into the flame shown in Figure II-D-1 are discussed in the following section.

(1) *Solution Transport.* In the initial step, the solution in the cuvette is transported into either a chamber of a chamber-type nebulizer burner or directly into the flame in a total consumption nebulizer burner. The transport rate F varies approximately with the fourth power of the capillary radius, with the pressure differential between the nebulizer tip and the solution in the cuvette, with the reciprocal of the solution viscosity and with the capillary length; that is, Poiseulle's equation for capillary flow is approximately valid as long as the solution viscosity does *not* exceed 2 centipoise (cp). In order to obtain precise and accurate results in all types of flame

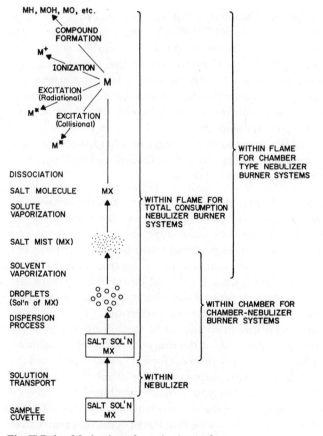

Fig. II-D-1. Mechanism of atomization in flames.

spectrometry, it is necessary to maintain the transport rate of solution constant or, if this is not possible, to at least monitor the transport rate to ascertain any change from some set value. If solutions vary considerably in physical characteristics, it may be necessary to use a solution pump.

(2) *Droplet Formation.* After the solution has been transported into the chamber or flame, the solution is broken up (dispersed) into droplets whose size d_o depends in a complex manner on solution viscosity η, surface tension γ, and density ρ on velocity of the spraying gas v on the solution transport rate F, and on the spraying gas flow rate Q. The value of d_o is approximately given by

$$d_o \approx \frac{k_1\sqrt{\gamma}}{v\sqrt{\rho}} + k_2 \left[\frac{\eta}{\sqrt{\gamma\rho}}\right]^{0.45} \left[1000\,\frac{F}{Q}\right]^{1.5} \qquad \text{(II-D-1)}$$

where k_1 and k_2 are constant characteristics of the sprayer. From this equation, the size of droplets increase from 10 to 65 μ as the gas flow rate decreases from 350 to 50 m sec^{-1}. The smallest droplets—a desirable situation in flame spectrometric methods—result when the velocity of the spraying gas is near the speed of sound. With total-consumption nebulizer burners, the average droplet diameter is of the order of 20 to 40 μ for aqueous and most organic and organic-aqueous mixture solvents, whereas with chamber-type systems the average droplet diameter reaching the flame is often 2 to 3 times smaller because of the removal of the larger droplets from the aerosol mist. (Removal is accomplished by means of flow spoilers and buffers, which serve to remove the large droplets to the drain.)

(3) *Solvent Evaporation.* Evaporation of the solvent from the aerosol mist to produce a salt particle mist may be complete before the analyte reaches the flame (true with some efficient chamber-type sprayers), or it may be quite incomplete (true with all total consumption nebulizer burners). The rate of solvent evaporation is nearly independent of the amount and characteristics of the salt particles and is quite dependent on three factors: the size of the solution droplet, the characteristics of the solvent, the number of solvent droplets per cubic centimeter and the flame characteristics. (The characteristics of the solvent are the diffusion coefficient of the solvent vapor in the flame atmosphere, the molecular mass of the evaporating solvent, the molecular volume of the solvent, the evaporation coefficient of the solvent, and the surface tension of the solvent.) The evaporation rate of solvent droplets in flames is probably given with good accuracy by the combined Fuchs-Fung-Okuyama equation[5] (an extension of the evaporation rate equation of Maxwell; the Maxwell equation is valid only for droplets of diameter greater than about 100 μ, whereas the Fuchs-Fung-Okuyama equation[5] is valid down to

0.1–0.01 μ). However, one point should be made about solvent evaporation for aqueous or nonaqueous solutions introduced into flames with a chamber-type sprayer, and for many solutions introduced at low transport rates into flames produced with total-consumption nebulizer burners, namely, that solvent evaporation is complete by the time the analyte aerosol rises in the flame gases to the measurement height. The efficiency of this process is characterized by the solvent aspiration efficiency ε_{solv} and is near unity for most nebulizer burner systems operated optimally. The chamber-type sprayers, however, waste most of the solution entering the chamber and the efficiency (yield) of the sprayer, represented by ε_{spray}, is about 0.03 to 0.1 for many commercial sprayers and many solvents.

(4) *Solute Evaporation.* Evaporation of the solute is the single most important factor affecting interferences and detection limits in flame spectrometry. Once the solvent has evaporated from the aerosol mist, a salt mist is left. The remaining salt mist evaporates at a rate dependent on the size of the salt particles resulting after solvent evaporation, the number of salt particles, the characteristics of the salt particles (similar to the characteristics for the solvent), and the flame characteristics.

The degree of solute vaporization ε_{sol} is given in general terms by

$$\varepsilon_{sol} = \frac{\int^{nT} \int^{\tau} \frac{dm_r}{dt} \, dn_r \, dt}{\int^{nT} \int^{\infty} \frac{dm_r}{dt} \, dn_r \, dt} \tag{II-D-2}$$

where the first integration is over the total number of particles, n_T, dm_r/dt is the rate of evaporation of a particles having a radius between r and $r + d_r$, and dn_r is the number of particles with a radius between r and $r + d_r$. The second integral is over the time period for solute vaporization (in the integral for the numerator the time period is τ, the time any particle spends on the flame from the time it leaves the burner to the upper most measurement height, and in the integral in the denominator, the time period is infinitely, i.e., long enough for all particles to become completely vaporized if they remained in the flame long enough). *Assuming that the Fuchs-Fung-Okuyama equation[5] is also valid* for evaporation rate of particles in flames, then

$$\frac{dm}{dt} = \frac{4\pi aDm(n_s - n_\infty)}{(D/\alpha v \alpha \phi) + (a/(a + \Delta)]} \tag{II-D-3}$$

where

$$\phi = \exp - \frac{3V_m \sigma}{akT} \tag{II-D-4}$$

$$v = \sqrt{kT/2\pi M_o} \qquad\qquad\qquad \text{(II-D-5)}$$

$$\Delta = \frac{2D}{\sqrt{8kT/\pi M_o}} \qquad\qquad\qquad \text{(II-D-6)}$$

where V_m is the molecular volume of the solute, σ is the surface tension of the solid, α is the evaporation coefficient, k is the Boltzmann constant, M_o is the molecular mass of the evaporating solute, D is the diffusion coefficient of the solute in the flame, n_s is the number of molecules per cubic centimeter at the solute surface, n_∞ is the number of solute molecules per cubic centimeter at an infinite distance to the solute surface, T is the absolute flame temperature, and a is the radius of the solute particle. If on the other hand, in the unlikely case that the particles[6] are moving (convective as well as diffusion-controlled) with respect to the flame gases, then equation II-D-3 must be multiplied by the Frossling factor;[6] if the particles interact with one another in the flame gases (also quite unlikely), then another correction factor[4] must be multiplied times the expression for dm/dt. These corrections factors, as well as the Fuchs-Fung-Okuyama equation, have been discussed by Zung.[5]

Lack of accurate values of physical parameters of solutes at flame temperatures makes it impossible to verify experimentally or use equation II-D-3 to estimate the influence of solute evaporation on shapes of analytical curves or on interference. However, it is apparent that to obtain the *least amount of analytical curve curvature at high analyte concentration and minimal interferences, a sprayer producing a small, average droplet distribution, dilute solutions if possible, and a hot flame with a low rise velocity* (e.g., C_2H_2/N_2O or C_2H_2/air) *is required.*

(5) *Atomization.* (a) *General Expressions for* β. After salt molecules of the analyte are produced, the final step in the production of an atomic vapor is an equilibrium* process (in contrast with the rate-dependent steps of solvent and solute evaporation) involving atomization. The degree of atomization β (also called the free-atom fraction) is simply the ratio of the concentration of analyte species in the form of atoms to the combination of analyte in all forms. For example, consider the case of a salt MX being introduced into the flame gases; for this case the most abundant species containing M would be M, M^+, MOH, MX, MH, and MO, and the chemical equilibria for the species are:

$$M \rightleftarrows M^+ + e \qquad K_1 = \frac{(n_{M^+})(n_e)}{n_M} \qquad\qquad \text{(II-D-7)}$$

* Sufficient time for approximate equilibrium exists in most flames.

$$M + O \rightleftarrows MO \qquad K_2^{-1} = \frac{n_{MO}}{n_M n_O} \tag{II-D-8}$$

$$M + OH \rightleftarrows MOH \qquad K_3^{-1} = \frac{n_{MOH}}{n_M n_{OH}} \tag{II-D-9}$$

$$MX \rightleftarrows M + X \qquad K_4 = \frac{n_M n_X}{n_{MX}} \tag{II-D-10}$$

$$M + H \rightleftarrows MH \qquad K_5^{-1} = \frac{n_{MH}}{n_M n_H} \tag{II-D-11}$$

where all equilibrium constants are written for the dissociation process and all n's are concentration of the species designated by the subscript. The β factor is therefore as general defined as

$$\beta = \frac{n_M}{n_M + n_M + n_{MO} + n_{MOH} + n_{MH} + n_{MX}} \tag{II-D-12}$$

If the flame gases and all analyte species are in *thermodynamic equilibrium*, then it is possible to *calculate β as a function of the equilibrium constants* (see Appendix 3 to relate equilibrium constants to gas temperatures) and of the concentrations of flame gas species (O, H, and OH), and of *species introduced* (in this case). Such an expression is complex and unwieldy for this discussion, and so several limiting cases are discussed later. However, it should be stressed that the *maximal value of β is unity* and that the sensitivity of measurement is a direct function of β (also of ε_{sol}).

(*b*) *Ionization of Analyte.* If only ionization is important (e.g., with cesium and rubidium), then the only equilibrium of importance for a metal M is given in equation II-D-7. The concentration of electrons n_e can arise by two means: (a) by ionization of M which produces a concentration of n_{M+} electrons and (b) by ionization of flame gas products (called free-flame electrons), which produces a concentration of n_f electrons. Thus the total concentration of electrons is given by

$$n_e = n_{M+} + n_f \tag{II-D-13}$$

Also, the total concentration of M (*material balance*) in all forms n_t is given by

$$n_t = n_M + n_{M+} \tag{II-D-14}$$

Combining the preceding expressions and solving for n_M gives

$$n_M = \frac{K_1 + n_f + 2n_t}{2} - \frac{1}{2}\sqrt{(K_1 + n_f + 2n_t)^2 - 4(n_t^2 + n_f n_t)}$$
$$\tag{II-D-15}$$

and so the free-atom fraction for ionization β is given by

$$\beta_i = \frac{n_M}{n_t} = \frac{K_1 + n_f}{2n_t} + 1 - \frac{1}{2}\sqrt{(K_1{}^2 + 2n_f K_1 + n_f{}^2 + 4K_1 n_f)/n_t{}^2}$$

$$(\text{II-D-16})$$

At *low concentrations* of *total M* (i.e., $K_1 \gg n_t$) and *for nonhydrocarbon flames*, (i.e., $n_t \gg n_f$), β_i reduces to

$$\beta_i = \frac{n_t}{K_1} \qquad\qquad (\text{II-D-17})$$

At high concentrations of total M in many flames (i.e., $n_t \gg K_1$ and $n_t \gg n_f$), the equation for β_i becomes simply

$$\beta_i = 1 \qquad\qquad (\text{II-D-18})$$

At low concentrations of total M (i.e., $K_1 \gg n_t$) and for hydrocarbon flames or nonhydrocarbon flames to which an electron donor has been added $(n_f \gg n_t)$, equation IV-D-16 for β_i again reduces to

$$\beta_i = 1 \qquad\qquad (\text{II-D-18})$$

In other words, at high concentrations of analyte in any flame or at low concentrations of analyte in any flame having a high free-electron concentration, ionization is negligible. Ionization is only really appreciable in hydrogen-supported flames or in very hot hydrocarbon flames (C_2H_2/N_2O). It is evident from equation II-D-17 that an increase in flame temperature will increase K_1 and decrease β_i. In such a flame, ionization can be minimized by also nebulizing an *ionization suppressant* (e.g., cesium chloride) into the flame. If an ionization suppressant is *not* added, deviations from linearity (a slope greater than unity and approaching 2 for analytical curves of log signal versus log analyte concentration) will occur at low analyte concentrations.

(c) Dissociation of Analyte Introduced into Flame. If the analyte MX is introduced into the flame and if no other equilibria are involved with either M or X, then the degree of dissociation $\beta_{\text{cpd MX}}$ is given by

$$\beta_{\text{cpd MX}} = \frac{K_4}{K_4 + n_X} \qquad\qquad (\text{II-D-19})$$

If the flame temperature is of the order of most analytical flames, $K_4 \gg n_X$ and so

$$\beta_{\text{cpd MX}} \sim 1 \qquad\qquad (\text{II-D-20})$$

for most analytes.

If the anion X (e.g., chloride, bromide) forms a stable HX species, then dissociation is even more apt to be incomplete. Assuming the only equilibria involving M that occur are equation II-D-10 and

$$H + X \rightleftarrows HX, \qquad K_6{}^{-1} = \frac{n_{HX}}{n_H n_X} \qquad \text{(II-D-21)}$$

and assuming the total concentration of M in all forms [i.e., $(n_t = n_M + n_{MX})$ and the total concentration of X in all forms except MX $(n_t = n_X + n_{HX})$], then the degree of dissociation $\beta_{cpd\ MX-HX}$ is given by

$$\beta_{cpd\ MX-HX} \cong \frac{K_4}{K_4 + [n_t K_6/(n_H + K_6)]} \qquad \text{(II-D-22)}$$

For most analytes characterized by the foregoing equilibria, $n_H K_4 > n_t K_6$, and so again

$$\beta_{cpd\ MX,HX} \sim 1 \qquad \text{(II-D-23)}$$

It is seldom expected that the compound introduced into the flame gases will limit the degree of atomization. It is much *more likely that the compound introduced is* "*completely*" *dissociated* and that the limiting compound formation is between the analyte atoms and flame gas molecules—especially O—to form stable *monoxides.* Such species are discussed in the next section.

(*d*) *Formation of Compound Between Analyte and Flame Gas Products.* Assuming only the formation of one compound MO (the most likely species for most elements in analytical flames) and assuming that *no other equilibria* are important for M, then the degree of atomization $\beta_{cpd\ MO}$ is given by

$$\beta_{cpd\ MO} = \frac{K_2}{K_2 + n_O} \qquad \text{(II-D-24)}$$

where n_O is the concentration of oxygen radicals in the flame gases. Thus we can see that $\beta_{cpd\ MO}$ is independent of n_t (i.e., independent of the concentration of M in all forms) and is only dependent on the temperature and composition of the flame and the stability of the MO species. Because n_o is often *much greater* than K_2 for many elements

$$\beta_{cpd\ MO} \ll 1 \qquad \text{(II-D-25)}$$

for many elements. To *increase* $\beta_{cpd\ MO}$, *it is necessary to raise the flame temperature and/or reduce the concentration of oxygen.* This explains the great success of flame spectrometric measurements within the interconal zone (red feather) of the C_2H_2/N_2O flame.

(*e*) *Dimerization of Analyte Atoms.* Dimerization, a rather minor process, can occur in low temperature flames; that is,

$$M + M \equiv M_2 \quad K_7 = \frac{n_{m_2}}{n_M{}^2} \tag{II-D-26}$$

For this rather trivial case, the degree of atomization is given by

$$\beta_{cpd,dim} = \frac{K_7}{4n_t} \left[\sqrt{[(K_7 + 8n_t)/K_7]} - 1 \right] \tag{II-D-27}$$

where n_t is the total concentration of M in all forms $(n_t = n_M + 2n_{M_2})$. At low values of n_t (i.e., $K_7/n_t > 16$)

$$\beta_{cpd,dim} \cong 1 \tag{II-D-28}$$

At very high total concentrations and for low temperature flames (i.e., $n_t/K_7 > 16$,

$$\beta_{cpd,dim} \sim \sqrt{K_7/2n_t} \tag{II-D-29}$$

Therefore, it is possible for analytical curves to deviate from the expected theoretical shape (from growth curve theory) at high concentrations of analyte. Fortunately dimeric species are seldom important in most analytical flames except for very low temperature diffusion flames.

(*f*)   *Experimental Values of Degree of Atomization.* No general solution for equations is given here because any expression for β depends specifically on the type of species present in flames. For example, with silver and copper, ionization and compound formation with flame gas species is minimal, whereas with aluminum, tungsten, thorium, zirconium, hafnium, beryllium, magnesium, calcium, strontium, barium, tantalum, and so on, numerous species are present. In Table II-D-1, experimentally measured β values for a number of elements in a variety of analytical flames are given. Note the wide variation in values of β in the C_2H_2/air flame and the smaller variation in the C_2H_2/N_2O flame.

(6) *General Expression Relating Atomic Concentration n_o in Flame Gases to Analyte Concentration C_o in Solution Sprayed into Flames.* The basis expression relating n_o (ground state atoms, cm^{-3}) to analyte concentration in the cuvette (moles cm^{-3} is given by

$$n_o = (6 \times 10^{23}) \frac{F \varepsilon \beta C_o g_o}{Q_t e_f Z_T} \tag{II-D-30}$$

where F is the solution transport rate (cm^3 sec^{-1}), Q_t is the flow rate of unburnt gases into the flame (cm^3 sec^{-1}), e_f is the expansion factor for the

Table II-D-1. Atomization Efficiencies (β value) of Atoms in Analytical Flames

| Element | Line(nm) | C_2H_2/Air | | | | C_2H_2/N_2O | |
		Ref. 1[a]	Ref. 2[c]	Ref. 3[c]	Ref. 4[d]	Ref. 4[d]	Ref. 5[e]
Ag	328.0	0.66	—	—	—	—	—
Al	394.4	$<10^{-5}$	—	—	—	0.42	0.59
Au	242.8	—	—	—	0.63	0.71	0.59
Ba	553.5	0.0011	0.009	0.0031	0.0026^f	0.30^f	0.15^f
Ca	422.7	0.14	0.066	0.067	0.070^f	1.4^f	0.69^f
Cd	228.8	0.50	—	—	—	—	—
Co	345.3	0.052	—	—	—	—	—
Cr	425.4	0.064	0.065	—	—	—	—
Cu	324.7	0.98	1.00	0.87	1.00	1.00	1.00
Fe	371.9	0.66	0.38	—	—	—	—
Ga	403.3	0.16	—	—	—	—	—
In	410.1	0.67	—	—	—	—	—
K	766.5[404.4]	[0.25]	—	0.45	0.28^f	—	—
Li	670.8[323.2]	[0.20]	—	0.21	0.26^f	0.44^f	—
Mg	285.2	0.59	0.64	—	0.84	1.5	2.3
Mn	403.0	0.45	0.59	0.93	0.70	0.76	0.80
Na	589.0	$0.50(1.00)^f$	—	1.00	0.53^f	0.33^f	0.65^f
Pb	283.3	0.44	—	—	—	—	—
Rb	780.0	—	—	—	0.16^f	—	—
Sn	286.3	$<10^{-4}$	—	—	0.078	0.76	0.71
Sr	460.7	0.13	0.075	—	—	—	—
Ti	399.9	—	—	—	—	0.3	—
Tl	377.6	0.36	—	—	—	—	—
V	437.9	—	—	—	—	0.91	—
Zn	213.9	0.45	—	—	1.10	0.91	1.00

[a] Used absolute integrated atomic absorption method to determine β's.
[b] Used relative ($\beta_{Cu} = 1$) integrated atomic absorption method to determine β's.
[c] Used atomic emission curve of growth to determine β's.
[d] Used relative ($\beta_{Cu} = 1$) integrated and peak atomic absorption methods to determine β's —only integrated values reported.
[e] Used relative ($\beta_{Cu} = 1$) peak atomic absorption method to determine β's.
[f] Ionization suppressor added.

References

1. L. De Galan and J. D. Winefordner, *J. Quant. Spectrosc. Radiat. Transfer,* **7**, 251 (1967).
2. P. J. T. Zeegers, M. P. Townsend, and J. D. Winefordner, *Spectrochim. Acta.,* **24B**, 243 (1969).
3. E. Hinnov and H. Kohn, *J. Opt. Soc. Amer.* **47**, 156 (1957).
4. J. B. Willis, *Spectrochim. Acta,* **25B**, 487 (1970).
5. S. P. Koirtyohann and E. E. Pickett, *Proceedings of XIII Colloquium Spectroscopicum International*, Ottawa 1967.

flame gases (no units—actually Qe_f is simply the burnt gas flow rate in cubic centimeters per second) and Z_T is the normalized electronic partition function (no units), which is defined by

$$Z_T = \sum_i g_i \exp \frac{E_i}{kT} \qquad \text{(II-D-31)}$$

where the g's are the statistical weights of the ground* (o) and excited (i) state, E_i is the excitation energy of state i, k is the Boltzmann constant, and T is the flame temperature (°K). The factors previously defined have the following units: F (cm³ sec⁻¹), ε (no units), and β (no units).

Because ε_{sol} F, and ε_{spray} (the chamber aspirator yield) decrease with analyte concentration at high analyte concentrations, analytical curves always curve downward toward the abscissa away from the theoretically predicted growth curves (log B versus log n)†. Because β_{ion} decreases with analyte concentration at low analyte concentration, the analytical curves of easily atomized elements have increased slope compared to the growth curves at low analyte concentrations. Finally, because $\beta_{cpd\ MO}$ is independent of the concentration of M, analytical curves do not change shape compared to the growth curves because of formation of compounds such as MO, MOH, and MH. However, if flame conditions are changed to increase atomization of MO, then the entire analytical curve (same shape) will move upward on the plot.

Nonanalyte species present in the analyte solution (i.e., interferences or concomitants), will generally affect (decrease) ε_{sol} to the greatest extent, thus resulting in chemical interferences. If such species are present at high concentrations, they can also affect (decrease) ε_{spray} and F.

Flame temperature affects ε_{sol}, β, and Z_T as discussed previously. Flame gas composition and temperature affect ε_{sol}, e_f, $\beta_{cpd\ MH}$, $\beta_{cpd\ MOH}$, and so on.

(7) *Diffusion of Analyte Atoms in Flames.* Different metals sprayed into flames diffuse laterally at different rates. Although this is not a large effect, it nevertheless is significant and can influence measurements of atomic fluorescence in flames by reducing the concentration of analyte atoms within the measured volume. Snelleman[7] has considered diffusion of metal atoms in flames assuming a point source of supply of atoms and a flame of uniform temperature. The concentration of metal vapor at any point due to diffusion is proportional to the mass of the metal vapor produced at the point source per second, inversely proportional to the square root of the product of diffu-

---

* If o is not the ground state, then all equations with g_o in the numerator should be replaced by $g_o e^{-E_o/kT}$ where E_o is the energy of state o and other terms are as previously defined.
† See Section II-C.

sion coefficient of metal vapor and flame gas rise velocity, and proportional to the reciprocal of the exponential of the flame rise velocity divided by the diffusion coefficient; and of course the concentration of metal vapor depends in a complex manner on the distance of the point of interest from the source of the vapor. In any event, diffusion can slightly affect the shape of the analytical curve and can be manifested as a small interference in some cases.

b. USE OF NONFLAME CELLS[8,11]

When this book was written, nonflame cells were only novel devices for obtaining great absolute sensitivities. However, the authors feel that such devices have considerable analytical utility because of their good atomization characteristics, low background flicker, and inefficiency of deactivation of excited atoms.

There are two basic approaches to the use of nonflame atomizers. In the first approach, the sample solution (it could be a solid) is accurately introduced into or onto the nonflame atomizer (e.g., a volume V_o of concentration C_o of analyte is pipetted into or onto the nonflame cell). The cell is first heated slowly to dry the introduced sample (sometimes a slightly higher heating rate is then used to char the sample) and then heated at a high rate to achieve a temperature of aboutf 3000°K to atomize the sample. The cells are usually made of some readily oxidizable substance, (e.g., graphite or tantalum), and so the furnace is maintained in an inert atmosphere, say argon, during heating. In the second approach, the sample solution is continuously and precisely introduced into the nonflame furnace at a rate of F cm³ sec⁻¹ by an efficient nebulizer which produces a small distribution of droplets. The furnace in the latter case is usually a tube furnace made of graphite or some metal and continuously heated electrically. The tube furnace is again maintained in an argon or other inert-gas environment. The factors affecting the concentration of atoms of analyte) n_o for both sampling approaches are considered below.

(1) *Discrete Sampling from a Cuvette.* Lʹvov[11] has shown that most metals, if present in weights of about 10^{-8} g, vaporize completely at a temperature of 3000°K in about 0.1 sec, resulting in a saturated vapor pressure of 0.1 mm or greater.* The exceptions to this are the refractory metals, such as hafnium,

---

*According to Lʹvov[11], the rate of vaporization R_v, (g cm⁻² sec⁻¹) of a substance from a free surface into a vacuum is given by

$$R_v = P_o \sqrt{M/2\pi RT} = 0.058 \, p_o \sqrt{M/T}$$

where p_o is the saturated vapor pressure of the substance of molecular weight M at temperature T.

tantalum, niobium, molybdenum, tungsten, chenium, and osmium. As long as the temperature of the furnace is sufficiently high, the vaporization time will be independent of the foreign gas pressure within the furnace. However, the vaporization time is quite dependent on the temperature of the furnace.

If the entire sample is vaporized in a cuvette (cylindrical cell with an open end) before any losses occur, then the peak atomic concentration n_o of analyte (cm^3) is given by

$$n_o \cong (6 \times 10^{23}) \frac{m_a \varepsilon \beta g_o}{e_g V_c M_a Z_T} \qquad \text{(II-D-32)}$$

where m_a is the mass* of the analyte (g), V_c is the cuvette volume (cm^3), Z_T is the electronic partition function (no units), M_a is the atomic weight of the analyte (amu), e_g is the gas expansion factor (no units) between the heated gas temperature and room temperature, and $\varepsilon \beta$ is the efficiency (no units) of producing atoms from the sample (i.e., ε is the efficiency of producing atoms, ions, and molecules from the solid sample and β is the free-atom fraction defined by equation II-D-12). The factor $\varepsilon \beta$ should be near unity for substances that are completely vaporized within the cuvette, assuming that the atmosphere of the cuvette is an inert gas. This is quite different from analytes in flames, where $\varepsilon \beta$ is often many orders of magnitude less than unity.

The resulting vapor in the cuvette may then be lost in four ways: diffusion of atoms through the walls and open ports of the cell; convection near the hot surface of the cell; escape of excess vapor from the relatively smaller volume of the cell; and removal by forced flow of inert gas through or by the cell during heating. With cuvette-type cells,† vapor losses are generally controlled almost exclusively by diffusion of vapor through the open ports and porous walls.‡ With filament-type cells, vapor losses are predominantly

---

* The mass of the analyte is related to the sample weight by

$$m_a = \gamma_s w$$

where w is the sample weight (g) and γ_s is the weight fraction of the sample which is the analyte. Of course w is simply given by

$$w = C_o V_o$$

where C_o is the concentration of the analyte (g/ml^{-1}) and V_o is the volume of sample (ml).

† Refer to Section III-B-4 for a discussion of nonflame cells.

‡ The removal of vapor by convective forces near the hot cuvette is negligible compared with the removal accomplished by diffusion. The escape of excess vapor from the smaller volume is negligible as long as the vapor volume is 10% or less of the cell volume; for example, for a cuvette of 4.5 mm diameter and 50 mm long at 2500°K and 1 atm argon, the limiting amount of an analyte of atomic weight 30 is $\sim 5 \times 10^{-5}$ g. Generally the cuvette-type cells are inside a housing through which a forced argon flow occurs; however, most of the argon losses are via cracks and openings within the housing and a relatively small flow is through the cuvette; for a cell 50 mm long and 5 mm in diameter, the actual flow through the cuvette is less than 0.025 cm^3 sec^{-1}.

controlled by the forced flow of inert gas by the filament with a minor contribution from diffusion from the filament into the atmosphere.*

If it is assumed that the cuvette contains just one opening (i.e., a cylinder), then under steady state conditions the mass of substance diffusing per unit time dm/dt is given by

$$\frac{dm}{dt} = -D_a \frac{d\rho_a}{dx} S \qquad \text{(II-D-33)}$$

where D_a is the diffusion coefficient, $d\rho_a/dx$ is the density gradient of analyte atoms in the direction of diffusion, and S is the cross-sectional area of the open port. The diffusion coefficient D_a (cm^2 sec^{-1}) is related to the diffusion coefficient D_o for the standard state of 1 atm pressure and 273°K by

$$D_a = D_o \left(\frac{T}{273}\right)^n \left(\frac{760}{p_o}\right) \qquad \text{(II-D-34)}$$

where n is about 1.5 to 2 for various gas mixtures. Now if it is assumed that the vapor density in the cuvette decreases linearly from its value of ρ_a at the bottom of the cuvette where the sample is being vaporized to zero at the cuvette ends, then

$$\frac{d\rho_a}{dx} = \frac{\rho_a}{l} = \frac{m_a}{lV_c} = \frac{m_a}{l^2 S} \qquad \text{(II-D-35)}$$

where the cuvette volume is V_c (cm^3) and m_a is the weight of the analyte (g). Substituting $d\rho_a/dx$ into equation II-D-33 gives

$$\frac{dm_a}{m_a} = -\frac{D_a}{l^2} dt \qquad \text{(II-D-36)}$$

Thus the relative loss of vapor is inversely proportional to the square of the cuvette length and is *independent* of its cross-sectional area. Integration of equation II-D-37 gives

$$m_{(t)} = m_o \exp -\frac{D_a t}{l^2} \qquad \text{(II-D-37)}$$

where $m_{(t)}$ is the mass of vapor at time t and m_o is the mass at time $t = 0$. Therefore, the atomic concentration of analyte varies in the same manner

$$n_{(t)} = n_o \exp -\frac{D_a t}{l^2} \qquad \text{(II-D-38)}$$

where $n_{(t)}$ is the concentration of atomic analyte at any time t and n_o is given by equation II-D-32 (assuming that the conditions preceding equation II-D-32 are valid).

* The relative contributions depend on the forced flow rate of inert gas by the filament.

If the cell walls are made of a porous material like graphite, the diffusion of vapor through them is given by an equation similar to II-D-37; that is,

$$m_{(t)} = m_o \exp - \frac{D_a' S' T}{l' V_c} \qquad \text{(II-D-39)}$$

where S' is the wall area, V_c is the cuvette volume, l' is the wall thickness of the cuvette, and D_a' is the diffusion coefficient of the gas in the porous material. Diffusion of vapor through the walls can be prevented by using metal (e.g., tantalum) cuvettes or, if graphite is used, then it is necessary to utilize either a metal lining such as tantalum or pyrolytic graphite.*

Therefore, the concentration of atomic analyte a time t is given by

$$n_{(t)} = n_o \exp - \frac{D_a' S' t}{l' V_c} \qquad \text{(II-D-40)}$$

where all terms are as defined previously.

The loss of vapor via diffusion through the open ends can be minimized by increasing the length of the cuvette, by using an inert gas of higher molecular weight, and by using a cover on the cuvette. An increase in cell length also necessitates more electrical power for heating. A higher molecular weight gas results in decreasing the diffusion coefficient D_a by $1/\sqrt{M_g}$, where M_g is the molecular weight of the inert gas. A cover is probably the simplest way to localize the atomic vapor for a longer time period.†

(2) *Discrete Sampling from a Hot Filament.* If the sample is vaporized from a heated filament (e.g., graphite or tantalum) into a continuous flowing stream of inert gas and if the analyte mass m_a (g) is vaporized as a plug in a time t, then the peak concentration n_p is given by

$$n_p = (6 \times 10^{23}) \frac{m_a \, \varepsilon \beta g_o}{t e_g \, Q_a \, M_a \, Z_T} \qquad \text{(II-D-41)}$$

where Q_a is the flow rate of inert gas (cm^3 sec^{-1}) past the filament, and all other terms are as defined previously. It is assumed in equation II-D-41 that the vaporized analyte mixes homogeneously with the inert gas.

---

* There is some danger of contamination when metal linings are used. Pyrolytic graphite, called pyrographite, is obtained by pyrolysis of hydrocarbons (e.g., CH$_4$, at a temperature of about 2300°K). Either the entire cuvette may be made of pyrographite or the cell may be lined with it. Pyrographite is much less permeable to gases, sublimes at a higher temperature (ca. 4000°K), has a higher thermal conductivity, and is less sensitive to air oxidation than normal graphite.

† For equation II-D-32 to be valid, it is probably necessary for many samples to use covered cuvettes to avoid losses .

Actually, the analyte concentration will vary with time $n_{(t)}$; the time-dependent concentration is a function of the amount of analyte on the filament, the type of filament, the area of coverage, the diffusion rate of analyte, the filament temperature, and several other factors. There is no simple relationship to describe the variation of $n_{(t)}$ with time for this dynamic process.

(3) *Continuum Sampling by a Nebulizer.* If a nebulizer is used to introduce continuously a sample solution into a suitable tubular furnace, the concentration of analyte n_0 is given by

$$n_o = (6 \times 10^{23}) \frac{FC_o \ \varepsilon\beta g_o}{e_g \ QZ_T} \qquad \text{(II-D-42)}$$

where F is the flow rate of analyte solution into the nebulizer, Q is the flow rate of inert gas carrier into the furnace (the inert gas carries the nebulized sample), and all other terms are as previously defined. In this case, there is no time dependence of analyte concentration because sample is continuously introduced into the furnace.

(4) *Comparison of Nonflame and Flame Atomizers.* The flame atomizer involves a more reliable, simpler means of sampling than the discrete sampling methods. However, the continuous sampling method should have the same reliability and simplicity of sampling that are found in flames. The environment of nonflame atomizers is simple to adjust for maximum $\varepsilon\beta$ and for maximum luminescence power yield (i.e., use of an inert gas). A nonflame atomizer, particularly the filament atomizer, should have a smaller background and background flicker than flames. The cuvette atomizer is generally viewed by the measurement system through a slit in the side (excitation is through the open port), and so thermal (blackbody) emission is appreciable whereas atomic fluorescence of vapors produced with the filament atomizer is observed above the filament where background should be small. On the other hand, there will be a considerably greater dilution of analyte with the filament atomizers than with the cuvette atomizers, and also $\varepsilon\beta$ should be greater (nearer unity) with the latter atomizers. Contamination of graphite and metal nonflame atomizers is generally greater than contamination of gases in flame spectrometry. Also, if the sample is not vaporized rapidly from nonflame cells, fractionation of elements occurs, resulting in lower peak concentrations. In conclusion, the major use of nonflame atomizers would seem to be greatest for analysis of small amounts (e.g., nanograms) of elements in small sample sizes (e.g., micrograms).

Other types of nonflame cells used in atomic fluorescence spectrometry include a heated metal loop with a small (1 μl) accurately pipetted sample and heated quartz containers. These devices now appear to have limited analytical use and are not discussed in detail here.

2. PRODUCTION OF MOLECULES IN MOLECULAR LUMINESCENCE SPECTROMETRY

a. USE OF FLAMES[1-4]

No analytical applications of molecular fluorescence spectrometry with analyte molecules produced by reaction and equilibration of the analyte atoms and flame gas products have been found at the time of writing. Because molecular fluorescence is less specific than atomic fluorescence and because molecular fluorescence should be less sensitive (only a small portion of band fluorescence would be measured by a spectrometric system), it is unlikely that molecular fluorescence spectrometry of species in flames will ever have great analytical use. Therefore, no discussion of the production of molecules in flames appears here, although the interested reader can refer to the section on atomization to gain some insight into molecularization (i.e., production of species such as MO, MOH, and MH).

b. USE OF CONDENSED PHASE[9,10]

In all types of solution photoluminescence spectrometry at room temperature (fluorimetry), the number of excited molecular species depends on both ground and excited state chemical equilibria involving that species. In phosphorimetry or in any low temperature luminescence studies, the concentration of the excited state species is determined by the ground state chemical equilibria (i.e., the chemical equilibria present prior to rapid freezing to $77°K$). Such processes as dissociation and association (e.g., dimerization) in the ground or excited states can be described by equations similar to equations II-D-10 and II-D-26 for similar processes in flames. If there are no equilibria, or if slow equilibration of excited state species occurs, then the number of excited state species is determined only by equilibrium processes regarding the ground state species and can be described by well-known expressions given in most elementary textbooks on chemical analysis. The influence of excited state equilibria (e.g., protonation or deprotonation processes in the excited state) are described by similar but more complex expressions involving the lifetime of excited state. An extensive treatment of these cases is not given here.

It should be stressed, however, that in all equations regarding the radiance of photoluminescence, the *concentration* of absorbing analyte is the actual number of analyte absorbers per unit volume of sample solution and not necessarily the concentration of analyte added. For example, if a weak monoprotic acid of ground state concentration C_t is in a sample cell and if the pH is made quite high, then the concentration of ground state monoprotic acid remaining at equilibrium is $C_t[H^+]/([H^+] + K_a)$, where K_a is the dissociation constant of the ground state weak acid; therefore, if the species absorbing radiation is the weak acid, a great decrease in photoluminescence radiance would result simply because of the ground state equilibrium. Also, excited state equilibria can greatly influence the concentration of the emitting species (see Section II-B-3).

The absorbing molecular species will, of course, depend on the molecular form (see Section II-B) of the species. For example, the native species may not fluorescence or phosphoresce, and so a reagent or reagents are added to produce a fluorescent or phosphorescent species. Similarly, a fluorescent or phosphorescent species may be produced from a nonluminescent species by a photochemical reaction. In any event, the final absorbing species can then undergo chemical equilibria as discussed previously and the final concentration of absorbers is dependent on the absorber, the concentration of the conjugate species, and the solution's temperature. No general treatment can be given for absorbing species produced via chemical reactions; but rather, the concentration of absorbers can *only* be determined once a specific system —with specific reactions—is defined, and then the concentration of absorbers can be estimated if all reactions and equilibria involving the species of interest are known.

3. GLOSSARY OF SYMBOLS

a	Radius of solute particles in flames, cm.
C_o	Concentration of species introduced into flame or nonflame, cm^{-3}.
C_t	Initial concentration of analyte molecules, moles liter^{-1}.
d_o	Diameter of droplets produced by sprayers in flame spectrometry, cm.
D	Diffusion coefficient of solute in flames, cm^2 sec^{-1}.
D'_a	Diffusion coefficient of vapor in walls of cuvette, cm^2 sec^{-1}.
D_a	Diffusion coefficient of analyte atoms in the atmosphere of a nonflame cell, cm^2 sec^{-1}.
D_o	Diffusion coefficient of analyte atoms under standard temperature (273°K) and pressure, cm^2 sec^{-1}.
e_f	Expansion factor of flame gases relative to room temperature, no units.
E_i	Excitation energy of state i, erg.
F	Solution transport rate into flame, cm^3 sec^{-1}.
g_o	Statistical weight of ground state o, no units.
g_i	Statistical weight of excited state i, no units.
k	Boltzmann constant, 1.38×10^{-16} erg °K^{-1} mol^{-1}.
k_1	Constant characteristic of sprayer (see equation II-D-1), $cm^{-1/2}$.

k_2 Constant characteristic of sprayer (see equation II-D-1), $cm^{-0.225}$.

K_i Equilibrium dissociation constant for equilibrium i.

l Length of nonflame cell cuvette, cm.

l' Wall thickness of cuvette, cm.

m Mass of solute particles in flame g.

m_a Mass of analyte in nonflame cell, g.

M_o Molecular weight of evaporating solute, g.

m_o Mass of analyte vapor in nonflame cuvette at time $t = 0$, g.

$m_{(t)}$ Mass of analyte vapor in nonflame cuvette at time t, g.

M_a Molecular weight of analyte, in amu.

M_g Molecular weight of inert gas in nonflame cells, amu.

n Constant in equation II-D-34, no units.

n_e Concentration of electrons in flame, cm^{-3}.

n_f Concentration of free electrons from flame gas products, cm^{-3}.

n_j Concentration of species j in flame, cm^{-3}.

n_o Concentration of analyte in proper state for absorption of light, cm^{-3}.

n_p Peak concentration of analyte in nonflame cells, cm^{-3}.

n_s Concentration of solute at solute surface, cm^{-3}

n_t Total concentration of analyte in all forms in flame, cm^{-3}.

$n_{(t)}$ Concentration of analyte at time t, cm^{-3}.

n_∞ Concentrations of solute at infinite distance from solute surface, cm^{-3}.

p_o Saturated vapor pressure of element, mm Hg.

Q Gas flow rate used in nebulizers, $cm^3\ sec^{-1}$.

Q_a Gas flow rate past filament, $cm^3\ sec^{-1}$.

Q_t Gas flow rate of unburnt gases into flame, $cm^3\ sec^{-1}$.

R_v Rate of vaporization of substance from surface into vacuum, $g\ cm^{-2}\ sec^{-1}$.

S Cross-sectional area of nonflame cell, cm^2.

S' Cross-sectional area of cuvette walls, cm^2.

S_a Area covered by sample, cm^2.

t Time analyte spends in nonflame cell, sec.

t Time particle spends in flame before measurement, sec.

T Gas temperature in nonflame cell, °K.

T Flame, temperature, °K.

V_c Cuvette volume, cm^3.

V_m Molecular volume, cm^3.

V_o Volume of solution pipetted into nonflame cell, cm^3.

w Weight of sample, g.

Z_T Electronic partition function, no units.

α — Evaporation coefficient, no units.

β — Degree of atomization (free-atom fraction), no units.

$\beta_{\text{cpd Mo}}$ — Degree of atomization when *only* equilibrium involving MO is important, no units.

$\beta_{\text{cpd MX}}$ — Degree of atomization when *only* equilibrium involving MX is important, no units.

$\beta_{\text{cpd MX HX}}$ — Degree of atomization when *only* equilibria involving MX and HX are important, no units.

β_i — Degree of atomization when *only* ionization is important, no units.

γ — Solution surface tension, erg cm^{-2}.

γ_s — Fraction of sample weight which is analyte, no units.

Δ — Correction factor (see equation II-D-6), cm.

ε_{sol} — Solute vaporization efficiency in flames or nonflames, no units.

$\varepsilon_{\text{solv}}$ — Solvent vaporization efficiency, no units.

$\varepsilon_{\text{spray}}$ — Sprayer efficiency, no units.

η — Solution viscosity, g cm^{-1} sec^{-1} (or poise).

ν — Correction factor (see equation II-D-5), cm sec^{-1}.

ρ — Solution density, g cm^{-3}.

ρ_a — Gas density of analyte atoms, g cm^{-3}.

σ — Solute surface tension, erg cm^{-2}.

τ — Time particle spends in flame from time particle leaves burner head to point of measurement, sec.

ϕ — Correction factor (see equation II-D-4), no units.

4. REFERENCES

1. R. Herrmann and C. T. J. Alkemade, *Flammenphotometrie*, 2nd ed., Springer, Berlin, 1960; *Chemical Analysis by Flame Photometry*, P. T. Gilbert, Transl. Wiley, New York, 1963.
2. R. Mavrodineanu, and H. Boiteux, *Flame Spectroscopy*, Wiley, New York, 1965.
3. E. Pungor, *Flame Photometry Theory*, Van Nostrand, New York, 1967.
4. C. T. J. Alkemade, Ph.D. Thesis, University of Utrecht, The Netherlands, 1954.
5. J. T. Zung, *J. Chem. Phys.*, **46**, 2064 (1967).
6. F. A. Williams, *Combustion Theory*, Addison-Wesley, Reading Mass., 1965.
7. W. Snelleman, Ph.D. Thesis, University of Utrecht, The Netherlands, 1965.
8. J. D. Winefordner, "Non-Flame Cells in Atomic Fluorescence Spectrometry," in *Proceedings of International Atomic Absorption Spectroscopy Conference*, Sheffield, England, July, 1969, Butterworths, London, 1970.
9. C. A. Parker, *Photoluminescence of Solutions*, American Elsevier, New York, 1968.
10. T. Förster, *Fluoreszenz Organischer Uerbindungen*, Vandenheok and Ruprecht, Göttingen, 1951.
11. B. V. Lʹvov, *Atomic Absorption Spectroscopy*, Israel Program for Scientific Translations, Jerusalem, 1969.

CHAPTER

III

INSTRUMENTATION AND METHODOLOGY

A. GENERAL LUMINESCENCE INSTRUMENTATION

1. INSTRUMENTAL LAYOUT AND REQUIREMENTS[1,2]

The basic instrumental components of atomic and molecular luminescence instrumentation are not unlike those used in other ultraviolet and visible methods, but the arrangement of these components is unique. Figure III-A-1 illustrates the general instrumental layout for all types of photoluminescence studies. The chemical system to be investigated (i.e., the sample) is contained in a sample cell. This may be an actual glass or quartz container, as in

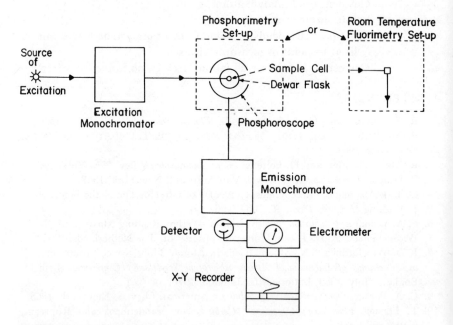

Fig. III-A-1. General instrumental layout for photoluminescence spectrometry. Note variation in sample compartment for phosphorimetry and room temperature luminescence spectrometry.

140

molecular luminescence spectrometry, or a flame cell, as in atomic fluorescence flame spectrometry. The sample is excited by ultraviolet or visible light from the source excitation, which is usually an arc or discharge lamp. The desired excitation wavelength region may be selected by means of the excitation monochromator if desired (this is seldom necessary or even desirable in atomic fluorescence experiments). The luminescence radiation is emitted by the sample in all directions; a portion of it is sampled by the detection system at some angle, traditionally 90°, from the axis of excitation. The detection system consists of the emission monochromator, which selects the wavelength region of luminescence emission to be measured; the photodetector, which converts the luminescence radiant flux to an electrical signal; and the amplifier-readout system, which amplifies and processes the electrical signal as required and displays it in a convenient fashion. Often one or more light choppers may be placed in the light path in order to separate the desired luminescence radiation from undesirable light emission from the sample. Simpler instruments use optical filters in place of monochromators and are called *filter fluorometers*. The more versatile system of Figure III-A-1 would be called a *spectrofluorometer* or *spectrophotofluorometer*, the prefix "spectro-" indicating that at least one monochromator is used. Such an instrument adapted for the measurement of phosphorescence would be called a *spectrophosphorimeter*. A more general term for all luminescence spectrometers might be *spectroluminometer*. The additional term "photo-" occasionally inserted into names of such instrumentation usually implies photoelectric detection of light intensity (rather than visual or photographic, now seldom used in luminescence studies). The spectroluminometer has the advantage over a filter instrument of being able to measure the spectral distribution of luminescence emission (the *emission spectrum*) and the variation in emission spectral radiance with excitation wavelength (the *excitation spectrum*).

The instrumental system is required to have reasonably *high spectral resolution, sensitivity*, and *signal-to-noise ratio* over the desired wavelength regions of excitation and emission. In addition, a certain degree of *convenience* and *speed* in obtaining data is desirable.

Spectral resolution is a measure of the ability of the instrument to isolate small wavelength regions of the excitation or emission radiation and to separate closely spaced lines or bands in the excitation or emission spectra. It is primarily a function of the design and adjustment of the monochromator(s). *Sensitivity* is a measure of the magnitude of the electrical output of the instrument which results from a given concentration of analyte in the sample, under a given set of experimental conditions. It is a function of almost every experimental variable in the instrumental system (e.g., the spectral radiance of the source emission; the design and adjustment of the monochromator(s); the nature and condition of the sample—i.e., temperature, matrix components (interferents); concentration and chemical form of analyte; photodetector sensitivity; and amplification (gain) of the readout system). More specifically, the *linear analytical sensitivity* of the entire system is defined as the slope of the analytical curve,* which is the plot of instrument response (e.g., electrical output, recorder deflection) versus analyte concentration. It is desirable that the analytical curve be linear (i.e., that the analytical sensitivity be independent of analyte concentration), but this is not always the case —especially at high analyte concentrations.

The *signal-to-noise ratio* is the ratio of the instrumental response (the signal) resulting from the presence of analyte in the sample to undesirable variations in that signal (noise). The noise may be in the form of a random fluctuation or an unpredictable drift in instrument response, and it is a function of many instrumental variables. Most of the noise in luminescence spectrometry is contributed by the source, the sample and sample cell, the photodetector, and (occasionally) the amplifier-readout system. The signal-to-noise ratio is important because it influences the precision with which the instrumental response may be measured and determines the minimum concentration of a particular analyte that may be detected under a given set of conditions (the *detection limit*).†

Additional factors influencing the design of luminescence instrumentation include the *convenience* and *speed* with which the instrument may be used to determine analytical curves, excitation and emission spectra, luminescence decay times, and the effect of sample temperature, polarization of excitation or emission radiation and other variables on luminescence radiances, spectra, and decay times.

2. SOURCES OF EXCITATION[3]

Sources of radiant energy may be classified according to the spectral distribution of the radiation emitted. *Continuum sources* are characterized by a broad spectral distribution covering a relatively large range of wavelengths (generally without sharp lines or bands). An ordinary tungsten filament light bulb is an example; it emits a continuum in the visible and infrared spectra. *Line sources* are characterized by spectra consisting of a number of relatively sharp lines or bands. An example is a low pressure mercury vapor lamp. The distinction between line and continuum sources is not always clear, however. In fact, one of the most common excitation sources used in luminescence spectrometry emits a characteristic line spectrum superimposed on a broad background continuum. The high pressure xenon and mercury arc lamps are of this type.

---

* Also called working curve and calibration curve.

† Also called limit of detection and minimum detectable concentration.

There are essentially two systems of units commonly used to express the intensity of light sources—the *radiometric* system, an absolute system based on the actual energy radiated by a source; and the *photometric* system, a relative system based on the apparent intensity of a source as viewed by the "average" human eye. The radiometric system is the more useful quantitatively and is used in this book.

The basic quantity is the *radiant flux*, which is the total energy radiated per unit time. Radiant flux is most commonly expressed in watts. In dealing with practical light sources, modification of these units is customary. In order to account for the finite radiating areas of real sources, intensities are expressed in watts square centimeter of radiating surface. Furthermore, a light source may not radiate equally in all directions because of the geometry of the device; or reflectors may be used to concentrate the light in a particular direction. Consequently, the intensity in a given direction may be more important for practical applications than the total light output, and intensities are often expressed as the power radiated into a unit solid angle, that solid angle being oriented in a specified direction. A unit solid angle, called a *steradian* (sr), is the solid angle subtended at the center of a sphere by the area on the surface equal to the square of the radius of the sphere. Because the area of a sphere of radius r is $4\pi r^2$, there are 4π steradians per sphere.

Source intensity expressed in watts per square centimeter per steradian is termed the *radiance* of the source. The intensity of a line in the spectrum of a line source is often expressed this way. Because the line is actually of finite spectral width, the intensity given in this way has been effectively integrated over all wavelengths in the line profile. For continuum sources, on the other hand, the total integrated intensity is usually less important than the intensity per unit wavelength interval at a particular wavelength. Consequently, intensities of continuum sources are most often expressed in watts per square centimeter per steradian per nanometer, which is called the *spectral radiance*.

a. INCANDESCENT SOURCES[3]

Probably the simplest type of light source, and the only one giving a smooth continuum spectrum without lines, is an *incandescent source*, such as a tungsten filament lamp. Although commonly used in visible and infrared absorption spectrophotometers, incandescent sources are not sufficiently intense, particularly in the ultraviolet region, to be useful in luminescence spectrometry. Nevertheless, the principles of black body emission are important.

An incandescent source emits radiation by virtue of its temperature, rather than by specific transitions between quantized energy levels characteristic of the particular incandescent substance. The radiation emitted is broad continuum whose spectral distribution is a function of the temperature of the emitter. The specific material of which the source is made may affect the overall intensity of the radiation, but has little effect on its spectral distribution. The spectral radiance as a function of wavelength is given by the *Planck black body distribution equation*:

$$B_{B\lambda} = \frac{5.8967\lambda^{-5}\,\varepsilon_\lambda}{\exp(14388/\lambda T) - 1} \qquad \text{(III-A-1)}$$

where $B_{B\lambda}$ = spectral radiance, watt $cm^{-2}\,sr^{-1}\,nm^{-1}$

$\quad\lambda$ = wavelength, nm

$\quad T$ = temperature, °K

$\quad\varepsilon_\lambda$ = spectral emissivity, typically 0.4 for tungsten at 3000°K, no units.

In Figure III-A-2, $B_{B\lambda}$ is plotted against λ for several different temperatures, according to equation III-A-1. The importance of this equation lies less in the applications of incandescent sources in luminescence spectrometry, which are few,, than in the continuum background emission component of high pressure arc lamps, which closely approaches the characteristic spectral distribution of high temperature black bodies. For instance, the continuum emission of a 500-watt xenon arc lamp corresponds closely to that of a 7000°K black body with an emissivity of 0.06.

The operating temperatures of the filament in an incandescent lamp depends on the power input. A typical 500-watt tungsten lamp operates at only about 3000°K and has an emissivity of 0.4. The intensity emitted by this lamp has a maximum at about 1000 nm and drops to one-hundredth of that value at about 300 and 5000 nm. Thus the tungsten lamp finds its greatest use in the visible and near infrared regions.

The intensity drops off very rapidly in the ultraviolet region. (In addition, the glass envelopes usually absorb strongly below 280 nm, but the intensity at that point is so low anyway that the use of quartz envelopes is of little practical value). The spectral radiance of a 500-Watt lamp at 300 nm is about 10^{-4} watt $cm^{-2}\,sr^{-1}\,nm^{-1}$, more than two orders of magnitude below that from a typical high pressure xenon arc lamp.

Incandescent lamps are important as sources in spectrometric applications because of their *excellent stability*, rather than because of their *spectral radiance*, which is low compared to continuum arc sources such as the high pressure xenon arc. Long and short term stabilities of the order of 0.01% can be obtained by powering the lamp from a programmable power supply controlled by a feedback signal obtained from a photocell monitoring the lamp output.[4] This order of stability is impossible to obtain with gas discharge and arc lamps.

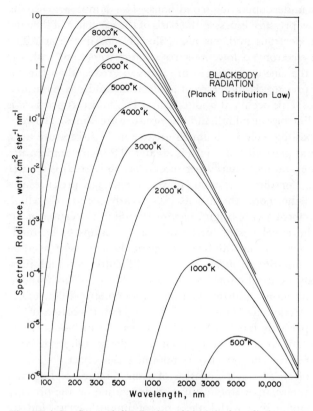

Fig. III-A-2.   Spectral distribution of black body radiation at several temperatures.

b. ARC LAMPS[3]

Most commercial molecular luminescence spectrometers use high pressure xenon, mercury, or mercury-xenon arc lamps as excitation sources because of their relatively high intensity, wide spectral range, and low cost. Arc lamps of this type are quite intense in the wavelength region from 200 nm through the near infrared.

An arc lamp consists of an enclosed arc discharge between two electrodes in a gas or metal vapor atmosphere. The metal vapor types are limited to the more volatile metals such as mercury, cadmium, zinc, gallium, indium, and thallium, and are capable of producing very intense line spectra of those elements. Gas-type arc lamps are constructed with hydrogen, nitrogen, and the rare gases.

A very important specification of an arc lamp is the *gas or metal vapor pressure* in the lamp under operating conditions. The lamps are usually classified as high or low pressure types on the basis of the existence of thermal equilibrium between electrons and gas molecules or ions in the arc. The kinetic energy of the electrons at low pressures is large compared with the gas energy, and thus the electron temperature is much greater than the gas temperature. As the pressure is increased, the electrons and gas molecules approach equilibrium. This occurs at about 10 mm Hg, and this pressure is taken as the dividing line between high and low pressure arc lamps. The voltage gradient in the positive column is much greater in high pressure arcs, so that the energy input per unit arc length is greater. Thus gas temperature and arc irradiance (i.e., radiant energy per second per unit area) is greater in high pressure arcs. For the same input power, then, high pressure arcs are made shorter. Furthermore, the arc itself is greatly *constricted* at high pressures and tends to form a very bright, narrow stream in the center of the tube. This enables the envelopes of high pressure arc lamps to be made smaller in diameter before serious wall heating occurs. Low pressure arcs, on the other hand, are comparitively *diffuse*. A serious difficulty with high pressure arcs is that they always exhibit some degree of instability and arc wander. The arc stream is convection stabilized; that is, the symmetrical flow of gas passing along the electrode axis holds the arc in the central portion of the gap. This flow is not perfectly laminar, however, and slight turbulences and instabilities cause the arc stream to shift about irregularly. Elliptical lamp envelopes and thoriated tungsten electrodes minimize this effect.

The *spectral output* of all arc lamps consists of some mixture of line and continuous emission, the relative amount of each depending on the pressure in the lamp. If sharp atomic lines with little or no continuous background are desired, low pressure lamps, operated at low current densities and temperatures, are used. This minimizes pressure and temperature broadening and self-reversal of lines. Lamps of this type may be useful in atomic fluorescence work. As the pressure and temperature is increased, the lines broaden and the background continuum increases in intensity. Very high pressures of many atmospheres are used if a continuous emission contaminated with a minimum of lines is desired. The background continuum radiation from arc lamps approaches that of a high temperature black body and is a result of strong interactions (i.e., collisions, etc.) between gas atoms, which occur more readily at high pressures. This type of lamp is preferred for molecular luminescence instrumentation.

Some types of arc lamps may be operated on either *AC or DC current*, and others are limited to DC operation only. Each mode has advantages and disadvantages. The design of AC power supplies is relatively simple, whereas DC supplies require a more complicated rectifier-filter design. An AC arc has a more uniform intensity distribution along the arc length, but the average arc width is larger. Most of the intensity of a DC arc is concentrated at the " cathode hot-spot," where the intensity may be several times the average value. A serious disadvantage of AC operation of high pressure arcs is that arc wander is greater than for DC operation. The rapid rise and fall of gas temperature during each cycle of the AC waveform disturbs the convection stabilization of the arc. Pressure waveforms caused by the AC modulation reflect off the lamp walls and may be focused back on the arc, further disturbing the convection equilibrium. These effects may be minimized by proper construction, but DC operation is almost always preferred for luminescence instrumentation. Low pressure arc lamps, on the other hand, are often AC operated, because the broad, diffuse discharge is less plagued by wander. Furthermore, the modulation effect of AC operation is often a convenience, since it eliminates the need for a mechanical chopper.

The *operation* of *arc lamps requires a number of precautions*. Some lamps intended for DC operation only have differing anode and cathode construction, and proper polarity must be observed. The anode of a high pressure arc lamp is made larger in order to more readily dissipate the heat caused by electron bombardment during operation. Vertical operation is preferred, because this minimizes deposition of vaporized electrode material onto the walls. Arcs more than 1 mm long tend to bow upward when operated horizontally and may cause excessive heating of the upper wall.

An important feature of all types of arcs is that they exhibit a *negative volt-ampere characteristic* and are, consequently, inherently unstable if operated from a constant voltage source. For this reason, arc lamps must be operated in series with a resistance, called the ballast, to ensure stability. The value of the ballast resistance is chosen so that the series combination of the lamp and the ballast has a slightly positive volt-ampere characteristic. For AC operation the ballast may take the form of an inductance, with the advantage that less power is dissipated in the ballast.

Source stability is particularly important in luminescence spectrometry because the measured luminescence radiance is usually directly proportional to the spectral radiance of the excitation source. Various means to increase arc lamp stability have been tried. Arc wander in a given lamp tends to decrease as the input power is increased. For this reason, arc lamps should not be operated far below their rated powers if maximum stability is desired. On the other hand, operation above the rated power may decrease lamp life. Feedback stabilization, used successfully with tungsten lamps, is expensive and of limited effectiveness. Ratio stabilization, which involves the measurement of the ratio of the luminescence intensity to the source intensity, is effective in most cases. Several commercial luminescence spectrometers now include a ratio fluorescence capability.

A high pressure mercury arc lamp is a very intense source of ultraviolet radiation. Its spectral output consists of a number of very intense mercury lines superimposed on a continuum background. A typical mercury arc spectrum appears in Figure III-A-3. Although more intense in the ultraviolet than an equivalent xenon arc lamp, the mercury arc lamp is far less smooth in spectral distribution and is therefore less suitable for instruments designed to record excitation spectra.

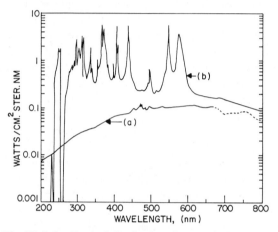

Fig. III-A-3.   Spectral distribution of two typical sources: curve *a*, PEK75 xenon arc lamp (with permission of PEK, Sunnyvale, Calif.); curve *b*, Osram HBO100 mercury arc lamp (with permission of Osram, Berlin).

The most widely used excitation source in molecular luminescence instrumentation is the high pressure xenon arc lamp.[5,6]. This type of lamp is constructed of two tungsten electrodes in a small quartz envelope. Most types have arc gaps of a few millimeters and are referred to as *short* or *compact arcs*. Xenon gas pressures under typical operating conditions range from 10 to 30 atm (see Figure III-A-3). Except for a few low intensity xenon lines around 450 nm, the spectral distribution in the visible and ultraviolet regions corresponds closely to that of a 600°K black body with an emissivity of 0.06.[7] A serious deviation from a black body distribution occurs in the near-infrared region, where there is a system of very strong lines between 800 and 1000 nm. Because most excitation spectra are obtained between 200 and 450 nm, the xenon lamp will not cause the appearance of false peaks. The ratio of line to continuum intensity depends on the current density in the arc. The spectral radiance of the continuum increases with approximately the 1.6 power of the current density, whereas the radiance of the line emission increases almost linearly. Thus a purer continuum is obtained at higher current densities.

Commercial xenon arc lamps are available in a wide range of input power ratings from a few watts to thousands of watts. The increased input power of the higher power lamps is due mainly to the increased total arc current, rather than arc voltage or arc current density. For lamps in the 100 to 5000-watt range, total arc currents range from 5 to more than 1000 A, but the arc voltage increases only slightly with lamp power and generally remains between 15 and 30 volts. This variation is owing primarily to the increased arc length, which varies typically from 1 to 9 mm. In general, the arc width also increases with arc power, so that the arc current density is not necessarily much greater in the high power lamps. The color temperature is nearly independent of lamp power. The increased *total light output* of *high power lamps ·is primarily due to* the *increased arc area*. However, the average spectral irradiance does increase somewhat with lamp power, because the spectral emissivity increases to some extent. The total radiative efficiency of a xenon arc lamp is typically 30 to 50%. The losses are primarily thermal and are the result of ions and electrons in the arc colliding with the envelope walls and giving up their kinetic energy. The heat is then carried from the walls by air convection or radiated as long wavelength infrared radiation. For a given current density, the energy loss by thermal conduction is expected to be proportional to the wall area. Thus a large-diameter, short-arc lamp would tend to have a higher total radiative efficiency than a long, small-diameter lamp of the same internal volume.

Compared with a 100-watt tungsten lamp, a 100-watt xenon arc has a lower radiative efficiency and therefore generates more heat, but its spectral radiance in the visible and ultraviolet is much higher because its color temperature is much more nearly optimum for this region. Calculating on the basis of the respective color temperatures and spectral emissivities, it is found that the spectral radiance of the xenon arc at 500 nm would be about 0.1 watt $cm^{-2} sr^{-1} nm^{-1}$ compared with about 0.005 watt $cm^{-2} sr^{-1} nm^{-1}$ for the tungsten lamp. In the ultraviolet, the difference is even more striking.*

---

* It is important to realize the distinction between emissivity and radiative efficiency when dealing with arc sources whose spectral distributions approximate those of a black body. The arc is not, of course, a real incandescent source, and trying to fit the arc into the black body scheme is somewhat artificial. The black body nomenclature serves merely as a basis of comparison of sources. The temperature is chosen for the best fit to the spectral distribution of the arc radiation, and then the emissivity is adjusted to reconcile the spectral radiance. A significant characteristic of a xenon arc is that the color temperature of the arc plasma (i.e., the black body temperature with the same spectral distribution) is close to 6500°K regardless of the input power, quite unlike a true incandescent filament lamp. The increased input power and light output of highpowered tubes is due more to the increased arc width than to increased color temperature, as mentioned previously.

Xenon arc lamps are almost always operated on DC for greatest stability and longest life. As with other high pressure DC arcs, polarity is important; and the lamp should be operated vertically with the anode (heaviest electrode) up. The starting voltage required is between 20 and 30 kV, considerably greater than that required for mercury arc lamps, because of the higher pressure in the xenon lamp before ignition. The pressure of xenon gas at room temperature before the lamp is started is relatively high (ca. 5 atm), whereas the vapor pressure of mercury at that temperature is very low (ca. 10^{-5} atm). The high voltage may be applied across the arc, to an auxiliary electrode, or to a wire placed close to the bulb. Immediately after ignition, the light output is about 80% of the final value, which is reached in a few minutes. Warm-up time is considerably less than that for mercury arcs.

Although the xenon arc, like the mercury arc, is convection stabilized, it is less sensitive to cooling conditions because there is no danger of xenon condensation. Forced air cooling is possible if care is taken to maintain a reasonably laminar air flow. Natural convection cooling is usually sufficient for all but the highest powered lamps.

The useful life of a xenon arc lamp is limited by the intensity loss due to deposition of electrode material on the bulb. Quoted values range from 200 to more than 1000 hours. Operation of lamps at powers in excess of the rated values increases the rate of electrode evaporation and decreases useful life. More starts and shorter average burning times also shorten life. Operation of a xenon lamp beyond its useful lifetime is not recommended, because absorption of radiated energy by the walls may generate excessive heat and cause the lamp to explode. As a rule, the average useful life of xenon arc lamps is somewhat longer than that of mercury arc lamps.

Arc lamps filled with a mixture of approximately 80% mercury and 20% xenon are often used. Such lamps have the ultraviolet lines of mercury and the infrared lines of xenon, as well as a strong continuous background. The electrical properties are similar to those of mercury arc lamps. The ultraviolet radiation of mercury-xenon lamps is considerably more intense than that of xenon lamps of the same power, but it is much less uniform in spectral distribution.

c. FLASH LAMPS[8,9]

In some applications, a very intense, short-duration pulse of exciting light may be desirable. This can be obtained from an arc lamp by applying a short high current pulse to the lamp, usually by discharging a high voltage capacitor through the lamp. Xenon arc lamps designed for this sort of service are called xenon flash lamps. They are used as sources of high intensity light pulses in the visible and ultraviolet region. Such lamps have been shown to have considerable analytical utility in luminescence spectrometry. They are particularly useful for the measurement of short luminescence decay times. In addition to xenon, other gases such as nitrogen and hydrogen may be used. Nitrogen flash lamps have been used to measure fluorescence decay times of the order of several nanoseconds.

A simple flash tube driving circuit is shown in Figure III-A-4. The DC power supply charges the energy storage capacitor C through resistance R to

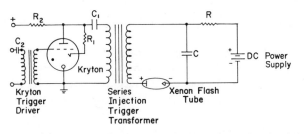

Fig. III-A-4. Simple flash tube circuit identical to the circuit in TM-12 Trigger Module, Technical Data, EG & G, Inc., Electronics Division, Boston, Mass. 02215 (with permission of EG & G, Inc.).

a voltage V, the operating voltage. This voltage must be below the self-flash voltage of the lamp, so that the lamp will operate only when triggered. The lamp is triggered by the application of a high voltage trigger pulse, which causes the fill gas in the tube to ionize. The energy stored in the capacitor C is then rapidly discharged through the lamp, producing an intense flash of light. The energy input (joule) to the lamp per flash is equal to the energy stored in the capacitor and is given by

$$J = \frac{1}{2} CV^2 \qquad \text{(III-A-2)}$$

where J is the input energy (joules), C is the value of the storage capacitor (μF),* and V is the operating voltage (kV). The value of C is typically between 0.1 and 500 μF, and V is usually between 0.5 and 3.0 kV. The maximum allowable value of J is limited by lamp construction and is specified by the manufacturer of the lamp.

The duration (sec) of the light flash t_f is approximately equal to the duration of the current pulse through the lamp

$$t_f = \frac{1}{2} R_f C \qquad \text{(III-A-3)}$$

* 1 microfarad (μF) $= 10^{-6}$ farad.

where R_f is the effective arc resistance during the flash and is approximately directly proportional to the arc gap length and inversely proportional to the $\frac{2}{3}$ power of the operating voltage V. Thus the shortest, most intense flashes are produced from short gap tubes operated from relatively high V and low C. The value of R_f is given by the manufacturer of the lamp for a specified set of operating conditions.

The *peak input power* (watt) per flash is given approximately by

$$P_i = \frac{J}{t_f} \qquad \text{(III-A-4)}$$

More important is the *peak light output* power per flash, which is P_i times the radiative efficiency, typically 25 to 50% for xenon flash tubes. The radiative efficiency is also a function of operating parameters, particularly V. For xenon tubes the peak output power per flash turns out to be roughly proportional to $C^{2/3}V^3$.

The *average input electrical power* $\bar{P}_i$, in watts, is

$$\bar{P}_i = Jf \qquad \text{(III-A-5)}$$

where f is the flash repetition frequency (Hz). A maximum value of $\bar{P}_i$ is also recommended by the manufacturer. The higher powered types may have to be forced-air cooled.

The average current drawn from the DC power supply is given by

$$i = VCf \qquad \text{(III-A-6)}$$

where the terms have the same definitions as previously. The value of the charging resistor R in Figure III-A-4 should be chosen so that

$$R = \frac{1}{5fC} \qquad \text{(III-A-7)}$$

Its power rating should be greater than $\bar{P}_i$. The value of f should never exceed about 0.1 t_f^{-1} to prevent continuous ionization of the lamp.

As an example of the application of the foregoing equations, consider the operation of a small commercial xenon flash lamp with maximum ratings of 5 joules per flash, 2-kV operating voltage, 10-watt average input power, and a quoted arc resistance of 5 ohms at 1 joule. Suppose that, in order to increase useful lamp life, the lamp is to be operated at 1.0 kV and 1 joule per flash, somewhat below the maximum ratings. The maximum allowable repetition frequency is that at which $\bar{P}_i$ is 10 watts and is equal to 10 Hz according to equation III-A-5. Equation III-A-2 gives the value of C as 2 μF. The peak input power per flash P_i is 2×10^5 watts (equation III-A-4) and the peak output power P_o will probably be about 5×10^4 watts (assuming

25% efficiency). The flash duration is 5 μsec (equation III-A-3). Equation III-A-7 gives the value of the charging resistor R as 10 kΩ. The average current drawn from the power supply is 20 mA by equation III-A-6.

The spectral distribution of flash lamps depends mainly on the nature and pressure of the fill gas. Xenon flash tubes with quartz envelopes have an effective color temperature of about 7000°K, similar to DC xenon arcs; but the emissivity is considerably greater and often approaches unity near the peak of the flash.

When a xenon flash tube is operated below its maximum peak input power per flash, not all the gas is ionized and the arc may take on a narrow, filamentary appearance. The position of the arc tends to shift from flash to flash, causing irregularities in the measured light output. If the energy input is increased so that all the xenon ionizes, the arc is stabilized. For this reason, it is usually not advisable to operate a flash tube very far below its maximum input. Certain tubes are specifically designed for stable, low energy operation.

3. MONOCHROMATORS[10-12]

In most modern commercial spectroluminometers, two monochromators are used, one for emission and one for excitation. Usually the two are identical in design.

A generalized diagram of the dispersive and optical portion of a monochromator is supplied in Figure III-A-5. Radiation from the sample is col-

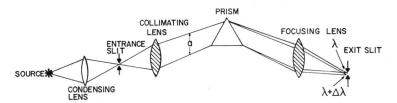

Fig. III-A-5. Generalized diagram of dispersive and optical portion of a monochromator.

lected by the condensing lens and focused onto the entrance slit, which defines the shape of the images at the focal plane. Radiation emerging from the entrance slit is collimated onto the dispersing element, a prism or grating, where it is separated into its spectral components. This radiation is focused onto the exit slit, which isolates a small region of the spectrum. In real instruments, the collimating and focusing elements may be lenses or mirrors; or, their functions are often combined into one element, but the schematic diagram shown will satisfy the requirements of the following general treatment of *dispersion, resolution,* and *spectral bandpass.* The focal lengths of the collimating and focusing elements are almost always equal in practice, and this is assumed in the following discussion.

a. DISPERSION

An important characteristic of a spectrometer is its *dispersion,* or its ability to separate different wavelengths of radiation so that they emerge from the dispersion element at different angles and come to a focus at different positions in the focal plane. The *angular dispersion D* is given by $d\theta/d\lambda$ where $d\theta$ is the angular separation of two dispersed beams separated by a wavelength difference of $d\lambda$. The value of D depends on the construction of the dispersing element (prism or grating). The *linear dispersion* is the separation between different wavelengths in the focal plane, and is given by $dx/d\lambda$, where dx (cm) is the linear distance between two wavelengths separated by $d\lambda$ (nm). The relation between angular and linear dispersion is easily determined by geometry. If F is the focal length of the lens which focuses a dispersed beam of light with angular dispersion $d\theta/d\lambda$, the linear dispersion at the focal plane will be

$$\frac{dx}{d\lambda} = F\frac{d\theta}{d\lambda} = FD \qquad \text{(III-A-8)}$$

A more useful measure of dispersion, and one which is most often specified by spectrometer manufacturers, is the *reciprocal linear dispersion* R_d, defined by

$$R_d = \frac{d\lambda}{dx} = \frac{1}{DF} \qquad \text{(III-A-9)}$$

The reciprocal linear dispersion is often expressed in Angstrom units per millimeter or nanometers per centimeter. Typical values of R_d for monochromators in a luminescence spectrometer range from 1 to 10 nm mm^{-1}.

b. SPECTRAL BANDPASS

The ability of a monochromator to produce a monochromatic beam of radiation from a polychromatic light source is expressed in terms of its *spectral bandpass s,* which is the range of wavelengths of radiation emerging from the exit slit when polychromatic radiation is incident upon the entrance slit. More precisely, the spectral bandpass is defined in terms of the *slit function* of the monochromator, which is the plot of intensity emerging from the exit slit versus wavelength when the monochromator is scanned through a monochromatic line from a line source evenly illuminating the entrance slit (see Appendix 3). The slit function will be a trapezoid or triangle whose half width (width at half-maximum) expressed in wavelength units is defined as the spectral bandpass s,

$$s = R_d W \qquad \text{(III-A-10)}$$

where we have R_d (nm cm^{-1}) and the width W (cm) of the entrance or exit slit, if they are equal, or of the larger of the two slits, if they are unequal. The base width $\Delta\lambda$ of the slit function is given by

$$\Delta\lambda = R_d(W_s + W_e) \qquad \text{(III-A-11)}$$

where W_s and W_e are the entrance and exit slit widths, respectively. If $W_s = W_e = W$,

$$\Delta\lambda = 2R_d W \qquad \text{(III-A-12)}$$

and the slit function is triangular. This is the most usual case.

c. RESOLUTION

The *resolution* $\Delta\lambda_R$ of a spectrometer may be defined as the minimum wavelength separation between two monochromatic spectral lines that are just separated. The exact definition depends on the criterion for separation. For essentially *complete* separation, the slit functions of the two lines may join but may not overlap at their adjacent bases. In this case, the resolution is the same as the basewidth of one slit function

$$\Delta\lambda_R = R_d(W_s + W_e) \qquad \text{(III-A-13)}$$

and

$$\Delta\lambda_R = 2R_d W = 2s \qquad \text{if} \qquad W_s = W_e = W \qquad \text{(III-A-14)}$$

neglecting diffraction and optical imperfections.

d. THEORETICAL RESOLUTION AND RESOLVING POWER

It might be thought, on the basis of the expression for linear dispersion given in equation III-A-8, that two adjacent spectral lines, no matter how close, could be resolved simply by increasing the focal length sufficiently. This would be so only if: (a) the slit widths could be made indefinitely small* and (b) the entrance slits were imaged in the focal plane as a geometrical image of the slit. In reality, of course, a finite slit width must always be used to allow a measurable amount of light to pass through. Even if an infinitely narrow slit could be used, however, the resolution would be limited by *Fraunhoffer diffraction* of the collimated beam by the effective aperture of the instrument. The effective aperture is the width of the collimated beam, or, in other words, the width of whatever aperture defines or limits the width of

* As $W \to 0$, the parallelism of the slit jaws would also have to be maintained.

the collimated beam (designated by the distance a in Figure III-A-5). The diffraction of the collimated beam by the edges of the effective aperture causes the entrance slit to be imaged at the focal plane as a diffraction pattern rather than as a geometrical image. The result is that, as the slit widths are decreased, the observed slit function approaches a Fraunhoffer diffraction pattern of finite width instead of an infinitely narrow line. The half width X of the central intensity maximum of the diffracting pattern is given by elementary diffraction theory.

$$X = \frac{\lambda F}{a} \qquad \text{(III-A-15)}$$

Thus, as the slit widths approach zero, the spectral bandpass approaches a minimum value $s_{\min}$ given by

$$s_{\min} = R_d X = \frac{R_d \lambda F}{a} \qquad \text{(III-A-16)}$$

By comparison with equation III-A-10, we see that in the common case* $W_s = W_e = W$, the effect of diffraction is to limit the *effective* slit width to a minimum value $W_{\min}$ as the mechanical slit width approaches zero.

$$W_m = \frac{\lambda F}{a} \qquad \text{(III-A-17)}$$

In addition to the diffraction effect, optical imperfections such as *spherical aberration*, *coma*, and *astigmatism* may cause the experimentally observed spectral bandpass to approach a value slightly higher than that calculated by equation III-A-16 as the slit widths approach zero. This additional aberration-limited spectral bandpass term s_a must also be included in $s_{\min}$. Thus equation III-A-16 should be replaced by

$$s_{\min} = \frac{R_d \lambda F}{a} + s_a \qquad \text{(III-A-18)}$$

It is not unreasonable to expect the slit width, diffraction, and aberration terms to add approximately linearly, and so the actual spectral bandpass for equal entrance and exit slits may be written as

$$s \simeq R_d W + s_{\min} \simeq R_d W + \frac{R_d \lambda F}{a} + s_a \qquad \text{(III-A-19)}$$

For purposes of calculation, all the length terms in this equation must be converted to units comparable with those of R_d; s_a must be determined

---

* Most monochromators have bilateral slits, but the slits may differ slightly (i.e., $W_s \neq W_e$). This is especially noticeable as W_s and W_e approach $W_{\min}$.

experimentally. An experimental plot of s versus W is a useful test of a monochromator. The spectral bandpass s is equal to the half width of the slit function obtained by scanning the monochromator through an isolated monochromatic line from a line source of very narrow true line width. The slope of the plot on linear coordinates is R_d, and the vertical intercept is $R_d \lambda F/a + s_a$. Because the diffraction term can be calculated, the optical quality can be assessed by the smallness of s_a. In such measurements, the entrance slit must be very evenly illuminated in order to obtain slit functions of good triangular shape.

The actual resolution expected from a monochromator should be calculated on the basis of equation III-A-19. For essentially *complete* resolution, two lines must be separated by $2s$ if the exit and entrance slit widths are equal. Somewhat less than complete resolution can sometimes be tolerated, however. In any case, $\Delta \lambda$ must always be greater than s.

The *resolving power* R of a spectrometer is defined by

$$R = \frac{\bar{\lambda}}{\Delta \lambda_R} \qquad \text{(III-A-20)}$$

where $\Delta \lambda_R$ is the wavelength difference (nm) between two monochromatic spectral lines which can just barely be distinguished as separate lines and $\bar{\lambda}$ is the mean wavelength of the two lines (nm). A value of 10^4 is not unusual for a medium-resolution monochromator.

The *theoretical resolving power* R_{theor} is the diffraction-limited resolving power for zero slit widths and no optical imperfections. Using the "Rayleigh criterion" of resolution, R_{theor} is given by

$$R_{\text{theor}} = \frac{\bar{\lambda}}{XR_d} = \frac{a}{R_d F} \qquad \text{(III-A-21)}$$

The extent to which R of a real spectrometer approaches R_{theor} is also considered as an indication of quality.

e. SLIT ILLUMINATION AND ENTRANCE OPTICS

The radiant flux of a spectrum at the focal plane of a spectrometer depends not only on the source radiance but also on the way in which the entrance slit is illuminated. Obviously, the highest possible radiant flux at the slit face is desirable in luminescence work to obtain high sensitivities. However, of all the light that passes through the entrance slit, only that part which is collected by the collimator lens is effective in producing the image at the focal plane. In other words, radiation falling outside the solid angle subtended at the entrance slit by the collimator lens will be wasted (and may, in fact, contribute to undesirable stray light in the instrument). Sometimes a condensing lens (mirror) is placed between the source and the entrance slit to collect and focus the light on the slit. The maximum radiant flux at the focal plane with a given source is obtained when the collimator is filled with light, regardless of the arrangement of the condensing lens. If the luminous area of the source is large enough to subtend a solid angle at the slit equal to that subtended by the collimator lens, then the collimator lens will be completely illuminated and a condensing lens will be of no help. This is often true of the emission monochromator in luminescence spectrometers.

If, on the other hand, the source area is too small to fully illuminate the collimator, then using a condensing lens to form an image of the source on the entrance slit will increase the image brightness at the focal plane. This is often true of the excitation monochromator. The lens should be placed so that it subtends a solid angle at the entrance slit equal to that subtended by the collimator lens. In this way, the collimator will be filled with light. The image of the source on the slit should just fill the slit.

It might be thought that the image brightness could be increased by using a condensing lens of large diameter and short focal length to form a small, high intensity image of the source on the entrance slit. This would indeed increase the radiant flux per unit area falling on the slit; but the solid angle at the slit would be proportionately decreased, and the spectral flux of the light emerging from the slit onto the collimator would remain unchanged.*

If the dimensions of the luminous area of the source are suitable dimensions for a slit, the entrance slit can be replaced by the source itself, thus simplifying the instrument and eliminating losses in the condensing lens. This is occasionally done in excitation monochromators when high pressure arc lamps are used as sources. A disadvantage, however, is that the effective slit width is not adjustable. Furthermore, arc wander causes unpredictable shifts in the monochromator calibration. In most cases, the use of a condensing lens and separate entrance slit is preferable.

f. SPEED

The *speed* of a spectrometer is an indication of the ability of the collimator lens to collect light emerging from the entrance slit. It is expressed by the "f number":

$$f = \frac{F_c}{D_c} \qquad \text{(III-A-22)}$$

* Because object area X solid angle = image area X image solid angle, an increase in image area compared to object area must be achieved with a decrease in image solid angle compared with object solid angle. The maximum radiant flux passing through the entrance slit and the maximum resolution results if the entrance slit is fully illuminated with source radiation. An array of lens and mirrors will only increase the radiant flux if more than *one* solid angle of radiation from the source fills the monochromator collimator.

where F_c and D_c are the focal length and diameter of the collimator lens (or mirror), respectively. In general, the term *speed* may be applied to any lens; the smaller the focal-length-to-diameter ratio (f number), the faster the lens, and the greater the light gathering ability. The solid angle, in steradians, of light collected by a lens from a point source is

$$\Omega = \frac{A_l}{F_c^{\,2}} \qquad\qquad \text{(III-A-23)}$$

where A_l is the area of the lens (cm^2) and F_c is its focal length (cm). Thus fast lenses (i.e., lenses with large diameters and short focal lengths) pick up a large solid angle of radiation.

g. LIGHT LOSSES

In a real spectrometer, the total flux of radiation at the focal plane is always less than that collected by the collimator lens, owing to unavoidable light losses in the instrument. The light transmitting ability of a spectrometer at a particular wavelength λ is indicated by its *transmission factor* T_f, which is the ratio of the monochromatic light flux of wavelength λ emerging from the exit slit (or incident on the focal plane) to that collected by the collimator lens. The primary causes of light losses in spectrometers are reflection, absorption, and scattering by the various optical components (e.g., prisms, lenses, gratings, and mirrors) in the instrument. It is obviously desirable to use lens and prism materials of high transmittance and mirrors of high reflectivity in the desired wavelength range. Furthermore, reflection losses at the surfaces of prisms and lenses can be reduced in a limited wavelength range by the use of special low reflectance coatings. Gratings should be blazed in the desired wavelength range. Finally, light losses can be reduced by reducing the total number of optical components in the light path. This can be done by combining the functions of two or more components into one (e.g., using a concave grating as both a dispersing element and a focusing element). In some designs, such as the Fery prism monochromator and the concave grating mountings, the only optical components in the light path is the dispersing element itself.

h. RADIANT FLUX EMERGING FROM SPECTROMETRIC SYSTEM[2]

All calculations involving the response of a spectrometric system to a "light source," as well as any predictions that can be made concerning the effect and optimization of the slit widths and other optical parameters, require that an expression be obtained for the light flux emerging from the exit slit of a monochromator as a function of the source radiance (line source) or source spectral radiance (continuum source) and monochromator optical variables. The radiance flux expressions for the excitation and emission monochromators in luminescence spectrometers are similar; the "source" radiance or spectral radiance refers to the excitation source radiance (or spectral radiance) and sample luminescence radiance (or spectral radiance), respectively.

To derive an expression for the radiant flux emerging from the exit slit, the following definitions are needed:

$B_N^o =$ radiance of a line source, watt cm^{-2} st^{-1}

$B_{C\lambda}^o =$ spectral radiance of continuum source, watt cm^{-2} sr^{-1} nm^{-1}

$W =$ Illuminated width of slit, cm

$H =$ Illuminated height of slit, cm

$T_f =$ transmission factor of optical system, no units

$\Omega =$ solid angle of light collected by collimator, sr

$s =$ spectral bandpass of monochromator, nm

$\Phi =$ total flux emerging from exit slit, watt

$K_o =$ optics factor $\equiv HT_f\Omega$, cm sr

Assuming that the exit and entrance slits are equal in W and H and that all light losses are included in T_f, then for a "line source"

$$\Phi = B_N^o WHT_f\Omega = B_N^o K_o W \qquad \text{(III-A-24)}$$

and for a "continuum source"

$$\Phi = B_{C\lambda}^o WHT_f\Omega s = B_{C\lambda}^o K_o sW \qquad \text{(III-A-25)}$$

Neglecting diffraction, $s = R_d W$; thus for a continuum source

$$\Phi = B_{C\lambda_o}^o W^2 HT_f\Omega R_d = B_{C\lambda_o}^o K_o R_d W^2 \qquad \text{(III-A-26)}$$

For a source of light containing both line and continuum components, the line-to-continuum ratio is inversely proportional to the slit width. This is of particular importance in luminescence spectrometry. In molecular fluorescence work, we must often attempt to measure relatively broad band emission spectrum of the sample in the presence of scattered light from the excitation source, often a mercury or mercury-xenon lamp. The relative contribution of the intense mercury lines can be reduced by using relatively large emission monochromator slit widths. In atomic fluorescence spectrometry with a continuum excitation source, the ratio of the desired fluorescence line emission to the continuum background due to scattering of the excitation radiation by flame particles and droplets is increased by using relatively narrow slits.

It is sometimes helpful to record an observed excitation or emission spectrum at several values of excitation and/or emission monochromator slit widths in order to determine which features of the spectrum are due to line emission (radiant fluxes directly proportional to slit width) and which are due to essentially continuum emission (radiant fluxes proportional to square of slit width).

It is of interest that the best balance between radiant flux* and resolution is obtained if the entrance and exit slits of a monochromator are equal.[2]

i. STRAY RADIATION

Stray radiation is particularly undesirable in luminescence instrumentation because of the very low luminescence intensities that must often be measured and because the very intense light from the excitation source may tend to "leak" into the emission monochromator, even in the absence of real luminescence of the sample.

Stray radiation in spectrometers may result from the following causes:

1. Leakage of room light into the instrument through unshielded slits or through light leaks in the instrument's case.

2. Reflection and scattering of radiation from walls, optics, mountings, slits, and baffles.

3. Reflection and scattering on optical surfaces.

4. Scattering of light within prisms and lenses.

5. Scattering of light by dust particles.

6. Unused orders in grating spectra.

7. Fluorescence of optical materials.

In monochromators, stray light causes contamination of the selected spectral bandwidth with light of other wavelengths. This undesirable effect may be minimized by proper design of the instrument and by the use of carefully selected optical materials. The amount of stray light in a monochromator is usually quoted by the manufacturer as a certain percentage of the total flux of desired radiation emerging from the exit slit. Typical values range from 0.1 to 1.0% for small commercial monochromators. Stray light in an existing instrument can often be reduced by painting the interior of the instrument flat back, by inserting baffles to obstruct unwanted rays, by carefully shielding slits from room light, and by using windows in front of the slits to seal out dust and corrosive fumes. The problem of stray radiation is

---

* Refer to Appendix 3 for a more detailed discussion of optical factors affecting the measured radiant flux and the measured instrumental signal.

most severe if the light intensity at the wavelength of interest is much less than that in neighboring regions. This situation occurs, for instance, when a monochromator is employed to select a wavelength in the ultraviolet region from a xenon arc lamp, which radiates most intensely in the visible. A filter that transmits only the wavelength region of interest may be placed in the light path to reduce interference by the unwanted radiation. A related problem arises when observing a weak atomic line close to a very strong one, as might occur in an atomic fluorescence experiment. Stray light from the strong line may seriously interfere with the weak line and may make resolution of the two lines impossible, even if the resolving power of the instrument is sufficient to separate the two lines if they were equally intense.

Probably the most satisfactory, and unfortunately most expensive, way of reducing interference from stray radiation is the use of the *double monochromator*, which is essentially a system of two monochromators connected in series, the exit slit of the first being the entrance slit of the second. With two identical monochromators, the dispersion and resolution are approximately doubled, and stray radiation is greatly reduced. For example, if the intensity of the stray radiation in each monochromator is 0.1% of that of the primary beam, then the double monochromator would reduce this to 0.0001% (i.e., 0.1% of 0.1%). A slight disadvantage, however, is that the transmission factor of the double monochromator would be about half that of either half by itself. A more serious difficulty is the necessity of keeping the wavelength settings of both monochromators exactly the same at all times during the scanning of a spectrum. This is generally accomplished by mounting both dispersing elements on the same rotating platform or shaft. In some double monochromators, different dispersing elements are used in each half; usually the first is a prism and the second a grating. It is also possible to get some of the effect of two dispersing elements by passing the light through a single dispersing element twice. This arrangement may be considerably less expensive than a true double monochromator, but it is also less effective in reducing scattered light.

At least one commercial luminescence spectrometer uses two double monochromators.

j. PRISM MONOCHROMATORS

Dispersion by a prism depends on the change of refractive index n of the prism material with wavelength. The angular dispersion D is given by

$$D \equiv \frac{\partial \Theta}{\partial \lambda} = \frac{\partial n}{\partial \lambda} \frac{\partial \Theta}{\partial n} \qquad \text{(III-A-27)}$$

where Θ is the deviation angle, that is, the angle through which an incident light beam is deviated in passing through the prism (Figure III-A-6a). The

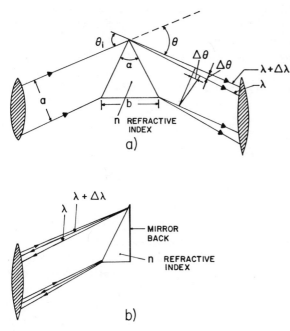

Fig. III-A-6. Dispersion of radiation by two typical prisms: (*a*) Cornu type, (*b*) Littrow type. (For simplicity, angles and distances do not appear in (*b*), but note that the littrow mounting is simply a folded Cornu mounting.)

first term in this equation is the *optical dispersion* of the prism material, and it depends only on the *wavelength* and the *nature* of the prism material and is independent of the geometry of the prism or its mounting. The second term $\partial\Theta/\partial n$ is the so-called *geometrical factor* and is a function of the geometry and mounting of the prism. In general, the prism mounting is arranged so that the ray internal to the prism is parallel to the prism base, this being the most symmetrical arrangement. In this case, the deviation angle Θ is a minimum, and astigmatism and reflective losses are minimized. The geometrical factor is then given by

$$\frac{\partial\Theta}{\partial n} = \frac{2\sin\alpha/2}{[1 - n^2 \sin^2(\alpha/2)]^{1/2}} = \frac{2\tan\Theta_i}{n} = \frac{b}{a} \qquad \text{(III-A-28)}$$

where α is the apex angle of the prism, Θ_i is the incidence angle at the first prism face, b is the thickness of the prism base, and a is the effective aperture as defined previously, assuming that the whole prism face is illuminated (usually true). By far the most common apex angle α is 60°, and the geometrical factor for this angle at minimum deviation is about 1.5. More common than a full 60° equilateral prism is the 30 × 60 × 90° Littrow prism with a reflective (aluminized) long side, shown in Figure III-A-6b. The reflection back through the prism gives the effect of a full 60° prism with a smaller amount of quartz. In addition, the birefringence of the quartz is canceled.

The theoretical resolving power of a prism may be shown to be

$$R_{\text{theor}} = b \frac{\partial n}{\partial \lambda} \tag{III-A-29}$$

Thus large prisms of high optical dispersion are desirable. Quartz prisms are used in the ultraviolet region, but glass is preferred in the visible because $\partial n/\partial \lambda$ is higher than quartz.

k. GRATING MONOCHROMATORS

All modern commercial luminescence spectrometers now have grating monochromators, principally because of their superior dispersion, resolution, and speed and because the dispersion of a grating is nearly independent of wavelength over a considerable range. The dispersion properties of a grating can be obtained from the basic grating equation. This equation, which is derived in elementary optics textbooks, states that maxima in the intensity of the diffracted light occur for those wavelengths λ for which

$$m\lambda = \Delta(\sin \Theta_i + \sin \Theta_d) \tag{III-A-30}$$

where m is a positive or negative integer called the *order*, Δ is the spacing between the grooves or rulings, Θ_i is the incidence angle, and Θ_d is the angle of diffraction, as in Figure III-A-7a. The angular dispersion of a grating is obtained by differentiating Θ_d with respect to λ in the previous equation:

$$\frac{d\Theta_d}{d\lambda} = \frac{m}{\Delta \cos \Theta_d} = \frac{m\delta}{\cos \Theta_d} \tag{III-A-31}$$

where δ is the number of lines (grooves) per centimeter ($=1/\Delta$). The reciprocal linear dispersion of a grating monochromator of focal length F is

$$R_d = \frac{10^5 \cos \Theta_d}{\delta F m} \tag{III-A-32}$$

where F is in meters and R_d is in nanometers per centimeter. In most monochromators, $\cos \Theta_d$ is nearly unity and does not change much over the wavelength range of the instrument. Thus R_d is nearly independent of wavelength. This fact greatly simplifies the calibration of grating monochromators. According to this equation, the highest dispersion (lowest R_d) is obtained

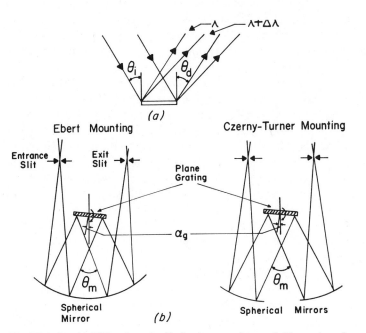

Fig. III-A-7. (*a*) Diffraction of radiation by a grating, and (*b*) rotation of a grating to vary wavelengths.

in monochromators of long focal length using finely ruled grating in a high order. Monochromators in luminescence spectrometers typically have focal lengths between 0.25 and 1.0 m and use gratings of 600 to 1200 lines mm^{-1} in the first order.

Reflection gratings may be plane or concave, the concave type being used mainly in photographic spectrographs. Most commercial luminescence spectrometers use plane reflection gratings in either the Ebert or the Czerny-Turner mountings (see Figure III-A-7*b*). In these mountings, the grating is rotated to change wavelengths while the angle between the incident and diffracted rays remains constant (Θ_m in Figure III-A-7). It is convenient to modify equation III-A-30 for this case. Defining α_g as the angle between the grating normal and the monochromator center line:

$$m\lambda = \left[2\Delta \sin\left(\frac{\Theta_m}{2}\right) \right] \sin \alpha_g \qquad (\text{III-A-33})$$

The factor in brackets is a constant for a given instrument and grating. For convenience in wavelength readout, a specially ground cam or "sine bar" mechanism is included in commercial monochromators to convert the linear rotation of the wavelength counter or dial to the sine rotation required by the equation, thus providing a linear wavelength scale. A " cosecant bar " can be used to provide a linear wavenumber (reciprocal wavelength) scale.

Note that according to equations II-A-30 and II-A-33, all grating monochromator wavelength dials read out in $m\lambda$, not simply λ; this must be considered if grating orders other than the first are used. For instance, the 546.074-nm green mercury line would be picked up in the second order at a dial reading of 2(546.074) or 1092.348 nm. On the other hand, the dial setting for a wavelength λ in the first order is also the proper setting for all wavelengths λ/m in correspondingly higher orders. This order overlap problem is particularly severe in spectrofluorimetry with a mercury arc excitation source, and the experimenter must understand quite well the operation of grating orders in order to avoid erroneous interpretation of data. For example, a fluorescence band at 507.4 nm could be interfered with by scattering of the intense 253.7-nm line from a mercury arc lamp, because this line would appear in the second order near a dial reading of 507.4 nm.

These problems of order overlap and ambiguity are the most serious disadvantages of grating spectrometers, but there are ways to avoid or reduce these difficulties. For instance, a filter may be placed in the monochromator light path to absorb unwanted wavelength regions, or a prism may be combined with the grating as an "order separator." The waste of precious light intensity in unwanted orders may be reduced by the use of " blazed " gratings,* whose rulings are shaped to throw most of the diffracted intensity into a particular $m\lambda$ region. In blazed gratings, the reflective flats are ruled at an angle β, the *blaze angle*, to the grating surface. The greatest diffracted intensities occur at those values of Θ_d most closely corresponding to the angle of specular reflection from the surface; this occurs when $(\Theta_i - \beta)$ equals $(\beta - \Theta_d)$ in Figure III-A-7a. For an Ebert or Czerny-Turner monochromator, this condition is met when $\alpha_g = \beta$; that is, at the value of $m\lambda$ satisfying

$$m\lambda = 2\Delta \sin \frac{\Theta_m}{2} \sin \beta \qquad \text{(III-A-34)}$$

It is again important to note that this equation tells us that gratings are blazed for a given $m\lambda$, not for a particular order or wavelength alone. A grating blazed at 500 nm in the first order also exhibits an intensity maximum at 250 nm in the second order. Thus second-order interference at 500 nm from ultraviolet radiation near 250 nm is *not* reduced by the use of a grating blazed at 500 nm in the first order.

---

* Blazed gratings are called echelette gratings.

The symmetrical path of light through Ebert and Czerny-Turner mono-chromators largely cancels out the adverse effects of spherical aberration, coma, and astigmatism. In addition, there is no chromatic aberration in any system with all-reflective optics. However, residual astigmatism does cause the spectral lines to be curved along their lengths; this effect can impair resolution if long, narrow straight slits are used. Special slits correspondingly curved along their lengths are used in some high resolution monochromators to correct this effect, but they are costly. Curved slits are seldom necessary in luminescence instrumentation, particularly if small slit heights are used ($H \leq 5$ mm).

1. FILTERS[2,13-15]

Filters may be used as substitutes for monochromators, as in filter fluoro-meters, or to decrease the stray light in spectrofluorometers. Filter fluoro-meters have the advantages of simplicity, low cost, and very high light gathering ability, but they are obviously less selective and cannot record spectra. Filters are often used in conjunction with grating monochromators to prevent interference from undesired orders or other stray light.

A large variety of glass and gelatine filters are currently manufactured by several firms (e.g., Corning Glass Works, Corning, N.Y.). Sill[13] has given a convenient summary of the transmission spectra of a large number of filters, both single and in combination. Both low-pass and high-pass filters are available. High quality band pass filters may be made by combining a low-pass and with a high-pass filter. Concentrated solutions of many substances may serve as very effective high-pass filters.[2]

Interferences filters[15] may also be useful in some cases. These have nar-rower pass bands than do absorption filters but are more expensive.

4. PHOTODETECTORS[16-20]

Most commercial luminescence spectrometers use multiplier phototubes for the measurement of luminescence intensity, although some also use other types of photodetectors to monitor the intensity of the source for ratio fluorescence measurements or for the measurement of corrected spectra or absolute quantum efficiencies. The principal requirements for measurement of luminescence radiation are high *sensitivity* and *signal-to-noise ratio* over the required spectral range, *high linearity* of response, *low dark current*, and *fast response time*. The *radiant sensitivity* of a detector γ is defined as the ratio of the output current (A) to the radiant flux on the cathode (watt). Ideally, the detector sensitivity should be independent of wavelength if undistorted luminescence spectra are to be measured. Unfortunately the last requirement is impossible to attain in the most sensitive detectors

a. GENERAL ASPECTS OF PHOTOEMISSIVE DETECTORS.*

The multiplier phototube is one type of photoemissive detector. The principle of operation of photoemissive detectors may be illustrated by the single-stage vacuum photodiode shown schematically in Figure III-A-8. This

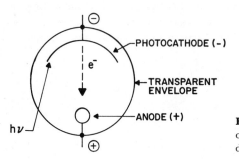

PHOTOCATHODE (-)

TRANSPARENT ENVELOPE

ANODE (+)

$h\nu$

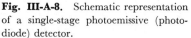

Fig. III-A-8. Schematic representation of a single-stage photoemissive (photodiode) detector.

device consists of two metal electrodes sealed in a glass or silica envelope and held at a potential difference of about 100 volts by means of an external battery or power supply. The negative electrode (photocathode) is coated with a photoemissive substance, usually a mixture of metals and metal oxides, of low work function. When light falls upon the photocathode. those photons of energy $h\nu$ greater than the work function Φ_a of the cathode material eject photoelectrons, which are attracted to and collected by the positive anode. The resulting current flowing in the external circuit is directly proportional to the rate of photoelectron emission, which is proportional in turn to the incident light radiant flux. Such a detector is stable, linear, and very fast, but unfortunately the sensitivity is comparatively low and is dependent on the wavelength. A photoemissive detector can not respond to light whose photons have an energy below the work function. This occurs at wavelengths above the *threshold wavelength* λ_{th} given by

$$\lambda_{\mathrm{th}} = \frac{hc}{\Phi_a} = \frac{12400}{\Phi_a} \, \text{Å} \qquad (\text{III-A-35})$$

where Φ_a is the cathode work function (eV).

Even below λ_{th} not all the photoelectrons may have enough kinetic energy $(h\nu - \Phi_a)$ to escape the surface, particularly if they originated some distance within the surface. At short wavelength, the sensitivity of the detec-

---

* For a discussion of the principles of nonphotoemissive detectors, refer to Appendix 4.

tor may be impaired by absorption by the envelope material. Most commercial tubes have λ_{th} between 600 and 1200 nm, and the short wavelength limit is about 350 nm with a glass envelope or 200 nm with silica. The shape of the spectral response curve (i.e., the plot of detector sensitivity versus wavelength) is thus a function of the cathode composition and envelope material and typically exhibits a broad maximum in the near-ultraviolet and visible region. There are a limited number of common combinations of cathode composition and envelope material; the corresponding response curves are denoted by "S" numbers such as S1, S5, and so on. A few are shown in Figure III-A-9.

A photoemissive detector is a "current source"; at a given radiant flux and wavelength, a given current is produced. This current flows through the load resistance R_L, developing the signal voltage $e_s = i_a R_L$. It is important to realize that the current is independent of the load resistance. Thus, for a given current, any desired signal voltage within reason may be obtained by

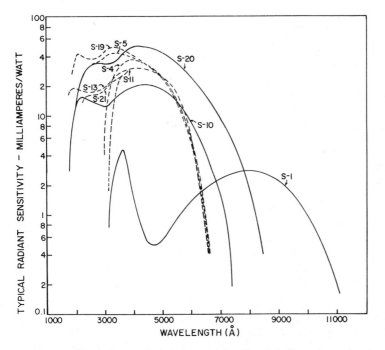

Fig. III-A-9. Typical spectral response curves for photoemissive surfaces taken from technical literature, Hamamatsu TV Co., Ltd., Hamamatsu, Japan. (With permission of Hamamatsu TV Co.)

selecting an appropriate load resistance. In general, a high value of R_L is desirable in order to reduce the required post detector electronic amplification. Two practical factors limit the magnitude of R_L: (a) the leakage resistance of the tube and associated circuit, which shunts the load resistance, and (b) the time constant $R_L C_S$, formed by the load resistance in parallel with stray capacitance C_S in the tube and wiring; this time constant degrades the effective response time of the detector. Values of R_L up to 10^{10} ohms are commonly used. If long detector leads are used, C_S may be 100 pF* or more, resulting in a time constant of up to 1 sec or more. Such a time constant would seriously degrade the accuracy of determination of the luminescence decay times (lifetimes) less than about 10 sec. Shorter leads, lower R_L, or combination of these would reduce the effect. In very high speed work, a miniature preamplifier is sometimes built right into the phototube base to minimize stray capacity.

All practical photoemissive detectors have a finite current output when no light is incident upon the cathode. This current, called the dark current i_d may interfere with the measurement of very low light intensities. The dark current may be the result of the following causes: (a) thermionic emission from the cathode or other internal parts of the device, (b) leakage currents over the envelope material or through tube base, socket, or circuit insulation, (c) positive ion current, and (d) field emission. The thermionic emission current i_{th} increases as Φ_a decreases (i.e., as λ_{th} increases). For this reason, near-infrared-sensitive (S-1) phototubes, which have long threshold wavelengths, tend to have higher thermionic emission currents than other tubes. In general, the phototube with the lowest acceptable λ_{th} should be used, to minimize thermionic emission. The temperature of the cathode has a great effect on the thermionic emission. In fact, the most generally satisfactory method of reducing thermionic emission in a given phototube is to cool the tube.

Another way to reduce thermionic emission current is to reduce the area of the cathode. However, provision must be made to focus all the light to be measured onto the small cathode; otherwise the signal current will be reduced proportionally. Some commercial phototubes are fitted with very small cathodes to help reduce the thermionic emission current.

The leakage current over the surface of the phototube envelope may often be reduced by carefully cleaning and drying the surface to remove conducting contaminents. Low-leakage base and socket materials should be used. Phototubes are often mounted in air-tight enclosures containing a dessicant to reduce moisture on the insulating surfaces. One excellent method of eliminating the effect of leakage currents in the output is to apply a circular band of

* 1 picofarad (pF) = 10^{-12} farad.

electrically conductive paint (called a *guard ring*) surrounding but not touching the point at which the anode lead passes through the envelope. When this guard ring is connected to the circuit ground, any leakage currents will then flow between the cathode and guard ring without interfering with the anode current. Leakage from the guard ring to the anode can be neglected because the potential difference between them is usually very low.

Positive ion current is the result of the impact of positive ions onto the cathode surface. It is usually a small contribution to the total dark current. Positive ions are produced in the phototube by the ionization of residual gas molecules by collisions with thermionically emitted electrons from the cathode acclerated toward the anode, by the field gradient between the cathode and anode, and by background radiation. By reducing the supply voltage to reduce the field gradient or cooling the tube to reduce the thermionic emission, residual gas ionization and resulting positive ion current will be reduced.

Field emission is the emission of electrons from the cathode caused by the electric field gradient at the cathode. Like the positive ion current, it is generally a small contribution to the total dark current in properly operated tubes, and can be reduced by lowering the supply voltage.

The total dark current i_d is the sum of the above-mentioned contributions and may range from 10^{-10} to 10^{-6} A for various types of photoemissive detectors.

The ideal output current of a photodetector is a smooth DC current proportional to the incident light flux. In real detectors, however, an AC or noise component is superimposed on the DC signal. This noise limits the ability of the experimenter to make accurate measurements of low light levels or to determine small differences between similar light levels.

The primary source of noise in photoemissive detectors at room temperature is generally shot noise. Shot noise is due to the fundamentally discontinuous (quantized) nature of light energy and electrical current. Photons arrive at the cathode randomly even though the overall intensity of the light beam is constant.* Thus photoelectrons are emitted from the cathode and arrive at the anode also randomly, and yet the long term rate of photoelectron pulses at the anode is constant and proportional to the light intensity. The same is true of thermionic electrons emitted from the cathode. Random, uncorrelated events like this are said to obey *Poisson statistics*, which has the fundamental property that the mean and the variance are identical; or in other words, the standard deviation is the square root of the mean. In terms

---

* Photon shot noise is often called simply photon noise. Photon noise is ultimately the limiting noise when other noises are eliminated. In the discussion here, shot noise includes both dark current shot noise and photon shot noise.

of a photoemissive detector, this means that the standard deviation σ_N of the numbers of electrons arriving at the anode in each of a series of equal time intervals of time t is given by

$$\sigma_N = \sqrt{N} = \sqrt{nt} \qquad \text{(III-A-36)}$$

where N is the mean number of electrons arriving at the anode per time interval t and n is the average rate of arrival. Because each electron carries a charge of e coulombs, the average anode current i_a is ne, and the standard deviation of the averages of the currents flowing during each time interval t is*

$$\sigma_i = \sqrt{ei_a/t} \qquad \text{(III-A-37)}$$

This σ_i is called the *rms shot noise current* and henceforth has the symbol $\overline{\Delta i_s}$. If the average current i_a is measured with a current meter which has an electrical noise bandwidth of Δf, in Hz, then the time interval t in the foregoing equation may be replaced by the effective averaging time of the current meter, which is

$$t = \frac{1}{2\Delta f} \qquad \text{(III-A-38)}$$

Thus

$$\overline{\Delta i_s} \equiv \sigma_i = \sqrt{2ei_a\,\Delta f} \qquad \text{(III-A-39)}$$

The average current i_a is the sum of the total photocurrent i_p and the thermionic current emission i_{th}

$$i_a = i_p + i_{th} \qquad \text{(III-A-40)}$$

Leakage currents are not included in the shot noise because they do not flow internally through the tube.

The most seriously detrimental effect of thermionic emission is not so much its contribution to the average anode current, which can be eliminated electrically, but rather its contribution to the total shot noise, particularly at low light levels where $i_p \leq i_{th}$.

Another possible source of noise in the output signal from a photodetector is Johnson noise in the load resistor. Johnson noise is a randomly varying potential occurring across any finite resistance. It is caused by the thermal agitation of electrons in the resistor material. It is not a function of the

---

* The reasoning is as follows: if σ_N is the standard deviation of the number of electrons, then $e\sigma_N$ is the standard deviation of their charge and $e\sigma_N/t$ is σ_i, the standard deviation of the average current in the time interval t. Combining this with equation III-A-36 and writing i_a for ne gives equation III-A-37.

current flowing through the resistance; thus Johnson noise in a phototube load resistance is constant at all signal levels. The rms Johnson noise current $\overline{\Delta i_j}$ and voltage $\overline{\Delta e_j}$ are given by

$$\overline{\Delta i_j} = \left(\frac{4kT\,\Delta f}{R_L}\right)^{1/2} \text{A} \qquad \text{(III-A-41)}$$

$$\overline{\Delta e_j} = (4kT\,\Delta f R_L)^{1/2} \text{V} \qquad \text{(III-A-42)}$$

where k is the Boltzmann constant, T is the temperature of the load resistance (°K), Δf is the electrical noise bandwidth of the circuit (Hz), and R_L is the load resistance (ohms). For instance, if $R_L = 10^8$ ohms, $T = 300°$K, and $\Delta f = 1$ Hz, $\overline{\Delta e_j} = 10^{-6}$ volt.

Since shot and Johnson noises are random and uncorrelated, they add quadratically. Thus the total rms noise current $\overline{\Delta i_p}$ due to shot and Johnson noise is

$$\overline{\Delta i_p} = (\overline{\Delta i_s}^2 + \overline{\Delta i_j}^2)^{1/2} = \left[\frac{2e\,\Delta f}{R_L}\left(iR_L + \frac{2kT}{e}\right)\right]^{1/2} \qquad \text{(III-A-43)}$$

Because $2kT/e$ equals about 0.05 volt at room temperature, Johnson noise is negligible as long as R_L is chosen high enough to make the signal voltage iR_L large compared to 0.05 volt.

b. CHARACTERISTICS OF SINGLE-STAGE PHOTOTUBES

Single-stage phototubes are not often used in luminescence instrumentation because of their low radiant sensitivities γ $(10^{-3}\text{--}10^{-1}$ A watt$)^{-1})$. They are, however, quite stable, have a wide linear dynamic range, and have very fast response times (typically 10^{-9} sec).

c. CHARACTERISTICS OF MULTIPLIER PHOTOTUBES

The most widely used photodetectors in luminescence spectrometry are the multiplier phototubes (called photomultiplier tubes by manufacturers), mainly because of their extremely high sensitivities.

A schematic diagram of a multiplier phototube and its circuitry is provided in Figure III-A-10. An evacuated tube of quartz or glass contains an anode A, a photocathode C, and a series of dynodes D_i. The resistor chain divides up the operating voltage V so that a potential difference of about 100 volts exists between adjacent electrodes. Photoelectrons emitted by the cathode are accelerated toward D_1 by the potential across R_1. The kinetic energy gained in this acceleration is several times the binding energy of the electrons

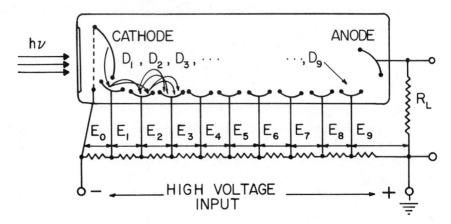

Fig. III-A-10. Schematic diagram of a multiplier phototube and its associated circuitry: D, dynodes; E, voltages applied to dynodes, and R_L, phototube load resistor.

on the surface of the dynodes. Each electron incident upon D_1 causes the emission of several secondary electrons, which are subsequently accelerated toward D_2 by the voltage across R_2. The same process is repeated for all the remaining electrodes. The resulting stream of electrons are collected by the anode. The anode current i_a flows through the load resistor R_L, generating the signal voltage $i_a R_L$, which is negative with respect to ground.

The *gain per stage g* is defined as the number of secondary electrons emitted by each dynode per incident electron and is typically 3 to 5. If z is the total number of dynode stages, the *total amplification* factor M is

$$M = g^z \qquad\qquad (\text{III-A-44})$$

For example, a typical commercial multiplier phototube with 10 dynode stages and a gain per stage of 4 would have a total amplification of $4^{10} = 10^6$. Thus the anode current i_a would be 10^6 times the cathode current i_c. Such a phototube would have a sensitivity about 10^6 times that of a single-stage phototube with the same cathode type and envelope material. A similar tube with 13 stages would have a total amplification of 10^8.

The gain per stage and amplification of a multiplier phototube depends very much on the operating voltage. Most commercial tubes with cesium–antimony dynode surfaces are operated at voltages from 80 to 120 volts per stage. It has been found experimentally that, in this range, the gain per stage is approximately proportional to the 0.7 power of the voltage.[19] The amplification of a 10-stage tube would then be proportional to the seventh power of the voltage. Because of this very sensitive dependence of the amplification on the operating voltage, multiplier phototubes require a very stable high voltage power supply with a total drift and ripple at least $\frac{1}{7}$ the maximum tolerable gain variation. A plot of the amplification of a typical 10-stage multiplier phototube versus the supply voltage V is shown in Figure III-A-11. The slope of the line on the logarithmic coordinate graph is about 7. The ability to change the sensitivity over such a wide range simply by changing the supply voltage is a significant and unique advantage of multiplier phototubes.

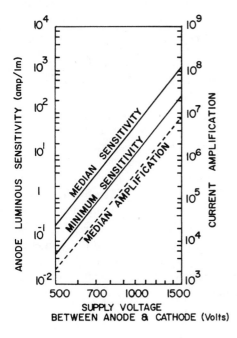

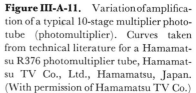

Figure III-A-11. Variation of amplification of a typical 10-stage multiplier phototube (photomultiplier). Curves taken from technical literature for a Hamamatsu R376 photomultiplier tube, Hamamatsu TV Co., Ltd., Hamamatsu, Japan. (With permission of Hamamatsu TV Co.)

The anode sensitivities γ_λ of multiplier phototubes (i.e., sensitivities expressed in terms of the anode current) are much higher than those of single-stage tubes and vary over a wide range of values. Radiant sensitivities generally fall between 400 and 80,000 A watt^{-1} at the wavelength of maximum response.

The spectral response of a multiplier phototube depends on the cathode type and envelope material and is not influenced by the multiplication process. Thus the response characteristics are similar to those of single-stage tubes and are catalogued by the same S number system.

At room temperature, the primary source of dark current in multiplier phototubes is almost always thermionic emission. Thermionic emission current from the cathode is amplified by the dynode secondary emission process in the same way as is the photoelectron current. (Any small differences between the average kinetic energies of thermal electrons and photoelectrons is easily swamped out by the large cathode-to-first-dynode potential). In addition, each of the dynodes exhibits some thermionic emission. Only the emission from the first few dynodes is significant, however, because that from latter dynodes is not sufficiently amplified. Leakage current, which is not amplified at all by the dynode multipliers, is much less important in multiplier phototubes than in single-stage tubes. Although the leakage current is directly proportional to the applied voltage, the thermionic emission current at the anode, like the photocurrent, is proportional to a high power of the voltage. At voltages below about 80 volts per stage, the gain may be so low that leakage current becomes significant. At 100 volts per stage, the operating voltage of most commercial photomultipliers, the thermionic emission current is always dominant, at least at room temperature. Above about 130 volts per stage, positive ion current and field emission may become significant. In addition, internal breakdown may occur at excessively high voltages, destroying the tube. It is therefore important never to operate a phototube above the manufacturer's recommended maximum voltage.

At room temperature the noise in multiplier phototubes is essentially all shot noise. Johnson noise in the load resistor is not amplified by the electron multiplier and hence is insignificant compared to the amplified shot noise. The equation for the shot noise $\overline{\Delta i_s}$ due to the total anodic current i_a (photocurrent plus thermionic current) must take into consideration the gain of the tube M as well as the shot noise contributed by the dynodes. This can be done by calculating the noise due to each electrode separately and then quadradically adding them to determine the total noise. The result is that the anodic shot noise $\overline{\Delta i_s}$ is

$$\overline{\Delta i_s} = \sqrt{2eBM\Delta f i_a} \qquad \text{(III-A-45)}$$

where B is the symbol for

$$B = 1 + g^{-1} + g^{-2} + \cdots + g^{-z} = \sum_{y=0}^{z} g^{-y} \qquad \text{(III-A-46)}$$

and where the other symbols have been defined previously. In general, B is slightly greater than one. For instance, if g is 4, then B is about 1.3. By comparison with equation III-A-39, it is seen that B may be considered a "correction factor," which is multiplied by the amplified cathode noise to account for the small amount of extra noise (15% or so) introduced by the dynode multiplication process.

As mentioned before, the thermionic component of dark current and dark current shot noise can be reduced greatly by cooling the tube. As a rule of thumb, each 10°K decrease in temperature halves the thermionic emission current. Liquid nitrogen and dry ice are commonly used coolants. In addition, thermoelectric cooling chambers are available which allow temperature adjustment. In many cases, it is possible to reduce the dark current by one or two orders of magnitude by cooling. Cooling may not, however, reduce the various other sources of noise significantly, so that these may become dominant at low temperatures. For each type of multiplier phototube, there is a minimum temperature below which it is not profitable to cool the tube. This occurs at about 253°K (−20°C) for most tubes with cesium–antimony cathodes. An additional noise source, which is important in some high grade multiplier phototubes operated at low temperatures, is cosmic ray scintillation noise.[21] This is the result of the passage of highly energetic cosmic particles (e.g., μ mesons) through the window of the tube, inducing Cherenkov radiation which may reach the cathode. Even though the average count rate of cosmic particles is low (ca. 1 or 2 min^{-1} cm^{-2}), each particle may produce hundreds of scintillation photons, resulting in a very large noise pulse at the anode. The effect is particularly evident in end-on tubes, because the cathode is so close to the window and because the window itself is relatively thick. Background radiation, as well as radiation from radioactive isotopes in the window material, may contribute to this source of noise.

The dynode voltages are usually obtained from a resistive voltage divider connected across a high voltage supply (as in Figure III-A-10). It is essential to observe polarity when connecting the phototube to the power supply; the *cathode must always be negative.* The current through the dynode resistor chain should be at least ten times the anode current so that the dynode potentials will not be reduced appreciably by the anode current. If the dynode chain current is excessively high, however, the heat dissipated by the resistors may warm the phototube, increasing the thermionic emission. For most applications, a dynode chain current of a few milliamperes is sufficient.

The most important interelectrode potential in a multiplier phototube is the cathode-to-first-dynode voltage, which should always be maintained at the value recommended by the manufacturer of the tube. This can be done by using a constant voltage (Zener) diode of the proper voltage in place of the cathode-to-first-dynode divider resistor. In this way, the voltage will remain constant even if the overall supply voltage or dynode chain current changes. The use of the recommended voltage ensures good collection efficiency and reduces the effect of stray magnetic fields.

Exposure of multiplier phototubes to high light levels may cause changes in the sensitivity of the tube (*fatigue*). Neither the cathode nor the anode currents should exceed their maximum ratings as specified by the manufacturer. A limitation on the cathode current density is set by the cathode resistivity. Photocurrent flowing through the cathode layer results in an electric field across the cathode which opposes the cathode-to-first-dynode field, lowering the collection efficiency and gain of the first stage. This can result in a large change in the total gain of the tube. The customary cesium–antimony cathode material has a relatively high resistivity; the current density in a cathode of this material should not exceed about 2×10^{-8} A cm^{-2}. In a multiplier tube with an M of 10^6 and a cathode area of 1.0 cm^2, this would correspond to an anode current of 20 mA. In this connection, the effect of phototube refrigeration should be considered. Most cathode materials are semiconductors and, as such, exhibit higher resistivity at low temperatures. Thus cooling a tube will reduce its maximum cathode current tolerance. This effect influences the selection of an optimum operating temperature.

Anode current limitations are set by the maximum tolerable anode power dissipation, heating of the dynodes, and space charge effects (fatigue). The maximum anode power dissipation is usually quoted by the manufacturer and generally falls between 0.1 and 1 watt. Maximum recommended long-term anode currents generally fall between 0.1 and 1 mA. For best stability, the anode current should not exceed one-hundredth of the maximum rating. A fatigued phototube may often be rejuvenated if operated in the dark for some time at the normal operating voltage. Multiplier phototubes must never be exposed to room light, for irreversible damage may result.

The response time of a multiplier phototube is usually less than 10^{-8} sec, sufficiently fast to allow the measurement of all but the fastest luminescence decay times.

5. AMPLIFIER-READOUT* SYSTEMS[22–24]

The function of the amplifier-readout system in spectrophotometric instrumentation is to amplify and process the small electrical signal from the photodetector and display it in a convenient and readable form. In luminescence spectrometry, the information to be displayed is most commonly luminescence signal due to luminescence radiance as a function of analyte concentration (quantitative analysis), wavelength (spectra), or time (luminescence decay times).

The amplifier-readout system normally determines the electrical noise bandwidth Δf of the entire system, which in turn influences the amount of random noise on the readout signal and determines the response time of the system.

---

* The reader is referred to Appendix 5 for a more detailed discussion on electronic signal processing.

a. DC AMPLIFIER SYSTEMS

The simplest and least expensive type of amplifier, and the one most commonly found on commercial luminescence spectrometers, is the DC amplifier.[25,26] The output signal of this type of amplifier (the *readout signal*) is an amplified and filtered version of the DC component of the photodetector signal. The DC amplifier is called a *low-pass* system; it passes and amplifies DC signals and all AC frequency components in the photodetector signal output below a certain frequency (approximately equal to the bandwidth Δf of the system) while attenuating all frequency components above Δf. When the light level seen by the photodetector is constant (i.e., not chopped), the signal of interest appears at zero frequency (DC) and is passed by the DC amplifier. Noise (e.g., shot noise) appears at nonzero frequencies (AC) and some of it (the portion above Δf) is attenuated. Thus the lower Δf, the lower the noise on the readout signal. The electrical noise bandwidth Δf_a of the amplifier is determined by the amplifier time constant τ_a, which is determined by an RC filter somewhere in the circuit of the DC amplifier.

$$\Delta f_a = \frac{1}{4\tau_a} = \frac{1}{4RC} \qquad \text{(III-A-47)}$$

Direct-current amplifiers are simple, relatively inexpensive, and easy to understand and operate. The biggest single disadvantage is that they pass and amplify *undesirable* DC signals as well as the desired photocurrent signal. *Undesirable DC signals include phototube dark current, stray light that falls on the detector even in the absence of sample luminescence, and offset (bias) current or voltage in the amplifier itself.* These signals may be balanced out electrically by adding an artificial DC signal equal in magnitude but opposite in polarity to the total undesirable DC signal, thereby canceling it out. However, drift in the dark current, stray light, or amplifier offset will cause errors in the output signal until the circuit is rebalanced. Furthermore, since the random AC noise components of these signals (e.g., thermionic emission shot noise) cannot be balanced out, they remain in the signal.

When using current-source photodetectors such as multiplier phototubes, it is usual to pass the detector anode current i_a through a load resistor R_L to generate a voltage signal $i_a R_L$ to be amplified by a voltage amplifier. The input impedence of the amplifier must be much larger than R_L in order to avoid a gain error. Electrometer amplifiers are commonly used. As mentioned previously, the effect of the input time constant $R_L C_S$ must also be considered, C_S being the shunt capacity from the signal lead to ground due to cable capacity, amplifier input capacity, and so on. An alternate approach is to feed the detector anodic current directly to the summing point (negative input) of an operational amplifier, using R_L as the feedback resistor. The output voltage is $I_a R_L$, as previously. If a field effect transistor (FET)-input operational amplifier is used, quite large values of R_L may be used in order to obtain high sensitivity. This arrangement has the advantage of reducing the effect of stray capacity and leakage paths across the input.[26]

For use with the usual nine-stage multiplier tubes, a DC amplifier should have a dynamic range of 10^{-5} to 10^{-10} A full scale. The noise and drift of the amplifier should always be negligible.

b. AC AMPLIFIER SYSTEMS

Alternating-current systems overcome the principal disadvantage of DC systems by reducing the response to dark current and the other DC-zero errors mentioned previously. There are several different types of AC systems, but they all require that the photosignal be *modulated* in some way; that is, a desirable and known variation in the signal is superimposed so that the amplitude of the variation contains some or all of the desired analytical information. In this way, the desired signal is moved up in frequency from DC to some AC frequency, leaving behind the DC errors. By far the most common type of modulation is *light intensity modulation*, sometimes called *chopping*. This can be achieved by passing the light beam to be measured through the cogs of a rotating sectored disk, called the *chopper*, which essentially switches the light beam off and on repetitively. The phototube signal thus alternates from dark current only to dark current plus photocurrent.

In luminescence spectrometry, it is usually advantageous to chop the excitation light beam, thereby modulating the luminescence selectively while not modulating constant emission from the sample (e.g., flame background emission) or any stray light that may leak into the phototube. In some cases, the *excitation source itself may be modulated* directly by operation from an AC or pulsed power supply. *Other types of modulation may also be used, including monochromator wavelength modulation and sample introduction modulation.* In any case, only the desirable portion of the photosignal is modulated; the DC-zero errors are not. The AC amplifier itself is designed to respond only to the "AC component" of the detector (i.e., the amplitude of the variation or modulation) while rejecting the "DC component," which is a function of the dark current.

Although AC amplifiers are similar in principle to DC amplifiers, they contain one or more capacitors in series with the signal. The capacitors block DC while passing AC. The system must then "demodulate" the AC component in order to provide a DC output signal proportional to the amplitude of the AC component at the input. This operation is performed by the *rectifier* stage. The output of the rectifier stage is amplified and filtered by a low-pass DC amplifier stage before it is set to the readout. The bandwidth of this filter stage must be considerably less than the modulation frequency to ensure complete filtering.

There are three main types of AC systems, differing in the type of amplification and demodulation.

Wide band (untuned) systems have amplifiers with relatively wide electrical bandwidths—usually extending from well below the modulation frequency to well above. Thus a relatively large amount of noise is passed along with the signal. The demodulator is a simple "rectifier" that either rejects all signal excursions of one polarity (half-wave rectifier) or inverts them to the other polarity (full-wave rectifier). The demodulator is followed by a filter stage, which is a low-pass DC amplifier that passes the DC (signal) component of the demodulator output and rejects or attenuates the AC (noise and modulation) component. The bandwidth of this filter stage determines the bandwidth of the amplifier system.

The principal advantage of this system is that unmodulated bias, offset, or drift errors do not affect the output signal. This includes dark current, unchopped stray or background light falling on the photomultiplier tube, and amplifier DC offset errors. The main disadvantage is that noise at the input, being an AC component, is amplified and rectified and contributes to the DC level at the output. This is called *noise offset*. Changes in dark current or unchopped stray light level cause changes in the shot noise, which in turn causes an error in the output signal which must be balanced out. Fortunately, shot noise is typically one or more orders of magnitude below the detector signal, so the effective error is much less than that in a DC system. At low light levels, however, where the signal-to-noise ratio is approximately 1, noise offset can be a serious problem.

Narrow band (tuned) systems reduce the last-mentioned difficulty by using a tuned amplifier with a narrow bandwidth centered on the modulation frequency. Only those noise components falling within this bandwidth are amplified. This reduces the amount of noise going into the rectifier, thus lowering the noise offset considerably. The system bandwidth is determined either by the bandwidth of the tuned AC amplifier stage or by the bandwidth of the output filter stage following the demodulator. Highly selective tuned amplifiers are not easy to make, however. Furthermore, if the modulation frequency drifts for any reason, then the amplifier must be retuned carefully. This difficulty is avoided in the synchronous systems.

c. SYNCHRONOUS (LOCK-IN) SYSTEMS

Synchronous systems, either wide or narrow band, are usually called *look-in* or *phase-sensitive* systems.[24,27] These systems differ from nonsynchronous AC systems in the demodulator stage, which, instead of being a simple rectifier whose operation is dictated by the *polarity* of the signal (*polarity-sensitive*), is controlled by the *frequency* and *phase* of the signal (*phase-sensitive*). The simplest type is a full-wave inverter (rectifier) whose inversion depends on a *reference signal*, a *noise-free signal* whose frequency and phase are synchronized with the sample signal modulation. This is easily obtained from a small lamp and photocell placed so that the chopper disk passes between them. Because noise is random and not frequency or phase synchronized with the signal modulation, the noise offset is completely eliminated, even with a wide band preamplifier stage. Such systems are relatively expensive, complex, and difficult to operate, however. The system bandwidth is determined by the filter stage.

d. PHOTON COUNTING* SYSTEMS[28-30]

Photon counting, a relatively new method for the measurement of low light intensities, is being used with increasing frequency in analytical spectrometry, particularly for Raman spectroscopy. Its use in luminescence spectrometry has thus far been limited to research applications.

In photon counting, a high speed electronic counter is used to count the photoelectron pulses at the anode of a multiplier phototube. Each photoelectron pulse contains, on the average, M electrons, for a total charge of eM coulomb. If M is of the order of 10^6, then the resulting charge is of sufficient magnitude to cause a distinct and easily observable voltage (and current) pulse across the phototube load resistor R_L. One such pulse occurs for each photoelectron ejected at the cathode, and hence the average pulse count rate $\bar{n}$ is proportional to the light level (under the condition that the cathode sensitivity is constant—which is a good assumption under normal conditions). Such photoelectron pulses are often of the order of several millivolts and are easily amplified by a wide band AC amplifier and counted by electronic counter circuits. This automatically provides a convenient, accurate, and easily read digital output. The technique is called *single-photoelectron counting*, or simply *photon counting* (even though not every photon is counted, of course).

The advantage of the photon counting system include the following:

1. A digital system is less subject to drift than an analog system.

2. Long counting times may be used at low light levels to accumulate the required number of total counts.

3. Many types of noise are discriminated against, particularly thermionic emission from the dynodes and leakage current.

4. The digital readout is easy to read and can be interfaced directly to computers.

* The basic principles of photon counting are given in Appendix 6.

Disadvantages include greater cost and complexity. In recording spectra and luminescence decay times, the counter can be replaced by a rate meter with analog output for connection to a recorder. In fact, the counter may not be necessary at all if chart readout is satisfactory.

Fast amplifiers, discriminators, rate meters, and counters designed for nuclear counting are commercially available in convenient modular form. These are easily adapted for photon counting. In addition, there are several commercial photon counting systems. The method of photon counting is further discussed in Appendix 6.

e. STROBOSCOPIC SYSTEMS[31]

Many luminescence studies involve the use of pulsed sources to study fast fluorescence or phosphorescence decay times. The measurement of low intensity, short duration light flashes poses a particular problem in photodetection. The simplest technique would be to amplify the signal from the phototube with a wide band amplifier and display the output on an oscilloscope. But the experimenter is faced with a dilemma. In order to preserve the waveform of a fast pulse, the bandwidth of the system must be high. For example, a 1-μsec rectangular pulse would require at least a 3-MHz bandwidth. However, the large bandwidth usually reduces the signal-to-noise ratio, preventing accurate measurements of low intensity light pulses.

This problem can be solved very effectively by employing a photomultiplier in the pulsed mode. Because the tube is sensitive to light only when the proper dynode voltages are applied, it can be gated on and off as desired by applying a rectangular high voltage pulse to the dynode chain. This gate pulse, which defines the " on " time of the phototube, is made much shorter than the duration of the light pulse to be studied. By means of suitable delay circuits, the gate pulse can be made to occur at any desired time during the light flash to be studied. If the light flash and phototube gate pulse are repeated periodically at a rate of a few pulses per second or greater, the average DC component of the train of photocurrent pulses can be measured with an average-reading DC meter. The bandwidth of the meter circuit can be made as small as desired to reduce the noise to a satisfactory level. The time variation (waveform) of the light flash is easily determined by measuring the average DC signal in this way for different delay settings (i.e., with the gate pulse occurring at different times during the flash). The measured signals are then plotted against the corresponding delay times to reconstruct the pulse waveform. A more convenient procedure is to sweep the gate pulse slowly through the light flash while recording the measured intensity as a function of time on a pen recorder. This eliminates hand plotting of data. If desired, the time variation of the spectrum of the light flash can be determined by scanning the monochromator at various fixed delay times.

The type of stroboscopic operation just described has still another advantage; the performance of the multiplier phototube may be considerably improved by the use of pulsed DC dynode voltages. Much higher gains can be achieved in the pulsed mode, since higher voltages are permissible. For example, a 1P21, whose maximum DC voltage rating is 1230 volts may be operated at pulsed voltages in excess of 1500 volts, increasing the gain approximately tenfold. The primary limitation on the DC voltage is the appearance of destructive positive ion regeneration. However, the positive ions, being ionized atoms of the residual fill gas, are relatively heavy and slow compared with electrons. Thus the positive ion current requires a certain amount of time, perhaps several microseconds, to build up. If the gate pulse width is less than a microsecond or so, the positive ion current simply does not have enough time to build up before the end of the gate pulse, even at pulse voltages much higher than that permissible for DC operation. This not only reduces the positive ion component of dark current and dark current noise but also permits the use of higher voltages, which yield higher gains. Thus the signal-to-noise ratio of the tube is considerably enhanced in the pulsed mode.

f. READOUT SYSTEMS

The function of the readout system is to display the signal from the amplifier system in a convenient and readable form. The most common types of readout in luminescence spectrometry are analog, such as meters, recorders, and oscilloscopes, although digital systems are showing considerable promise.

Meters are simple and inexpensive, but they provide no permanent record and are difficult to read if the signal is noisy. Accuracy is limited to about 1 to 2%. The best types have mirrored scales to eliminate parallax errors.

Pen recorders are much more expensive than meters, but they provide a permanent record and they permit noise measurements and the averaging of noisy signals. Servomechanism (potentiometric) types are the most accurate (0.1–0.5%) and sensitive (1–10 mV full scale), but they are expensive and relatively slow. Recorder bandwidth is typically 1 or 2 Hz, and the slew rate—the maximum rate of change of the pen position (usually in. sec^{-1}) — is also limited. This is important in the measurement of luminescence decay times. Galvanometric (millimeter) recorders are generally faster and less expensive but also less accurate and sensitive than servo recorders.

Most pen recorders record voltage (or current) as a function of time, but *x-y* servo recorders permit the recording of two electrical variables against each other.

Oscilloscopes offer the widest readout bandwidth (up to 10^8 Hz). Such large bandwidths may be required for the study of very fast luminescence decay times. Accuracy is typically 1%. An oscilloscope camera can provide a permanent record if required. Storage oscilloscopes are also available.

Digital readouts usually involve the electronic conversion of the analog readout signal to digital form by means of a digital voltmeter. Photon counting systems provide a digital readout naturally. Digital displays are easy to read and very accurate (0.1–0.01%) but are more expensive than a simple meter. Printing of tapes or pages and interfacing to a computer is also possible.

g. SPECIAL SPECTROSCOPIC TECHNIQUES

Spectroscopic methods involving interferometry and Fourier spectrometry and correlation and multiplex spectrometry have not been utilized in analytical luminescence spectrometry. However, these methods should have considerable potential in future luminescence instrumentation, and so the reader is referred to Appendix 7 for a brief discussion of the principles and uses of these novel techniques.

6. GLOSSARY OF SYMBOLS

a	Linear aperture of monochromator, cm.
A_l	Lens area, cm^2.
b	Thickness of prism base, cm.
$B_{B\lambda}$	Spectral radiance of black body, watt cm^{-2} sr^{-1} nm^{-1}.
$B_{C\lambda_0}^o$	Spectral radiance of continuum light source, watt cm^{-2} sr^{-1} nm^{-1}.
$B_N{}^o$	Radiance of line light source, watt cm^{-2} st^{-1}.
c	Speed of light, cm sec^{-1}.
C	Capacitance, farad.
C_S	Stray capacitance, farad.
D	Angular dispersion, angular degrees nm^{-1}.
D_c	Collimator diameter, cm.
e	Charge on electron, coulomb.
e_s	Signal voltage, volt.
f	f number of optical component, no units.
F	Frequency of measurement system, Hz.
F_c	Focal length of collimator lens, m.
g	Gain per stage of multiplier phototube, no units.
H	Monochromator slit height, cm.
$\bar{\imath}$	Average power supply current, A.
i_a	Average anodic current of photoemissive detector, A.
i_c	Average cathode current of photoemissive detector, A.
i_d	Average dark current of photoemissive detector, A.
i_p	Average photoanodic current, A.

i_{th} Average thermionic emission anodic current of photoemissive detector, A.

J Stored energy in condenser of flash tube current, joule.

k Boltzmann constant, 8.0×10^{-5} eV $^\circ$K^{-1}.

K_o Optics factor, $HT_f\Omega$, cm sr.

m Grating diffraction order, no units.

M Multiplier phototube multiplication factor, no units.

n Refractive index, no units.

n Rate of electron arrival at surface, sec^{-1}.

$\bar{n}$ Average counting rate, sec^{-1}.

P_i Peak input power of flashtube, watt.

$\bar{P}_i$ Average input power of flashtube, watt.

P_o Peak output power of flashtube, watt.

r Radius of sphere for calculation of unit solid angle, cm.

R Resistance of resistive device, ohm.

R Resolving power, no units.

R_d Reciprocal linear dispersion, nm cm^{-1}.

R_f Flashtube arc resistance, ohm.

R_L Phototube load resistor, ohm.

R_{theor} Theoretical resolving power.

s Spectral bandpass, nm.

s_a Aberration limited spectral bandpass, nm.

$s_{\min}$ Minimum spectral bandpass corresponding to $W_{\min}$, nm.

t Time interval, sec.

T Temperature, $^\circ$K.

T_f Optical transmission factor, no units.

V Power supply voltage, volt.

W Monochromator slit width, cm.

W_e Exit slit width, cm.

$W_{\min}$ Minimum resolving power slit width, cm.

W_s Entrance slit width, cm.

x Linear distance at focal plane, cm.

X Half width of diffraction pattern, cm.

y Index for summations, no units.

z Number of dynodes in multiplier phototubes, no units.

α Prism apex angle, angular degrees.

α_g Grating rotation angle, angular degrees.

β Blaze angle of grating, angular degrees.

γ_λ Phototube anodic sensitivity, A watt^{-1}.

δ Number of grooves per centimeter for grating, cm^{-1}.

Δ Grating constant, nm (or cm).

Δf Electrical noise bandwidth, Hz.

$\overline{\Delta f_a}$ Amplifier noise bandwidth, Hz.

$\overline{\Delta i_a}$ Rms amplifier noise anodic current, A.

$\overline{\Delta i_j}$ Rms Johnson noise current, A.

$\overline{\Delta i_p}$ Rms phototube noise anodic current, A.

$\overline{\Delta i_s}$ Rms phototube shot noise anodic current, A.

$\Delta\lambda$ Base width of monochromator slit function, nm (or cm).

$\Delta\lambda_R$ Wavelength resolution, nm (or cm).

ε_λ Spectral emissivity of a black body, no units.

Φ Deviation angle (prism), angular degrees.

ϕ_a Work function of photoemissve surface, eV.

Θ_d Diffraction angle (grating), angular degrees.

Θ_i Incident angle on prism or grating surface, angular degrees.

Θ_m Angle between incident and diffracted rays on grating, angular degrees.

λ Wavelength of radiation, nm (or cm).

λ_{th} Threshold wavelength for photoemissive surface, nm (or cm).

ν Frequency of radiation, Hz.

π 3.1416 ...

σ_i Standard deviation of average current, appropriate units, a.

τ_a Amplifier time constant, sec.

Ω Solid angle of radiation collected by optical device, sr.

7. REFERENCES

1. D. W. Ellis, "Luminescence Instrumentation and Experimental Details," Chapter 2 in *Fluorescence and Phosphorescence Analysis*, D. M. Hercules, Ed., Wiley-Interscience, New York, 1966.
2. C. A. Parker, *Photoluminescence of Solutions*, Elsevier, Amsterdam, 1968.
3. L. R. Koller, *Ultraviolet Radiation*, 2nd ed., Wiley, New York, 1965.
4. L. A. Rosenthal, *Rev. Sci. Instr.*, **36**, 1329 (1965).
5. N. Z. Searle et al., *Appl. Opt.*, **3**, 923 (1964).
6. C. M. Doede and C. A. Walker, *Chem. Eng.*, **62**, 159 (1955).
7. W. A. Baum and L. Dunkelman, *J. Opt. Soc. Amer.*, **40**, 782 (1950).
8. Edgerton, Germeshausen, and Grier, Inc., *Xenon Flash Tube Handbook*, Boston, Mass.
9. D. E. Perlman, *Rev. Sci. Intr.*, **37**, 340 (1966).
10. R. A. Sawyer, *Practical Spectroscopy*, 3rd ed., Dover, New York, 1963.
11. E. J. Meehan, *Optical Methods of Analysis*, Wiley-Interscience, New York, 1964.
12. F. A. Jenkins and H. E. White, *Fundamentals of Optics*, McGraw-Hill, New York, 1957.
13. C. W. Sill, *Anal. Chem.*, **33**, 1584 (1961).
14. E. J. Bowen, *The Chemical Aspects of Light*, Clarendon Press, Oxford, 1946.
15. H. A. Strobel, *Chemical Instrumentation*, Addison-Wesley, Reading, Mass., 1960.

16. C. G. Cannon, *Electronics for Spectroscopists*, Wiley-Interscience, New York, 1960.
17. Radio Corporation of America, *RCA Phototube and Photocell Manual*, Harrison, N.J.
18. S. Rodda, *Photoelectric Multipliers*, MacDonald, London, 1953.
19. Electra Megadyne, Inc., *EMI Photomultiplier Tubes Bulletin*, Los Angeles, Calif.
20. K. S. Lion, *Instrumentation in Scientific Research*, McGraw-Hill, New York, 1959.
21. A. T. Young, *Rev. Sci. Insts.*, **37**, 1472 (1966).
22. J. J. Brophy, *Basic Electronics for Scientists*, McGraw-Hill, New York, 1966.
23. E. J. Bair, *Introduction to Chemical Instrumentation*, McGraw-Hill, New York, 1962.
24. Brower Laboratories, Inc., *A Practical Guide to the Measurement of Signals Buried in Noise*, Westboro, Mass.
25. L. P. Morgenthaler, *Basic Operational Amplifier Circuits for Analytical Chemical Instrumentation*, McKee-Pedersen Instruments, Danville, Calif., 1967.
26. T. C. O'Haver and J. D. Winefordner, *J. Chem. Educ.*, **46**, 241 (1969).
27. S. L. Ridgway, "The Use of Lock-in Amplifiers for the Detection and Measurement of Light Signals, *Signal Notes*, **1**, 1, (Princeton Applied Research Corp. 1967).
28. G. A. Morton, *Appl. Opt.*, **7**, 1 (1968).
29. R. Foord et al., *Appl. Opt.*, **8**, 1975 (1969).
30. M. L. Franklin, C. Horlick, and H. V. Malmstadt, *Anal. Chem.* **41**, 2 (1969).
31. M. L. Bhaumik, *Rev. Sci. Instr.*, **36**, 37 (1965).

B. ATOMIC FLUORESCENCE INSTRUMENTATION[1-3]

1. GENERAL INSTRUMENTAL REQUIREMENTS

The instrumental layout for atomic fluorescence studies is essentially that of Figure III-A-1. The sample cell is usually a chemical flame, into which is sprayed the sample solution containing the analytical element. The flame simply provides a high temperature gaseous environment, which produces an atomic gas of the analytical element. Other types of cells have also been used. The source of excitation is a high intensity line or continuum source (*without* an excitation monochromator).

The monochromator, detector, and readout electronics are conventional, and DC, AC, and synchronous amplifier systems have been used. Modulation for the AC and synchronous systems is usually performed by means of a rotating sector disk placed in the excitation beam; this arrangement allows discrimination against atomic thermal emission. Direct electrical modulation of the source has also been attempted.

Vickers and Vaught[4] have described a simple atomic fluorescence system that dispenses with the emission monochromator. By using a solar-blind photomultiplier and a cool, low background flame, they were able to obtain useful results with a very simple experimental setup.

At the time of this writing, only one instrument specifically designed for atomic flame fluorescence spectrometry is available commercially, although some atomic absorption instruments are easily convertible for fluorescence studies. The principal modification is that the source-flame axis must be pivoted to some angle, traditionally 90°, to the flame-monochromator axis. For that reason, a slot burner is not practical; a flame emission burner head (e.g., chamber-type or total-consumption nebulizer burner) may be used. Atomic fluorescence experiments usually require a more intense excitation source than those customarily provided in atomic absorption units, although some high intensity hollow cathode discharge tube (HCDT) sources may be useful. Furthermore, the solid angle of excitation radiation incident upon the flame may and should be made as large as practical in atomic fluorescence studies, whereas this solid angle is limited by the monochromator in atomic absorption work.

2. SOURCE OF·EXCITATION

a. REQUIREMENTS FOR A SOURCE IN ATOMIC FLUORESCENCE SPECTROMETRY

The source of excitation in atomic fluorescence spectrometry must be very intense over the absorption line width of the analyte and must be stable. Narrow line sources (e.g., metal vapor arc lamps, hollow cathode discharge lamps, and electrodeless discharge lamps), have been used by most investigators, although the line width of the source is not as critical as in atomic absorption spectrometry. Many different types of line sources have served successfully in atomic fluorescence studies. Continuum sources, such as xenon arc lamps have been less helpful analytically.

b. METAL VAPOR ARC LAMPS

Metal vapor arc lamps (see Figure III-B-1a) of the type made by Osram and Philips have been used by several investigators. Metal vapor arc lamps have a high radiant output but are often considerably self-reversed when operated under recommended conditions. Therefore many workers have cooled these lamps or operated them at lower than recommended currents. Although manufactured for cadmium, zinc, thallium, gallium, indium, mercury, and the alkali metals, metal vapor arc lamps have found only limited use for atomic fluorescence flame spectrometry. These lamps are *not* recommended for analytical use in atomic fluorescence spectrometry.

c. HOLLOW CATHODE DISCHARGE TUBES

Hollow cathode discharge tubes have been used in atomic fluorescence work. An HCDT (see Figure III-B-1b) is a type of low pressure metal vapor discharge lamp widely used as a line source in atomic absorption spectroscopy. The envelope is usually made of glass, with a quartz window sealed in

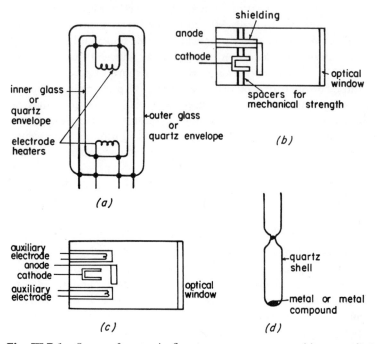

Fig. III-B-1. Sources for atomic fluorescence spectrometry: (*a*) vapor discharge lamp, (*b*) shielded hollow cathode discharge tube, (*c*) high intensity hollow cathode discharge tube, (*d*) electrodeless discharge tube.

one end. The cathode is a cuplike hollow cylinder containing the metal whose spectrum is to be produced, and a simple wire electrode serves as an anode. The interior is filled with argon or neon at a pressure of a few torr. The lamp is operated from a constant current AC or DC power supply capable of delivering a current of several milliamperes. A high voltage is applied to the tube to initiate the discharge. Positive ions of the inert fill gas bombard the cathode and "sputter" excited metal atoms from the surface. If the inert gas pressure is adjusted properly, the discharge is contained entirely on the inside of the cup-shaped cathode. The spectral output consists of a very sharp, unreversed line spectrum of moderate intensity. Both atom and ion lines are observed in some lamps.

An HCDT can be constructed for almost any of the metallic elements in the periodic table. If the metal is difficult to machine or is very expensive, a small amount of the metal can be placed inside a cathode made of aluminum, which has a simple spectrum and does not sputter easily. In some cases, the spectra of two or more metals can be obtained from one HCDT by making the cathode of a suitable alloy, sintered powders, or intermetallic compounds.

The main disadvantage of conventional HCDTs in atomic fluorescence applications is that the radiance of the emitted lines is only moderate. If the operating current is increased in an attempt to increase the radiance, the lines tend to become broadened and self-reversed. In addition, the increased rate of cathode sputtering causes blackening of the window and eventual destruction of the cathode. However, the line radiance of HCDTs can be increased by suitable modifications of the construction of the tubes. The so-called *high intensity HCDTs* (see Figure III-B-1c) have a pair of auxiliary electrodes mounted in front of the hollow cathode. A separate discharge across the auxiliary electrodes further excites the metal atoms sputtered from the hollow cathode, increasing the emitted radiance. A separate power supply is needed for the auxiliary electrodes. High intensity HCDTs have been found to be especially good sources for copper, magnesium, silver, nickel, cobalt, iron, arsenic, and lead. Detection limits with these high intensity HCDTs is similar to values obtained with electrodeless discharge tubes. Despite the high cost of such sources and the inconvenience of two power supplies, these lamps are quite useful for atomic fluorescence spectrometry.

Demountable hot hollow cathode lamps (i.e., lamps operated at high currents) have been used in atomic fluorescence spectrometry but with less success than either the high intensity HCDTs or the electrodeless discharge lamps.

d. ELECTRODELESS DISCHARGE TUBES

Probably the most useful atomic line source for fluorescence applications is the electrodeless discharge tube (EDT). A typical EDT (see Figure III-B-1d) is constructed of a sealed quartz tube containing a few torr of an inert gas and a small amount of the metal (or salt of the metal) whose spectrum is desired. The lamp has no electrodes (hence the name) and is operated by being placed in an intense field of Rf or microwave radiation. The application of a Tesla coil to the tube serves to start the discharge by ionizing some of the inert gas atoms. The electrons so produced are accelerated by the high frequency electric component of the Rf field and acquire enough energy to excite and ionize atoms. The resulting heat vaporizes the metal compound, which is subsequently excited (and ionized to some extent) by collisions, producing the desired spectrum. The discharge is confined by the "skin effect" to the outer surface of the plasma, thus minimizing self-absorption. Furthermore, since the pressure and gas temperature in the discharge is low,*

---

* The electron temperature is high, however (e.g., $10,000^2 K$).

the atomic lines are very sharp. The radiance of the emission is quite high for many metals. Electrodeless discharge tubes can be constructed for a large variety of elements (most of the periodic table). If the metal itself is not sufficiently volatile, a volatile salt, such as the iodide, can be used. The operating characteristics of the EDT depend on numerous parameters, such as the form and amount of metal or metal salt, type and pressure of fill gas, frequency of excitation, and tube dimensions. Care should be taken to optimize these parameters for best results.

Specific preparations of EDTs used in atomic fluorescence spectrometry have been described by Dagnall, West, and co-workers and Winefordner and co-workers in their extensive papers on atomic fluorescence spectrometry. Table III-B-1 provides a listing of the elements (and their form) for which EDTs have prepared.

Table III-B-1. Elements for which Electrodeless Discharge Tubes Have Been Prepared and Reported to be Analytically Useful in Atomic Fluorescence Spectrometry[a]

Prepared as Pure Element

H, He, N, O, Ne, P, S, Cl, Ar, Zn, Se, Br, Cd, I, Xe, Hg

Prepared as Element—Chloride

Li, Be, Na, Mg, K, Ca, Rb, Sr, Cs, Ba, W, Pd, Ag, Pt, Au

Prepared as Element—Iodide

Ti, Zr, Nb, Hf, Mn, Fe, Co, Ni, Ca, Rh, Ir, B, Al, Si, Ga, Ge, As, In, Sn, Sb, Te, Tl, Pb, Bi, U

Prepared as Element—Bromide

Mo

[a] Only the most useful form of the element is listed.

Most EDTs have been operated at microwave frequencies by means of a resonant cavity (three-quarter wave coaxial Broida cavity or quarter wave coaxial Evenson cavity) or an "A" antenna (see Figure III-B-2). With the "A" antenna, the EDT is often inserted into an evacuated quartz chamber to maintain the tube at a fairly high temperature; more intense, stable radiation results with such thermostating devices. An EDT prepared to operate on one of the devices will usually not operate optimally on the other devices. Although EDTs can also be operated at Rf frequencies (capacitative or inductive coupling), Rf operation is seldom used because of the reduced lifetimes of tubes at lower frequencies.

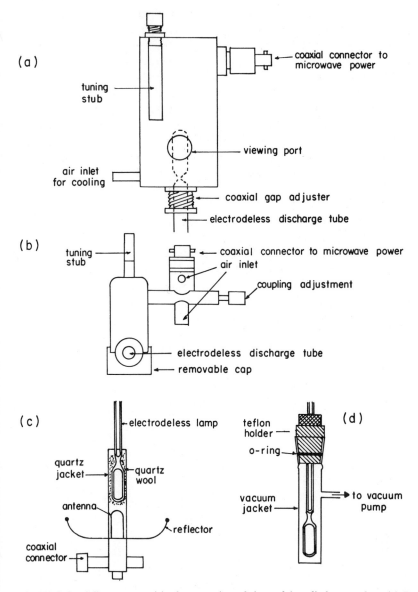

Fig. III-B-2. Microwave cavities for operation of electrodeless discharge tubes: (*a*) Broida-type cavity, (*b*) Evenson-type cavity and (c) microwave "A" Antenna for operation of EDTs, (*c*) open quartz insulator jacket, (*d*) quartz vacuum jacket.

193

When EDTs are operated at microwave frequencies, the tubes emit only the fill gas spectrum upon initiation of the discharge. As the tube heats up, the fill gas spectrum is reduced in radiance and the spectrum of the element (analyte) within the tube increases. An EDTs operated with coaxial cavities reaches its peak radiance faster than one operated on the "A" antenna. However, EDTs operated with coaxial cavities are more difficult to tune properly and often have a tendency to drift. On the other hand, the "A" antenna system with thermostat is less convenient and a less efficient coupling device. Since EDTs will probably become the accepted line sources for atomic fluorescence spectrometry, more research is needed on preparation of these sources and an optimization of coupling devices.

e. CONTINUUM SOURCES

The only continuum source that has served successfully in atomic fluorescence spectrometry is the xenon arc lamp (see Figure III-B-3). The obvious advantage of a continuum source is that only one source is needed for all elements. However, xenon arc lamps are much less intense over the narrow absorption line. A high resolution emission monochromator is required to reduce noise from scattering of the excitation light by droplets and particles in the flame and to minimize spectral interferences. An excitation monochromator, however, is generally unnecessary. Detection limits by atomic fluorescence with a continuum source are substantially poorer in most cases than those obtained using an intense line source; even so, several features

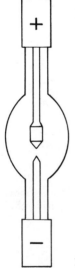

Fig. III-B-3. Typical xenon arc lamp.

indicate the continuum sources should be of considerable analytical use in atomic fluorescence spectrometry—their general use for all elements, their well-defined and predictable growth and analytical curves, their excellent stability, and their use for qualitative analysis (wavelength scanning). In addition, continuum sources are useful in *fundamental studies* in atomic fluorescence spectrometry because of the well-defined and predictable growth curves.

3. THE FLAME CELL

a. REQUIREMENTS FOR A FLAME IN ATOMIC FLUORESCENCE SPECTROMETRY

The flame should be sufficiently hot and reducing and of low burning velocity to provide a good aspiration and atomization efficiency (see Section II-D) and sufficiently cool to minimize ionization and flame background radiation. Actually the foregoing requirements are mutually exclusive. The C_2H_2/N_2O flame has a low burning velocity and is very reducing and hot but yet has a high background in certain spectral regions and produces considerable ionization even if ionization suppressors are used for some elements. On the other hand, the $H_2/O_2/Ar$ flame has a low flame background, results in less ionization, and is a poorer quenching medium but has a high burning velocity and is not as reducing as the C_2H_2/N_2O flame. Therefore, there appears to be *no optimum* flame for atomic fluorescence spectrometry but rather a " compromise " flame.

There are two basic types of flames—laminar premixed and turbulent—produced by chamber-type nebulizer burners or by total-consumption nebulizer burners (see Figure III-B-4). The turbulent flames produced by the mixing of gases (oxidant and fuel) above the burner result in considerable turbulence, audible noise, and air entrainment, whereas the laminar flames are produced by mixing the gases at room temperature within a chamber prior to ignition. Laminar flames are less turbulent and less audibly noisy and have less flame flicker than turbulent flames.

b. PREMIXED FLAMES

The most important premixed flames used in atomic fluorescence spectrometry are listed in Table III-B-2. The most useful low temperature premixed flames have been the H_2/air and $H_2/O_2/Ar$ flames. The latter flame produces greater power yields because the quenching ability of argon is poor; it is also slightly hotter than the H_2/air flame. Hydrogen-supported flames are ideal for easily atomized elements (e.g., arsenic, selenium, bismuth, thallium, and antimony) in simple solutions (i.e., no appreciable concentrations of matrix components).

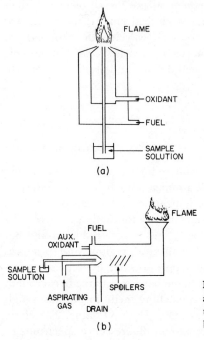

(a)

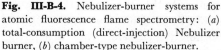

(b)

Fig. III-B-4. Nebulizer-burner systems for atomic fluorescence flame spectrometry: (a) total-consumption (direct-injection) Nebulizer burner, (b) chamber-type nebulizer-burner.

Some workers have used the propane/air flame for volatile elements with good success. The H_2/N_2O flame appears to be of use primarily for the volatile elements.

The C_2H_2/O_2 and C_2H_2/air flames are of little analytical use because of their high backgrounds and appreciable burning velocities. On the other hand, the C_2H_2/N_2O flame has found considerable use for elements that are difficult to atomize in flames (e.g., beryllium and germanium).

C. UNMIXED TURBULENT FLAMES

Unmixed turbulent flames have also been used by numerous investigators for atomic fluorescence spectrometry and the most important ones are listed in Table III-B-2. The $H_2/Ar/entrained$ air and the H_2/air flames have been most used. Unfortunately, such flames are rather prone to chemical (solute vaporization) interferences and are not good atomizers (low temperature and not very reducing). The turbulent H_2/O_2 and C_2H_2/O_2 have been used but are generally less satisfactory than the two-above mentioned turbulent flames. The $H_2/entrained$ air flame has also served with great success for easily atomized elements, as has the H_2/N_2O flame.

Table III-B-2. Flames of Interest in Atomic Fluorescence Spectrometry[a]

Gas Mixture	Flame Temperature (°K)
Premixed Flames	
Argon hydrogen	550–800[e]
Nitrogen/hydrogen	
Air/propane	2198
Air/hydrogen	2160
Argon/oxygen/hydrogen[b]	2350
Air/acetylene	2548
Nitrous oxide/hydrogen	2823
Oxygen/hydrogen	2933
Nitrous oxide/acetylene	3050
Oxygen/acetylene	3373
Turbulent Flames	
Hydrogen/entrained air	ca. 2100
Argon/hydrogen	ca. 2000
Air/hydrogen	ca. 2130
Oxygen/hydrogen[c]	2643
Oxygen/hydrogen[d]	2570–2260
Oxygen/acetylene[d]	2590

[a] Taken from R. Smith, " Flame Fluorescence Spectrometry," in *Spectrochemical Methods of Analysis*, J. D. Winefordner, Ed. Wiley, New York, 1971.
[b] Argon-to-oxygen ratio = 4:1.
[c] Rotational temperature of dry flame.
[d] Temperature with 1–3 ml H_2O min^{-1}.
[e] Temperature for innermost part of flame.

d. PREMIXED TURBULENT FLAMES

A hybrid flame of the previously named types is the premixed turbulent flame produced by premixing oxidant/fuel prior to introduction into the oxidant and fuel ports of a total-consumption nebulizer burner. The resulting flame is quite laminar in appearance and has lower flame flicker than the comparable unmixed turbulent flame. The premixed turbulent flame is also a more efficient atomizer than the comparable unmixed flame.

4. NONFLAME CELLS

Flames are plagued with several serious disadvantages. Flames have appreciable flame background and noise in certain spectral regions and are rather poor atomizers of elements forming stable monoxides in the flame gases (e.g., titanium, hafnium, zirconium, tungsten, molybdenum, beryllium, germanium) or of elements in a sample solution containing a concentrated matrix that is difficult to vaporize (see Section II-D). In addition, flames contain many molecular species (e.g., CO, CO_2, N_2) that are efficient quenchers of excited atoms. Therefore, many workers have attempted to overcome these disadvantages by using nonflame cells with controlled atmospheres. However, most nonflame cells used up to the time of writing this book were less convenient to work with than the simple flame cell.

Most nonflame cells have been based on the graphite tube furnace approach.[5] When a sample is placed in a graphite tube heated to about 3000°K by ohmic heating (electrical current) in an argon atmosphere, the sample vaporizes and atomizes within the reducing atmosphere. The atoms produced are then excited by means of a modulated line or continuum source, and the resulting atomic fluorescence is measured by means of a photodetector-phase-sensitive-amplifier system. The hot atomic vapor diffuses out of the graphite tube. (Also, convection due to flow of argon through the graphite tube removes atomic vapor—refer to Figure III-B-5a.)

Other nonflame cells used in atomic fluorescence spectrometry have included the graphite filament furnace[6] in which the sample solution is placed on the filament, which is heated in an inert atmosphere and the platinum loop cell[7] in which the sample solution adhering to a platinum loop is vaporized and atomized into an argon stream by heating the loop (see Figure III-B-5b and III-B-5c).

All nonflame cells up to the time of writing have utilized discrete sample sizes (hypodermic syringe) and so have had limited accuracy and precision. However, it certainly seems possible to utilize nebulization methods to eliminate this disadvantage of nonflame cells.

Another application of atomic fluorescence of vapors in nonflame cells is the use of the so-called *atomic resonance monochromator* for atomic absorption spectrometry.[8] Radiation from a hollow cathode lamp passes through the flame containing the atomized analyte and then enters a cell containing a vapor of the element of concern. The metal vapor is produced by a hollow cathode sputtering device within the resonance monochromator cell. On absorbing the radiation from the flame, atomic fluorescence of the atomic vapor results, and only the fluorescence characteristics of the element of interest is measured by a photomultiplier at right angles to the incident beam. Any radiation from impurities in the hollow cathode, nonresonance

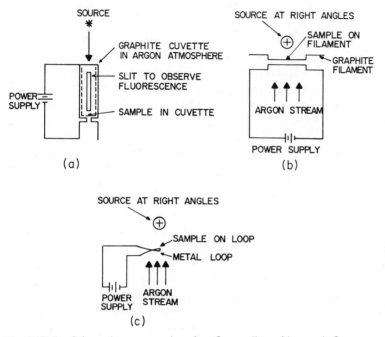

Fig. III-B-5. Schematic representation of nonflame cells used in atomic fluorescence spectrometry: (*a*) graphite cuvette furnace (cf., Ref. 5), (*b*) graphite filament furnace (cf. Ref. 6), (*c*) metal loop atomizer (cf. Ref. 7).

lines, atomic emission in the flame, flame background, and so on, does not give rise to the fluorescence process; therefore, it is not observed by the photomultiplier. The resonance detector is thus a selective detector for a given element.

5. REFERENCES

1. R. Smith, "Flame Fluorescence Spectrometry," in *Spectrochemical Methods of Analysis*, J. D. Winefordner, Ed., Wiley, New York, 1971.
2. J. D. Winefordner and T. J. Vickers, *Anal. Chem.*, **42**, (5), 206R (1970).
3. J. D. Winefordner and J. M. Mansfield, *Appl. Spectrosc. Rev.*, **1**, 1 (1967).
4. T. J. Vickers and R. M. Vaught, *Anal. Chem.*, **41**, 1476 (1969).
5. H. Massmann, *Spectrochim. Acta*, **23B**, 45 (1968).
6. T. S. West and X. K. Williams, *Anal. Chim. Acta*, **45**, 27 (1969).
7. M. P. Bratzel, R. M. Dagnall, and J. D. Winefordner, *Appl. Spectrosc.*, **24**, 518 (1970).
8. J. V. Sullivan and A. Walsh, *Spectrochim. Acta*, **22**, 1843 (1966).

C. MOLECULAR LUMINESCENCE INSTRUMENTATION[1-3]

1. SAMPLING DEVICES

The most common cell for condensed phase (solution) fluorimetry is similar to the 1-cm square cell that is standard for absorption spectrophotometry.[2] Usually all four faces of the cell are polished. Glass is suitable for excitation wavelength above 350 nm; quartz is used below 350 nm. Nonfluorescent synthetic silica is preferred to natural quartz in careful work. Larger cells are used for gas samples.

In fluorescence spectrometry, the design and geometry of the cell and cell holder greatly influence the amount of scattered excitation radiation seen by the detector, particularly if the wavelengths of excitation and emission are the same or very close. The excitation geometry is characterized by the angle between the axes of excitation and observation. (Section II-C-7 furnishes a consideration of the influence of angle between exciting and measuring beams and of the means of excitation in measurement, i.e., front surface versus right angle.) The common 90° angle is most suitable for dilute solutions or gaseous samples (fraction of radiation absorbed by analyte plus interferences is less than 5%) and is compatible with the square cell geometry. It is important that the detection system "see" only the fluorescing solution and not the cell walls through which the excitation light enters, but yet the entire sample should be illuminated and the entire luminescence emitted towards the detector should be measured. (See Section II-C-6 for an extensive theoretical treatment of the influence of incomplete illumination and/or incomplete measurement of the luminescence). For solid samples, or solutions that are very concentrated or turbid, the fluorescence may be viewed at an acute angle from the excitation axis. This is called the frontal or surface configuration.[1,2]

Sample cells for use in phosphorimetry and low temperature fluorescence studies are usually small (1-mm i.d.) tubes made of nonluminescent synthetic silica.[3] The tube is positioned in a silica Dewar flask filled with a liquid coolant (commonly liquid nitrogen). Right angle illumination is most common, although the frontal configuration is useful for solutions that form opaque "snows" at low temperature (cf, the rotating sample cell discussed later).

The small sample tubes are somewhat more difficult to fill, empty, and clean than fluorescence cells, but the smaller sample volume (typically 0.1 ml) means much lower absolute detection limits at a given concentration. The tubes can be emptied with a long, small diameter polyethylene tube connected to a water aspirator. They may be cleaned by successive rinses with nitric acid, distilled water, and solvent or sample solution. Very dirty tubes can be stored under concentrated nitric acid to remove the last traces of organic matter.

Precision of measurements of low temperature luminescence of samples in small sample tubes is often limited by sample tube positioning errors. Hollifield and Winefordner[4] have described a sample tube spinning apparatus (see Figure III-C-1) that averages out optical inhomogeneities and minimizes the effect of variations in sampling position. They observed a ten-fold decrease in relative standard deviation when the spinning apparatus was used. A commercial nuclear magnetic resonance (NMR) sample tube spinner may also be used.[5]

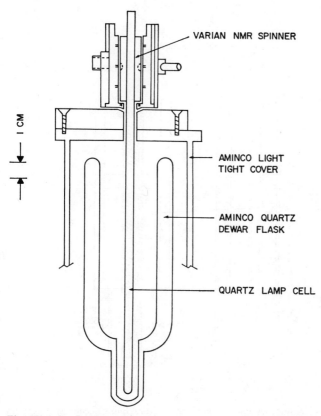

Fig. III-C-1. Representation of rotating sample cell immersion method of Refs. 4 and 5. (With permission of the authors.)

Parker[2] has described several elaborate specimen containers and compartments for variable temperature luminescence studies, including an apparatus for sample deaeration to reduce the quenching effect of oxygen and a frontal illumination system for low temperatures.

2. SAMPLE COOLING METHODS[2,3]

The simplest method of cooling a sample for low temperature studies is to immerse the sample tube directly into a transparent liquid coolant in a Dewar flask. A portion of the Dewar is left unsilvered to allow passage of the light. This system, illustrated in Figure III-C-2a, is the one most common in analytical applications. It has the advantage of rapid sample cooling and a stable, known sample temperature. However, there are several disadvantages:

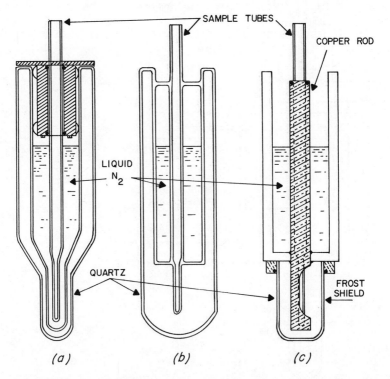

Fig. III-C-2. Types of cooling systems for low temperature luminescence spectrometry (predominantly phosphorimetry): (a) simple immersion method, (b) conduction cooling via quartz tube and sample, (c) conduction cooling via copper rod. (Drawings are reproductions of those found in Ref. 3—with permission of the authors.)

1. The excitation and emission light beams must each pass through three quartz layers. Even if the quartz is quite transparent, there are reflective losses at each of the twelve interfaces (typically 4% per interface), amounting to as much as 50% total light loss.

2. The coolant must be transparent and nonluminescent.

3. The convective motion and refractive index changes in the coolant cause flickering of both the incident and luminescence light beams; this effect may make a significant contribution to the total noise.

4. Occasional bubbling of the coolant may cause extremely unstable readings.

Other methods of sample cooling minimize these disadvantages by reducing the number of quartz layers through which the light beams must pass and by eliminating direct contact of the coolant with the end of the sample tube containing the sample. In Figure III-C-2b the sample tube is made an integral part of the Dewar flask. The reduction of reflective losses is only minor, but convection and bubbling noise are eliminated. The sample is cooled by conduction through the quartz sample tube walls and through the rigid solution itself. The disadvantages are the uncertainty and inhomogeneity in the sample temperature and the considerable inconvenience involved in changing samples. The system in Figure III-B-2c reduces thermal gradients by means of a highly conductive copper rod machined to fit the sample tube snugly. The frost shield reduces icing and fogging of the viewing area. However, the space inside the shield cannot be simply evacuated, and so condensation of air onto the sample tube and fogging of the shield itself by conductive cooling through the internal air space cannot be avoided. The copper rod conduction system has also been used for frontal illumination of solid samples.

Both conductive cooling systems can use solid-liquid coolants, but the immersion system is limited to boiling liquids. Liquid nitrogen, liquid air, and liquid nitrous oxide have been commonly used in phosphorimetry, although many other coolant systems could be used for other temperatures.[3] Minkoff[6] has given a helpful review of low temperature techniques.

3. MODULATION METHODS

The observation of long-lived ($\bar{\tau} > 10^{-3}$ sec) luminescence without interference by prompt fluorescence and scattered light is most easily accomplished by the use of a mechanical phosphoroscope. The phosphoroscope is a rotating mechanical shutter that allows periodic out-of-phase excitation and observation of the sample. Two basic forms are in common use.

The rotating can phosphoroscope is shown in Figure III-C-3a. It consists of a hollow metal cylinder with two or more equally spaced apertures along

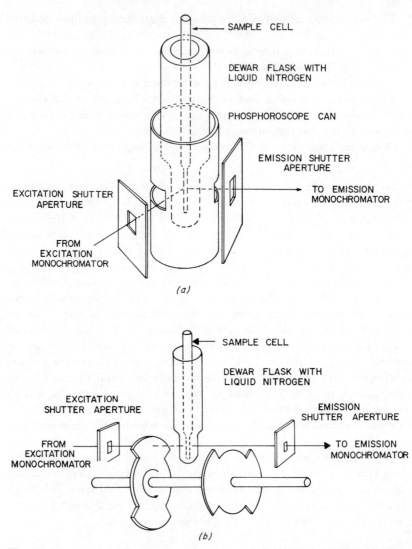

SAMPLE CELL

DEWAR FLASK WITH
LIQUID NITROGEN

PHOSPHOROSCOPE CAN

EMISSION SHUTTER
APERTURE

TO EMISSION
MONOCHROMATOR

EXCITATION SHUTTER
APERTURE

FROM
EXCITATION
MONOCHROMATOR

(a)

SAMPLE CELL

DEWAR FLASK WITH
LIQUID NITROGEN

EXCITATION
SHUTTER APERTURE

EMISSION
SHUTTER APERTURE

FROM
EXCITATION
MONOCHROMATOR

TO EMISSION
MONOCHROMATOR

(b)

Fig. III-C-3. Schematic representation of phosphoroscope interaction with radiation: (*a*) rotating can device, (*b*) Becquerel-type device. (Drawings are reproductions of those found in Ref. 7—with permission of the authors.)

its circumference. The cylinder is rotated by a variable speed motor. The apertures allow the radiation from the excitation monochromator to strike the sample periodically and allow luminescence emission to reach the entrance slit of the emission monochromator between the excitation periods. The intensity of prompt fluorescence and scattered incident light decays very rapidly after termination of the excitation period, and so only long-lived luminescence (phosphorescence and delayed fluorescence) will remain when the phosphoroscope has turned from the point at which excitation is terminated to the point at which the observation period is initiated.

The Becquerel or rotating disk phosphoroscope (Figure III-B-3*b*) achieves the same effect with two sectored disks mounted on a common motor-driven axle. This system is somewhat more versatile than the rotating can type because the phase relation between the two disks may be adjusted more easily.

Either type of phosphorocopic system produces a train of approximately trapezoidal pulses of light. The resulting photocurrent pulses from the detector are usually measured with a DC readout system in commercial instruments. The response of the DC system is proportional to the average area under the current pulses. An AC or lock-in detection system could also be used, because the light is conveniently modulated by the phosphoroscope.

The operation[7] of both the can and disk phosphoroscopes is characterized by three time periods which are a function of the design and speed of the rotating components: the cycle time t_C, which is the period of one cycle of excitation and observation; the exposure time t_E, which is the time during which the sample is excited or observed (assumed equal); and the delay time t_D, which is the time between the end of the excitation period and the beginning of the observation period. The magnitudes of these time periods, relative to the lifetime τ of the luminescent species, influences the measured radiant flux of luminescence. It is apparent that the presence of the phosphoroscope reduces the measured radiant flux because the sample is being excited (or observed) for only a fraction t_E/t_C of each measurement cycle. It can be shown[7] that the ratio α of the average DC photocurrent observed using a phosphoroscope to that which would be observed without the phosphoroscope if prompt fluorescence and scattered light were not to interfere, is given by

$$\alpha = \frac{\tau(\exp - t_D/\tau)(1 - \exp(t_E/\tau))^2}{t_C[1 - \exp(-t_C/\tau)]} \qquad \text{(III-B-1)}$$

For long-lived luminescence ($\tau \gg t_C$), α is independent of τ and of the speed of the phosphoroscope and is given by

$$\alpha = \left(\frac{t_E}{t_C}\right)^2 \qquad \text{(III-B-2)}$$

Because t_E must be less than one-half of t_C, this α must be less than 0.25. For shorter decay times, or slower phosphoroscope speeds, where $\tau < t_C$, α begins to drop off. This occurs at about $\tau = 10^{-3}$ sec for the Aminco phosphoroscope operating at maximum speed (7000 rpm).

It is usually desirable to operate the phosphoroscope fast enough so that $\tau \gg t_C$, because in that case undesirable changes in the phosphoroscope speed or changes in the sample luminescence lifetime (due to temperature variations, quenching by dissolved oxygen, etc.) will have little effect on the α factor. On the other hand, some degree of resolution of a mixture of two spectrally similar compounds can be accomplished if their decay times are significantly different; t_C is made approximately equal to the larger decay time, therefore attenuating the response due to the shorter lived species. The degree of attenuation may be calculated using equation III-B-1.

Equation III-B-1 may be modified and generalized to apply to a stroboscopic system (see Section III-A-6-d) using a pulsed excitation source and a gated multiplier phototube.[8,9] If the half-intensity width t_F of the flash tube excitation source is much less than the lifetime τ of the luminescent species, then the corresponding α value is

$$\alpha' = \frac{f t_F (\exp{(-t_D/\tau)}(1 - \exp{-t_P/\tau})}{1 - \exp{-1/f}} \qquad \text{(III-B-3)}$$

where $\alpha' =$ ratio of average luminescence radiant flux observed per unit time to the hypothetical luminescence radiant flux per unit time produced by steady state luminescence.

$f =$ repetition frequency of light source and photomultiplier pulses, sec^{-1}.

$t_D =$ delay time between end of excitation pulse and beginning of phototube pulse, sec.

$\tau =$ lifetime of luminescent species, sec.

$t_P =$ duration of phototube pulse, sec.

It may be shown that[8] the use of a pulsed system of this type would allow enhancement of the signal of one luminescent species with respect to other species of longer *or* shorter lifetime. The lifetime of maximum enhancement and the extent of relative enhancement is determined by the adjustable parameters t_D, t_P, and f.

4. COMMERCIAL INSTRUMENTS

Fluorescence spectrometers are manufactured by several firms. The Aminco-Bowman Spectrophotofluorometer (American Instrument Co., Silver Spring, Md.) contains two Czerny-Turner grating monochromators.

Either may be scanned automatically for recording excitation and emission spectra. A high pressure xenon arc lamp is normally supplied as the excitation source. A 1P28 multiplier phototube and a DC amplifier system are used to detect the luminescence emission. Meter readout is supplied, although output connections for a recorder and oscilloscope are provided. Available accessories include adjustable slits, interchangeable gratings, photomultiplier cryostat, polarization accessory, multiple sample turret, and a phosphorescence attachment.

The Aminco SPF-125 is a smaller, simpler, less expensive spectrofluorometer intended primarily for quantitative clinical tests.

The Farrand Mark-1 Spectrofluoremeter (Farrand Optical Co., Inc., Mount Vernon, N.Y.) is similar in principle to the Aminco-Bowman device. It also uses two grating monochromators. Electrically operated light shutters for decay time studies of phosphorescence are provided.

The Baird-Atomic Fluorispec (Baird-Atomic, Cambridge, Mass.) is unique in that it uses two double grating monochromators for high resolution and extremely low scattered light. A phosphoroscope attachment is available as an accessory.

None of the foregoing instruments gives "true" excitation or emission spectra corrected for variations in source radiance, monochromator transmission, and detector sensitivity with wavelength. Compensated or "corrected spectra" instruments include additional components to provide this correction. The Perkin-Elmer model 195 linear energy spectrofluorometer (Perkin-Elmer Corp., Norwalk, Conn.) uses two silica prism monochromators. A beam splitter samples a small portion of the light leaving the excitation monochromator and directs it onto a thermocouple detector*. The thermocouple responds to the total energy striking it and is insensitive to the wavelength of the light (in contrast to the more sensitive photoemissive detectors). The thermocouple signal controls a slit servomechanism which keeps the energy emerging from the excitation monochromator constant. In this way, energy-corrected excitation spectra are obtained. Emission spectra are corrected by means of an adjustable electrical cam which controls a slit-servo in the emission monochromator.

The Turner Model 210 Absolute Spectrofluorometer (G. K. Turner Associates, Palo Alto, Calif.) records corrected excitation and emission spectra, as well as absorption spectra compensated for fluorescence errors. Excitation correction is accomplished by means of a time-shared bolometer detector, which compares the energy emerging from the excitation monochromator to that from a small reference lamp and adjusts the reference lamp to provide equal energy. The multiplier phototube is time-shared between a

* Refer to Appendix 4.

beam from this reference lamp and the fluorescence radiant flux emerging from the emission monochromator. The ratio of the fluorescence signal to the reference signal is recorded. Emission correction is provided by a cam-driven wedge attenuator in the reference beam. The cam is operated from the emission monochromator wavelength drive and is shaped to compensate for the spectral response of the multiplier phototube and emission mono-chromator.

The Aminco energy-corrected luminescence spectrometer uses a reference thermopile to obtain a wavelength-independent measure of the excitation intensity passing through the sample. Excitation spectra are corrected by di-viding the uncorrected phototube signal by the thermopile signal. Emission spectra are corrected by means of an adjustable cam, linked to the emission monochromator wavelength drive, which drives a plunger on a potentio-meter wired into an operational amplifier circuit.

Phosphoroscope attachments are available for the Aminco, Farrand, and Baird-Atomic instruments. These include a rotating mechanical shutter and a liquid nitrogen Dewar flask.

5. GLOSSARY OF SYMBOLS

f Frequency of pulsing, Hz.

t_C Phosphoroscope period, sec.

t_E Exposure time of analyte to exciting light per cycle, sec.

t_D Delay time between end of excitation and beginning of observation, sec.

t_F Duration of light flash, sec.

t_P Phototube pulse (gate) width, sec.

α Phosphoroscope factor, ratio of average DC pnotocurrent observed using a phosphoroscope to the DC photocurrent observed with no phos-phoroscope, no units.

α' Phosphoroscope factor with flash tube excitation, ratio of average luminescence radiant flux observed per unit time to hypothetical lumiminescence radiant flux produced by steady state luminescence, no units.

τ Luminescence decay time (lifetime), sec.

6. REFERENCES

1. D. W. Ellis, "Luminescence Instrumentation and Experimental Details," in *Fluorescence and Phosphorescence Analysis*, D. M. Hercules, Ed., Wiley Interscience, New York, 1966, Chapter 3.
2. C. A. Parker, *Photoluminescence of Solutions*, Elsevier, Amsterdam, 1968.
3. J. D. Winefordner, W. J. McCarthy, and P. A. St John, "Phosphorimetry as an Analytical Approach in Biochemistry," in *Methods of Biochemical Analysis*, Vol. 15, D. Glick, Ed., Wiley-Interscience, New York, 1967.

4. H. C. Hollifield and J. D. Winefordner, *Anal. Chem.*, **40**, 1759 (1968).
5. R. Zweidinger and J. D. Winefordner, *Anal. Chem.*, **42**, 639 (1970).
6. G. J. Minkoff, *Frozen Free Radicals*, Wiley-Interscience, New York, 1960.
7. T. C. O'Haver and J. D. Winefordner, *Anal. Chem.*, **38**, 602 (1966).
8. M. L. Bhaumik, *Rev. Sci. Instr.*, **36**, 37 (1965).
9. T. C. O'Haver and J. D. Winefordner, *Anal. Chem.*, **38**, 1258 (1966).

D. SIGNAL-TO-NOISE RATIO THEORY

1. GENERAL ASPECTS

a. NOISE[1-3]

The readout (meter reading, pen position, scope deflection, digital number, etc.) of a spectrometric system consists of a desirable signal component, which is related in some way to the nature and concentration of the sample, and an undesirable signal component, which interferes with measurements of the sample. This undesirable component may consist of:

1. A *constant DC offset* or *bias* error, which might include background, dark current, amplifier offsets, zero-adjust errors, and so on. These signals are easily eliminated or compensated for.

2. A *drift* component, due perhaps to phototube fatigue, power supply drift, photodecomposition of a sample, or monochromator wavelength drift. Drift can be a serious source of error in many cases.

3. An AC *noise* component—it may be *periodic*, such as that due to pickup of 60-Hz power line frequency, 120-Hz power supply hum, or stray pickup of radio signals; or it may be *aperiodic* (random), such as shot or Johnson noise. Periodic noise is usually easy to eliminate by means of an electrical notch filter tuned to the proper frequency. Random noise, however, appears at not just one frequency but is distributed over a wide band of frequencies and is therefore much more difficult to filter out completely.

The frequency distribution of noise components is characterized by a power density spectrum, which is a plot of noise power per unit frequency interval (watt Hz^{-1}) versus frequency, rather analogous to the spectrum of a light source. Noise whose power density is independent of frequency is called *white* noise (see Figure III-D-1). This is a very common type of noise; shot noise and Johnson noise are white under most conditions. *Nonwhite* noise may also be observed (see Figure III-D-1). One particularly common and troublesome type of nonwhite noise is called "pink" or $1/f$ *noise* because its power density is inversely proportional to frequency.

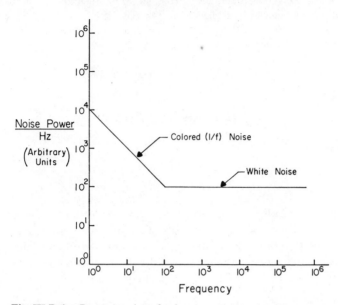

Fig. III-D-1. Representation of noise power spectrum.

b. ELECTRICAL BANDWIDTH

The *electrical bandwidth* of a system is a measure of how quickly the output (pen position, needle reading, oscilloscope trace, etc.) follows changes in the input signal (current or voltage). A system with high bandwidth (wide band) faithfully displays rapid changes in the input signal, whereas a system with low bandwidth (narrow band) smoothes out and/or attenuates input signal changes. As a result, systems with finite bandwidth exhibit a finite *response time, an averaging time, and a frequency response.*

(1) *Response Time.* If an input *step function* is applied to a system (i.e., an instantaneous change in the input signal), the output will not change instantly but will reach about 98% of its final reading in a time called the *response time t_r*, related to the electrical bandwidth Δf by[2]

$$t_r \cong \frac{1}{\Delta f} \qquad\qquad \text{(III-D-1)}$$

(2) *Averaging Time.* Any input signal changes are averaged over an effective time t_a :

$$t_a = \frac{1}{2\Delta f} \qquad\qquad \text{(III-D-2)}$$

Thus if a direct measurement of a white noise source is made with an rms AC voltmeter of electrical bandwidth Δf, then the observed rms voltage will be equal to the standard deviation of a number of successive time averages of the amplitude of the noise source, where each average is taken over a time period equal to $1/2\Delta f$.

(3) *Frequency Response.* If the input signal to the readout system is varied in amplitude sinusoidally at some frequency f, then the system response to that amplitude change will be a function of f. A plot of the system response to the sinusoidal variation in signal amplitude versus the frequency of the amplitude variation is called the *frequency response* curve of the system. The range of frequency Δf over which the system responds is called the *passband*, or more commonly, the *electrical bandwidth*, in frequency units. If the input signal is DC then a *low-pass* readout system can be used. A low-pass system responds equally to DC and to all AC frequencies up to approximately Δf, and it attenuates frequencies above Δf. If some form of modulation is used to produce an AC signal, then a *bandpass* system can be used. A bandpass system responds only to frequencies between certain frequency limits f_l and f_u, where $f_l < f_u$. The bandwidth Δf is given by $f_u - f_l$. In real systems, the rate of attenuation of response to frequencies outside of Δf is gradual, and this makes the determination of the exact bandwidth limits less direct.

Most readout systems are characterized by more than one bandwidth. For instance, AC systems have both a predetection bandwidth (the bandwidth of the AC amplifier before the demodulator) and a postdetection bandwidth (the bandwidth of the low-pass DC filter stage following the demodulator). Both AC and DC systems may use a chart recorder readout which is a low-pass system itself and has its own bandwidth. The total electrical bandwidth of the entire system Δf_t is given by

$$\Delta f_t = \left(\frac{1}{\Delta f_1{}^2} + \frac{1}{\Delta f_2{}^2} + \cdots + \frac{1}{\Delta f_j{}^2} \right)^{-1/2} \tag{III-D-3}$$

where $\Delta f_1 \cdots \Delta f_j$ are the bandwidths of each individual stage or component of the system. Obviously, if one Δf is much lower than the other, it will effectively determine the system bandwidth.

C. EFFECT OF ELECTRICAL BANDWIDTH ON NOISE

The response of a readout system to noise at the input depends on what portion of the noise spectrum is passed by the system. Noise frequency components beyond (above or below) the passband are attenuated. For white noise, the noise power per unit frequency interval (watt Hz^{-1}) is constant. Therefore, the total noise power passed by a system is the product of the input noise density (watt Hz^{-1}) and the electrical noise bandwidth Δf of the system (Hz). Thus the measured noise voltage or current at the output is proportional to the square root of total noise power and thus to $\sqrt{\Delta f}$, which is why $\sqrt{\Delta f}$ appears in the equations for shot and Johnson noise. If the noise is not white, the relation between noise and bandwidth is different; but in general reducing Δf reduces the noise at the output.

d. SIGNIFICANCE OF RESPONSE TIME

As a rule, random noise in the output signal of an analytical system may be reduced by reducing the electrical bandwidth of the system. However, the response time t_r increases as Δf decreases. In many cases, the increase in response time will introduce systematic errors or other undesirable side effects which effectively limit the extent to which Δf may be profitably reduced.

(1) *Analysis Time.* After changing a sample, it is necessary to wait at least one response time, and preferably longer, before taking a reading, so that no systematic error will occur because of incomplete response of the instrument to the change. If Δf is reduced to reduce noise, the response time t_r increases, so each reading takes longer. If multiple readings are desired for a statistical check, the total analysis time may become excessive if the response time is too long (Δf is too low). The proper balance between Δf and t_r must be judged for the particular application.

(2) *Sample Consumption.* In the flame techniques, such as atomic fluorescence spectrometry, the sample is consumed at some rate* continuously during the analysis. If we attempt to increase the signal-to-noise ratio, the increase in response time that is produced will require the consumption of more sample for each reading, with the result that the absolute (weight) detection limit may be increased.

(3) *Drift Errors.* Instrumental drifts resulting from source spectral radiance drift, photomultiplier fatigue, amplifier drift, and so on, have a greater effect on analytical accuracy the longer it takes to get a reading. Decreasing Δf to reduce noise errors may increase drift errors.

(4) *Sample Stability.* Some samples in molecular luminescence work are photolyzed by the high excitation radiance or are otherwise unstable. Excessively long readout response time may lead to cumulative errors and irreversible damage to the sample.

---

*Actually one of the authors (JDW) has recently utilized discrete sampling (1 μ 1) in atomic fluorescence flame spectometry. With discrete sampling, a large Δf and a smaller t_r are required.

(5) *Spectrum Scanning.*[4,5] Excitation and emission spectra are usually obtained by scanning the appropriate monochromator wavelength drive at a constant rate and recording the emission signal versus time. Excessively long t_r, fast scan speeds, or both may distort spectral peaks and blur out important details and fine structure. The optimum scan speed depends on Δf and on the structure of the spectrum.

For recording line spectra, the monochromator scan speed r (nm sec^{-1}) should be approximately

$$r \leq \frac{s'}{t_r} = s'(\Delta f) \tag{III-D-4}$$

where s' is the spectral bandpass of the monochromator, t_r is the system response time (sec), and Δf is the electrical noise bandwidth (sec^{-1}). For recording band spectra, r should be

$$r \leq \frac{\Delta \lambda'}{t_r} = (\Delta \lambda') \, (\Delta f) \tag{III-D-5}$$

where $\Delta \lambda'$ is the half width (nm) of the most narrow band or spectral feature to be observed. If the electrical bandwidth Δf is adjustable, rather than the scan speed, then the foregoing relations may be used to determine the smallest permissible value of Δf.

(6) *Luminescence Decay Time Measurements.* There is a conflict between random noise errors at large Δf and systematic errors at small Δf due to the long response time in luminescence decay time measurements, also. The response time must be much smaller than the decay time to be measured.

In many of the previously described situations, an optimum value of electrical bandwidth may exist for a particular set of experimental conditions. This optimum value must usually be determined experimentally. In those cases in which excessively long response time introduces systematic errors, a series of measurements made at various values of Δf and extrapolated to $\Delta f = \infty$ ($t_r = 0$) will give an indication of the significance of t_r. However, those readings taken at high Δf may be imprecise because of the increased noise.

e. NOISE SOURCES IN SPECTROMETRIC INSTRUMENTATION

The most important sources of random noise in luminescence spectrometry include the following ones.

(1) *Source of Excitation.* Drift and random fluctuations* in source radiance may be due to line voltage variations, lamp aging, arc wander, sputtering and

---

*A $1/f$ noise component also exists in source fluctuation.

evaporation of electrode material onto lamp walls, thermal turbulence and convection in the arc gases, and so on. The noise is a constant fraction of total source radiance and thus increases linearly as the measured photosignal is increased by, for example, opening the monochromator slits or increasing detector or amplifier gain.

(2) *Sample and Sample Cell.*   The sample and the sample cell constitute a noise source highly dependent on the particular model. It includes such factors as flame noise in flame fluorescence work and coolant convection and bubbling in low temperature luminescence spectrometry of molecules in the condensed phase.

(3) *Photodetector.*[6-10]   Detector noise, mainly dark current shot noise in multiplier phototubes, is attributable to the fundamental quantum nature of electrons and photons and is often the fundamental limiting noise when other noise sources have been reduced or eliminated.

(4) *Amplifier-Readout System.*[3]   Electronic noise in the amplifier and readout system is usually insignificant with modern instrumentation. However, some types of noise may still cause trouble, particularly stray pickup of 60-Hz hum, radio and TV signals, radar, and so on. Drift in DC systems may be significant.

f. ADDITION OF NOISE SOURCES

In analytical applications we are usually interested in the total effect of the various noise sources on the readout signal. It is convenient to express noise amplitudes in terms of rms current noise at the output of the photodetector. Each noise source is evaluated in terms of its contribution to the total noise $\overline{\Delta i_t}$ in the photodetector output current. The way in which the various noise sources combine is important. Offsets and drifts add linearly and algebraically (i.e., with regard to sign). On the other hand, independent random noises add quadratically:

$$\overline{\Delta i_t} = (\sum_n \overline{\Delta i_n}^2)^{1/2} \qquad \text{(III-D-6)}$$

where $\overline{\Delta i_t}$ is the total rms current noise in the photodetector output current and the $\overline{\Delta i_n}$ are the individual rms current noises due to each of the various noise sources. In some cases the relative contribution of the various noise sources to the total noise may be determined by investigating the dependence of total noise at the output on the detector current; this may be accomplished by varying the monochromator slit width or height to change the radiant flux reaching the detector. Source noise is a constant fraction of the total source radiance (spectral radiance), and thus its noise contribution to the output noise increases linearly with the radiant flux reaching the detector (and thus with the detector current). Shot noise increases as the square root of detector current. Electronic noise is constant, independent of any variable before the electronics. It may be possible to determine, for instance, whether any one noise source is dominant.

g. SIGNAL-TO-NOISE RATIO AND THE OPTIMIZATION OF EXPERIMENTAL CONDITIONS[11-15]

The signal-to-noise ratio* $i_L/\overline{\Delta i_T}$, must be considered when experimental conditions for any spectrochemical method are being optimized. It is generally desirable to obtain the largest possible signal-to-noise ratio in order to obtain the most precise results and the lowest limit of detection. If analytical expressions for i_L and $\overline{\Delta i_T}$ could be obtained, then the optimum value of any experimental parameter X could be found by differentiating $i_L/\overline{\Delta i_T}$ with respect to X, setting the resulting derivative equal to zero, and solving for X_{opt}. It is usually necessary to do this graphically, however. A plot of $i_L/\overline{\Delta i_T}$ versus each parameter X, obtained experimentally, will reveal which parameters do in fact exhibit maxima.

h. EFFECT OF RANDOM NOISE ON ANALYTICAL PRECISION AND DETECTION LIMITS[16,17]

In many instances the precision of analytical measurements is limited by random noise. This is particularly true in trace analysis near the limit of detection. In such cases, statistical methods may be used to evaluate the effect of random noise on analytical *precision*.

Analytical determinations invariably involve the measurement of at least two signals, one or both of which may be noisy. Most commonly, the *difference* between two signal levels is sought (e.g., the difference between a sample signal and a blank signal). It is necessary to consider separately the noise on the blank signal $\overline{\Delta i_b}$ and the noise on the sample signal $\overline{\Delta i_s}$. Consider two noisy current signals i_s and i_b having rms noises $\overline{\Delta i_s}$ and $\overline{\Delta i_b}$, respectively, such that $i_s > i_b$ (these are sample and blank signals, respectively). Let us define the difference $i_s - i_b$ as the analytical signal S (i.e., $S = i_L$)

$$S = i_s - i_b \qquad \text{(III-D-7)}$$

The respective rms noise σ_s is the standard deviation of the difference; that is, $\sigma_{i_s - i_b}$, and can be shown to be

$$\sigma_s = \sqrt{\overline{\Delta i_s}^2/n_s + \overline{\Delta i_b}^2/n_b} \qquad \text{(III-D-8a)}$$

---

* The i_L represents the detector photocurrent due to analyte luminescence.

where n_s and n_b are the number of measurements of i_s and i_b, respectively, which constitute each measurement of S. In practical work n_s and n_b are made equal. Thus if $n_s = n_b = n$:

$$\sigma_s = \sqrt{(\overline{\Delta i_s}^2 + \overline{\Delta i_b}^2)/n} \qquad \text{(III-D-8b)}$$

The magnitude of σ_s has an effect on several important specifications of analytical measurements.

(1) *Relative Standard Deviation*[15]   In a random-noise-limited system, the relative standard deviation *RSD* of a series of successive readings of S is given by

$$RSD = \frac{100\sigma_s}{S} = \frac{100}{S}\sqrt{(\overline{\Delta i_s}^2 + \overline{\Delta i_b}^2)/n} \qquad \text{(III-D-9)}$$

There are two useful limiting cases for this equation. For large sample concentrations, $i_s \gg i_b$. In luminescence spectrometry the most important sources of noise increase in amplitude with the analytical signal, and thus $\overline{\Delta i_s} \gg \overline{\Delta i_b}$ at large sample concentrations; therefore,

$$RSD \cong \frac{100\,\overline{\Delta i_s}}{\sqrt{n}\,S} \qquad \text{(III-D-10)}$$

At low sample concentrations, $i_s \approx i_b$ and $\overline{\Delta i_s} = \overline{\Delta i_b} = \overline{\Delta i}$. Thus

$$RSD \simeq \frac{100\sqrt{2}\,\overline{\Delta i}}{\sqrt{n}\,S} = \frac{141\overline{\Delta i}}{\sqrt{n}\,S} \qquad \text{(III-D-11)}$$

(2) *The Limit of Determination.*   Often we want to know the smallest analytical signal S_l that can be *measured* with a given maximum acceptable *RSD*. This is easily found from equation III-D-9 or its two limiting cases:

$$S_l = \frac{100}{RSD}\sqrt{(\overline{\Delta i_s}^2 + \overline{\Delta i_b}^2)/n} \qquad \text{(III-D-12)}$$

The sample concentration corresponding to this value of S is called the *limit of determination.*

(3) *Limiting Detectable Sample Concentration.*[16, 17]   In some cases it may be useful to know the smallest analytical S_m that can be *detected* with a given level of confidence. This is derived from small sample statistical theory by the use of the "Student t" statistic. The Student t is defined as the ratio of the

smallest detectable difference between two means to the standard deviation of those differences, or in our terms:

$$t = \frac{(i_s - i_b)_m}{\sigma_s} = \frac{S_m}{\sigma_s} \qquad \text{(III-D-13)}$$

where S_m is the minimum detectable S. The value of t (no units) is available in tables and is a function of the number of degrees of freedom and the confidence level desired. The number of degrees of freedom is equal to $2n - 2$, where n is the number of measurement pairs, each measurement pair consisting of one measurement each of i_s and i_b. Combining equations III-D-8 and III-D-13, we have

$$S_m = t\sigma_s = \frac{t\sqrt{\overline{\Delta i_s}^2 + \overline{\Delta i_b}^2}}{\sqrt{n}} \qquad \text{(III-D-14)}$$

In general, S_m is small, so that $i_s \simeq i_b$ and $\overline{\Delta i_s} \simeq \overline{\Delta i_b} = \overline{\Delta i}$. In that case

$$S_m = \frac{\sqrt{2t}\,\overline{\Delta i}}{\sqrt{n}} \qquad \text{(III-D-15)}$$

The sample concentration giving a signal equal to S_m is called the *limiting delectable sample concentration*, sometimes shortened to *limit of detection* or *detection limit*.

2. ATOMIC FLUORESCENCE FLAME SPECTROMETRY

a. SIGNAL AND SHAPE OF ANALYTICAL CURVES[18,19]

The average photodetector signal current i_L due to atomic fluorescence of the analyte (assuming the atomic fluorescence line half width is considerably less than the spectral bandpass of the measuring monochromator) is given by

$$i_L = mW'K_o'\gamma B_L \qquad \text{(III-D-16)}$$

where m is a factor that accounts for the light loss when a mechanical source intensity chopper is used (m equals 0.5 for most choppers and 1.0 if no chopper is used), W' is the monochromator* slit width (cm), K_o' is the optics factor* (cm sr) discussed in Section III-A-4-h, γ is the radiant sensitivity of the photodetector (A watt^{-1}), and B_L is the radiance (watt cm^{-2} sr^{-1}) of fluorescence (cf. Section II-C-10). The magnitude of B_L was seen to depend on the atomic concentration of n_o atoms in the flame gases, according to

---

* The primes designate emission monochromator. Actually in atomic fluorescence studies only an emission monochromator is used (i.e., no excitation monochromator).

equations II-C-24 through II-C-35, and n_o was shown to be related to the sample solution concentration C_o by

$$n_o = (6 \times 10^{23}) \frac{F \varepsilon \beta C_o g_o}{Q_t e_f Z_T} \qquad \text{(III-D-17)}$$

where F is the solution transport rate (cm^3 sec^{-1}), ε is the aspiration efficiency (no units), β is the atomization efficiency (no units), C_o is the sample concentration (mole cm^{-3}), e_f is the flame expansion factor (no units), Q_t is the flow rate of gases into the flame (cm^3 sec^{-1}), and Z_T is the electronic* partition function (see Section II-D-1-b).

If ε and β do not vary with sample concentration C_o, n_o will vary directly with C_o. In that case a plot of i_L versus C_o (called the *analytical curve*) should have the same shape as the growth curves given in Section II-C. Unfortunately $\varepsilon\beta$ decreases significantly both at very low analyte concentrations (because of change in the degree of ionization) and at high analyte concentrations (because of reduced solution transport rate, increased solute vaporization interference, and decreased aspirator yield). Compound formation of the analyte atoms with flame gas products (e.g., O, H, OH) results in displacement of the entire analytical curve. In practice, therefore, the slope of the *logarithmic analytical curve* is greater than that of the growth curve at very low sample concentrations and less than that of the growth curve at high sample concentrations. At very high concentrations, the fluorescence power efficiency Y_L also decreases.

For a more extensive discussion of the shapes of growth and analytical curves, respectively, refer to Sections II-C-10 and II-D-1-b. In Section II-D nonflame cells are also discussed.

b. NOISE

In atomic fluorescence flame spectrometry, the total rms random noise current $\overline{\Delta i_T}$, referred to the input of the amplifier-readout system, is the quadratic sum of six important and independent noise sources:

$$\overline{\Delta i_T} = \sqrt{(\overline{\Delta i_D}^2 + \overline{\Delta i_B}^2 + \overline{\Delta i_S}^2 + \overline{\Delta i_L}^2 + \overline{\Delta i_E}^2 + \overline{\Delta i_A}^2)} \qquad \text{(III-D-18)}$$

where $\overline{\Delta i_D}$ is the total rms detector noise† current, $\overline{\Delta i_B}$ is the rms flicker noise in the flame background, $\overline{\Delta i_S}$ is the rms flicker noise due to scattering of incident radiation by unevaporated solvent and droplets and solute particles

---

* The g_o is the statistical weight of the ground state. If o is not the ground state, then replace g_o by $g_o e^{-E_o/kT}$ (see Section II-D).

† For phototubes, $\overline{\Delta i_D}$ is primarily determined by detector shot noise.

in the flame gases, $\overline{\Delta i_L}$ is the rms flicker noise of the fluorescence radiation, $\overline{\Delta i_E}$ is the rms flicker noise in the thermal radiation emitted by the analyte atoms at the analytical wavelength λ_o, and $\overline{\Delta i_A}$ is the rms noise in amplifier-readout system referred to the input. All the foregoing noises are referred to the amplifier input as rms noise currents, and all are bandwidth-limited by the electrical bandwidth of the amplifier-readout system. No theoretical analytical expressions are known for source noise and flame flicker noise, so no general equation for $\overline{\Delta i_T}$ can be given. It is necessary to determine empirically the effect of various experimental parameters on $\overline{\Delta i_T}$.

C. SIGNAL-TO-NOISE RATIO AND OPTIMIZATION OF EXPERIMENTAL CONDITIONS FOR LOW ANALYTE CONCENTRATIONS[20]

In atomic flame fluorescence spectrometry, the experimental parameters that have been of concern in signal-to-noise ratio optimization at low analyte concentrations ($\overline{\Delta i_L}$ is negligible compared with other noise sources) include the following:

(1) *The Monochromator Slit Width W.* The variation of $i_L/\overline{\Delta i_T}$ with monochromator slit width W exhibits a maximum, indicating that an optimum slit width exists. This is expected intuitively; $i_L/\overline{\Delta i_T}$ is reduced by increased $\overline{\Delta i_D}$ and $\overline{\Delta i_A}$ at small slit widths and by increased flame background noise $\overline{\Delta i_B}$ at large slit widths. If $\overline{\Delta i_S}$ becomes appreciable (i.e., intense sources, wide slits, and scatterers in flame), then the peak of the bell-shaped curve of $i_L/\overline{\Delta i_T}$ versus W will shift toward smaller values of W.

(2) *Electrical Bandwidth of Amplifier Readout System* Δf. The value of $i_L/\overline{\Delta i_T}$ increases as Δf decreases, but drift errors may predominate at low Δf as discussed in Section III-D-1. The optimum value of Δf is usually of the order of 1 Hz.

(3) *Source Radiance (Spectral Radiance).* The value of $i_L/\overline{\Delta i_T}$ increases with source radiance (spectral radiance) up to a point and then decreases as $\overline{\Delta i_S}$ and $\overline{\Delta i_L}$ become the dominant noise sources. At low source radiance (spectral radiance), $\overline{\Delta i_D}$, $\overline{\Delta i_B}$, and $\overline{\Delta i_E}$ may be dominant. Thus the highest useful source radiance (spectral radiance) is approximately the value of the radiance (spectral radiance) resulting in $\overline{\Delta i_S}$ that is approximately equal to or greater than all other noise sources combined.

(4) *Flow Rates of Flame Gases,* Q_t. Gas flow rates affect both i_L and $\overline{\Delta i_T}$ by influencing flame temperature ($\overline{\Delta i_E}$ and $\overline{\Delta i_B}$), solution flow rate, aspiration

efficiency, and so on. Many workers have demonstrated that optimum values do exist and must be found experimentally.

(5) *Flame Height and Burner Position.*   Several workers have plotted atomic concentration contour plots of a variety of flame types and burner configurations. No theoretical treatment is practical; the signal-to-noise optimization must be done experimentally. Position variables affect not only the fluorescence signal i_L but also the noise due to flame background and source scattering (i.e., $\overline{\Delta i_B}$ and $\overline{\Delta i_S}$, respectively).

(6) *Fluorescence Power Yield,** Y_L.*   The value of $i_L/\overline{\Delta i_T}$ increases with Y_F, but Y_L is limited to a maximum value of unity; Y_L may be increased by diluting the flame gases with an inert gas (argon, for example).

(7) *Multiplier Phototube Supply Voltage.*   The variation of $i_L/\overline{\Delta i_T}$ with phototube voltage exhibits a very broad flat maximum, indicating that this parameter is not particularly critical over a considerable range. At very low voltages, however, amplifier noise, reading errors, and leakage currents may predominate; and at excessively high voltages regenerative ionization may occur (see Section III-A-5).

(8) *Phototube Load Resistance R_L and Amplifier Gain.*   Since both i_L and $\overline{\Delta i_T}$ are amplified by the same extent R_L and amplifier gain have little or no effect on $i_L/\overline{\Delta i_T}$. However, R_L must be large enough so that Johnson noise and amplifier noise are a negligible contribution to the total noise, and the amplifier gain must be high enough to eliminate reading errors.

All the above-mentioned parameters may be optimized theoretically if all the required spectral, flame compositional, and instrumental variables are known. Interdependent experimental variables may also be optimized by proper statistical experimental design.

d. ESTIMATION OF LIMITING DETECTABLE SAMPLE CONCENTRATIONS[21]

Once the various experimental parameters have been optimized, the limiting detectable atomic concentration n_M can be estimated theoretically by combining equations III-D-15 and III-D-16, substituting for B_L (using appropriate equation from Section II-C), and solving for n_M. The limiting detectable sample concentration C_m is then calculated by means of equation III-D-17.

---

* Recall from Section II-C that the designation Y_L is used to represent the fluorescence power yield.

3. MOLECULAR LUMINESCENCE SPECTROMETRY

a. SIGNAL AND SHAPE OF ANALYTICAL CURVE

The average photodetector signal current i_L due to molecular luminescence of the analyte (assuming that the spectral bandwidth s' of the emission monochromator is smaller than the luminescence band half width) is given by an equation similar to equation III-D-14:

$$i_L = \alpha W' K_o' s' \gamma B_{L\lambda} \qquad \text{(III-D-19)}$$

where α is the phosphoroscope factor (see Section III-C-3; no units), s' is given by $R_d' W'$ (nm), R_d' and W' are the reciprocal linear dispersion and slit width of the emission monochromator, and $B_{L\lambda}$ is the luminescence spectral radiance (watt cm^{-2} sr^{-1} nm^{-1}) and is approximately given by

$$B_{L\lambda} = \frac{B_L}{\Delta\lambda_G} \qquad \text{(III-D-20)}$$

where $\Delta\lambda_G$ is the half width of the luminescence band assuming a triangular profile (nm) and B_L is the luminescence radiance (watt cm^{-2} sr^{-1} nm^{-1}; see Section II-C). Equation II-D-19 is valid for either right angle or front surface illumination measurement modes (assuming that the proper equation for B_L is used). It should be stressed that in all commercial luminescence instruments a continuum source-excitation monochromator is used, and so B_L is given by equations II-C-30 or II-C-31 at low absorber concentrations or by equations II-C-33 or II-C-35 at high absorber concentrations. Thus B_L is directly proportional to the excitation radiance $B_N{}^o$, which is given by $B_{C\lambda_o}^o s$, where $B_{C\lambda_o}^o$ is the spectral radiance of the continuum source at the wavelength setting of the excitation monochromator and s is the spectral bandpass of the excitation monochromator (i.e., $R_d W$). Also, the cell dimensions are generally greater than the slit width W and the slit height of the excitation monochromator. Thus i_L depends on W^4 if it is assumed that the entrance and exit slits of both excitation and emission monochromators are identical (the normal case in luminescence spectrometers).

Furthermore, because B_L depends linearly on absorber concentration (at low absorber concentrations) so does i_L; this dependence illustrates the analytical usefulness of the molecular luminescence method for quantitative analysis. At high absorber concentrations, B_L becomes independent of analyte concentration (for the line source case of a continuum source excitation monochromator—see Section II-C) and so does i_L. Therefore, analytical curves [log (luminescence signal, i_L) versus log (analyte concentration)] have a slope of unity at low analyte concentrations and a slope of zero at high analyte concentrations. The latter region (slope of zero) is not analytically useful except instrumentally for quantum counters.

b. NOISE

The total rms current noise $\overline{\Delta i_t}$ in condensed phase luminescence spectrometry may be written[22]

$$\overline{\Delta i_T} = \sqrt{\overline{\Delta i_D}^2 + [(\zeta^2 + \varepsilon^2)^{1/2} + \varepsilon]^2[\Delta f(i_L + i_B)] + \overline{\Delta i_A}^2} \quad \text{(III-D-21)}$$

where $\overline{\Delta i_D}$ = total rms detector noise current, as given by equation II-A-45

ζ = source flicker factor (i.e., the ratio of rms fluctuation in the source radiance to the mean source radiance, measured at unit electrical bandwidth—$\Delta f = 1.0$ Hz)

ε = coolant flicker factor (i.e., the ratio of the fluctuation in the emitted radiance due to convection currents and bubbling in the coolant to the mean radiance of emission, measured at unit electrical bandwidth)

Δf = electrical bandwidth of amplifier-readout system, Hz

i_L = photocurrent due to analyte luminescence, A

i_B = photocurrent due to background, A

$\overline{\Delta i_A}$ = rms electronic noise generated by amplifier-readout system, referred to the input, A.

In this equation it is assumed that the source and coolant flicker noises are white, which may not be quite valid in general. For room temperature work, ε is probably close to zero, simplifying the equation considerably.

c. SIGNAL-TO-NOISE RATIO AND CHOICE OF EXPERIMENTAL CONDITIONS[23,24]

The experimental parameters that have been of concern in the optimization of experimental conditions in molecular luminescence spectrometry include the following.

(1) *Monochromator Slit Width W.* The signal-to-noise ratio $i_L/\overline{\Delta i_T}$ increases with W up to a point and then levels off. The best choice of slit width would then be the smallest value that does not significantly reduce $i_L/\overline{\Delta i_T}$ below the value achieved at large slit widths. A larger slit width is undesirable because of the loss of spectral selectivity. Rather large slit widths ($W > 1$ mm) are generally used in quantitative work. Smaller slits may be used in spectrum scanning to achieve better resolution, because it is possible to choose a somewhat more concentrated solution and thus be able to obtain the desired signal-to-noise ratio at a smaller slit width.

(2) *Electrical Bandwidth.*   Electrical bandwidth Δf has an effect similar to that in atomic fluorescence; namely, reducing Δf generally reduces random noise but at the expense of increasing the importance of several systematic errors. The total random rms noise $\overline{\Delta i}_T$ is proportional to $(\Delta f)^{1/2}$ only if $\overline{\Delta i}_T$ is white, which may not be the case in real systems. In particular, source flicker and coolant noise would not be expected to be white except at higher frequencies (e.g., above 100 Hz).

(3) *Phototube Voltage.*   Phototube voltage has the same effect as in atomic fluorescence systems.

(4) *Source Radiance.*   The value of $i_L/\overline{\Delta i}_T$ increases with $B_N{}^\circ$ (the radiance emerging from the exit slit of the excitation monochromator and striking the sample—see Section II-C) until source noise predominates, and at this point $i_L/\overline{\Delta i}_T$ levels off. At even higher values of $B_N{}^\circ$, photodecomposition of the sample may become a problem. Nevertheless, higher $B_N{}^\circ$ allows us to use narrower slit widths, thus increasing spectral resolution.

(5) *Power Yield Y_L.*   Just as for atomic fluorescence, $i_L/\overline{\Delta i}_T$ increases with Y_L. Some means of increasing Y_L are discussed in Section II-B-3-c.

(6) *Amplifier Gain.*   Amplifier gain has no effect on $i_L/\overline{\Delta i}_T$ as long as it is sufficient to avoid serious reading errors.

d. ESTIMATION OF LIMITING DETECTABLE SAMPLE CONCENTRATION

The limiting detectable sample concentration C_m may be estimated by substituting equations III-D-19 and III-D-20 into equation III-D-15, setting $S_m = i_L$, and solving for sample concentration. Experimental estimations of the flicker factors ζ and ε and of the background signal current i_B must be made. Winefordner, McCarthy, and St. John,[22] who have compared values of C_m calculated in this way to experimentally measured detection limits, found generally good agreement.

4. GLOSSARY OF SYMBOLS

a_s Area of luminescing sample, cm^2.
B_A Radiance absorbed by analyte, watt cm^{-2} sr^{-1}.
B_L Radiance luminesced by analyte, watt cm^{-2} sr^{-1}.
$B_N{}^\circ$ Radiance emerging from exit slit of excitation monochromator and striking analyte, watt cm^{-2} sr^{-1}.
C_m Limiting detectable analyte concentration (limit of detection), mole $liter^{-1}$.
C_o Sample (analyte) concentration, moles $liter^{-1}$.

e	Charge of electron, coulomb.
e_f	Flame gas expansion factor, no units.
f	Frequency of measurement system, Hz.
F	Solution transport rate with nebulizer in flame spectrometry, cm^3 sec^{-1}.
f_l	Lower frequency limit of measurement system, Hz.
f_u	Upper frequency limit of measurement system, Hz.
g_o	Statistical weight of lower (ground) state involved in atomic absorption transition, no units.
i_b	Photoanodic signal due to blank, A.
i_B	Photoanodic current due to background, A.
i_L	Photoanodic current due to luminescing analyte, A.
i_p	Photoanodic current (signal due to sample), A.
i_s	Photoanodic current due to sample, A.
K	Linear analytical sensitivity, $\mu g\ cm^{-3}\ A^{-1}$.
K_o	Optics factor, cm sr.
L	Emission (luminescence) path length, cm.
l	Absorption path length, cm.
m	Chopper factor, no units.
n	Number of measurement pairs, no units.
Q_t	Flow rate of gases into flames, $cm^3\ sec^{-1}$.
r	Monochromator scan speed, nm sec^{-1}.
R_d	Reciprocal linear dispersion of optical system, nm cm^{-1}.
RSD	Relative standard deviation, appropriate units.
s	Spectral bandpass of excitation monochromator, nm.
s'	Spectral bandpass of emission monochromator, nm.
S	Analytical signal, A.
S_m	Minimum detectable signal, A.
t	"Student" t, no units.
t_a	Effective averaging time, sec.
t_A	Sample aspiration time, sec.
t_r	Response time of measurement system, sec.
T_s	Transmittance of complete optical system, no units.
W	Excitation monochromator slit width, cm.
W'	Emission monochromator slit width, cm.
X_m	Minimum detectable amount of analyte, μg.
Y_L	Power yield of luminescence, no units.
Z_T	Electronic partition function, no units.
α	Phosphoroscope factor, no units.
β	Atomization efficiency in flames, no units.
γ	Photoanodic sensitivity of multiplier phototube, A $watt^{-1}$.
Δf	Electrical noise bandwidth, Hz.

D. SIGNAL-TO-NOISE RATIO THEORY   225

| Δf_j | Electrical noise bandwidth of stage j, Hz. |

Δf_j Electrical noise bandwidth of stage j, Hz.
Δf_t Electrical noise bandwidth of total system, Hz.
$\overline{\Delta i}$ Signal rms noise current, A.
$\overline{\Delta i}_A$ Amplifier rms noise current, A.
$\overline{\Delta i}_b$ Blank rms noise current, A.
$\overline{\Delta i}_B$ Background rms noise current, A.
$\overline{\Delta i}_D$ Detector rms noise current, A.
$\overline{\Delta i}_E$ Flame emission rms noise current, A.
$\overline{\Delta i}_F$ Atomic fluorescence rms noise current, A.
$\overline{\Delta i}_n$ Individual rms noise current, A.
$\overline{\Delta i}_p$ Phototube rms shot noise current, A.
$\overline{\Delta i}_s$ Sample signal rms noise current, A.
$\overline{\Delta i}_S$ Scatter rms noise current, A.
$\overline{\Delta i}_T$ Total rms noise current, A.
$\Delta \lambda'$ Half width of most narrow band or spectral feature, nm.
$\Delta \lambda_G$ Half width of luminescence band (Gaussian shape), nm.
ε Coolant rms flicker factor, $Hz^{-1/2}$
ε Aspiration efficiency in flame spectrometry, no units.
ζ Source rms flicker factor, $Hz^{-1/2}$
π 3.1416...
σ_s Standard deviation of difference of means, appropriate units.
Ω Solid angle of luminescence collected by emission monochromator, sr.

5. REFERENCES

1. E. J. Bair, *Introduction to Chemical Instrumentation*, McGraw-Hill, New York, 1962.
2. Brower Laboratories, Inc., *A Practical Guide to the Measurement of Signals Buried in Noise*, Westboro, Mass.
3. L. Smith and D. H. Sheingold, *Analog Dialog*, **3**, (1), Analog Devices, Inc., Cambridge, Mass. March, 1969.
4. I. G. Williams and H. C. Bolton, *Anal. Chem.*, **41**, 1755 and 1762 (1969).
5. J. Rolfe and S. E. Moore, *Appl. Opt.*, **9**, 63 (1970).
6. C. G. Cannon, *Electronics for Spectroscopists*, Wiley-Interscience, New York, 1960.
7. Radio Corporation of America, *RCA Phototube and Photocell Manual*, Harrison, N.J.
8. Electra Megadyne, Inc., *EMI Photomultiplier Tube Bulletin*, Los Angeles, Calif.
9. S. Rodda, *Photoelectric Multipliers*, MacDonald, London, 1953.
10. K. S. Lion, *Instrumentation In Scientific Research*, McGraw-Hill, New York, 1959.
11. G. Czerlinski and A. Weiss, *Appl. Opt.*, **4**, 59 (1965).
12. C. A. Nittrouer, *Elec. Inst. Digest*, October, 1968.
13. O. C. Chaykowsky and R. D. Moore, *Research/Development*, p. 32, April, 1968.
14. T. Coor, *J. Chem. Educ.*, **45**, A533 and A583 (1968).

15. J. D. Winefordner, W. J. McCarthy and P. A. St. John, *J. Chem. Educ.*, **44**, 80 (1967).
16. P. A. St. John, W. J. McCarthy and J. D. Winefordner, *Anal. Chem.*, **39**, 1495 (1967).
17. M. L. Parsons, *J. Chem. Educ.*, **46**, 290 (1969).
18. J. D. Winefordner, M. L. Parsons, J. M. Mansfield, and W. J. McCarthy, *Spectrochim. Acta*, **23B**, 37 (1967).
19. P. J. T. Zeegers, R. Smith, and J. D. Winefordner, *Anal. Chem.*, **40**, (*11*), 26A (1968).
20. M. L. Parsons, W. J. McCarthy, and J. D. Winefordner, *J. Chem. Educ.*, **44**, 214 (1967).
21. J. D. Winefordner, M. L. Parsons, J. M. Mansfield, and W. J. McCarthy, *Anal. Chem.*, **39**, 436 (1967).
22. J. D. Winefordner, W. J. McCarthy, and P. A. St. John, " Phosphorimetry as an Analytical Approach in Biochemistry," in *Methods of Biochemical Analysis*, Vol. 15, D. Glick, Ed., Wiley-Interscience, New York, 1967.
23. J. J. Cetorelli, W. J. McCarthy, and J. D. Winefordner, *J. Chem. Educ.*, **45**, 98 (1968).
24. P. A. St. John, W. J. McCarthy, and J. D. Winefordner, *Anal. Chem.*, **38**, 1828 (1966).

E. METHODOLOGY

1. ATOMIC FLUORESCENCE FLAME SPECTROMETRY[1-4]

a. MEASUREMENT OF ATOMIC FLUORESCENCE SPECTRA

Although it is seldom done, it is possible to measure atomic fluorescence spectra by using a continuum source, an automatic wavelength scanning emission monochromator, and a strip-chart recorder synchronized with the scan rate. To obtain a true fluorescence spectrum with peak heights independent of the instrumental system, it is necessary to multiply each line height (signal) by the product $(\gamma_\lambda \cdot B_{C\lambda} \cdot T_{s\lambda'})^{-1}$, where γ_λ is the photoanodic sensitivity of the multiplier phototube at the wavelength of the line of concern, $B_{C\lambda}$ is the spectral radiance of the continuum source at the wavelength of the line of concern, and $T_{s\lambda'}$ is the transmittance of the entire spectrometric system at the wavelength of concern. To avoid the confusing admixture of fluorescence spectra among any DC incident light scattering and flame background, it is best to modulate the source and use a tuned (AC) detector sytsem. If a DC measurement system is employed, it is best to choose a small monochromator slit width to maximize the line-to-continuum ratio.

Often when a line source is used, wavelength scanning is performed only over a small wavelength interval about the fluorescence line; this procedure isolates the fluorescence line from emission lines and emission bands due to the flame gases (when using a DC measurement system).

b. MEASUREMENT OF ATOMIC FLUORESCENCE POWER YIELD

Pearce, de Galan, and Winefordner[5] have described a method to measure fluorescence power yields of atoms in flames, and have given an expression for the fluorescence power efficiencies Y_L as a function of measurable experimental variables. A suitable experimental set-up includes a 900-watt xenon arc lamp excitation source mounted on a radial arm to allow accurate measurements of both the absorption and luminescence signals. The power yield Y_L is given by

$$Y_L = \left(\frac{i_L}{\Delta i}\right)\left(\frac{WH}{W'H'}\right)\left(\frac{4\pi}{\Omega}\right)\left(\frac{A_L}{A_I}\right) \qquad \text{(III-E-1)}$$

where i_L = readout signal due to atomic fluorescence

Δi = difference signal observed in the absorption measurement

W and H = monochromator slit width and height for absorption measurement

W' and H' = monochromator slit width and height for fluorescence measurement

Ω = solid angle of radiation incident on the monochromator entrance slit

A_L = flame area from which fluorescence is observed

A_I = area of image of xenon arc on the entrance slit of monochromator.

Power yields vary considerably from one spectral line to another. As a result of entrainment of ambient air into the turbulent flames, the values obtained in different flames are approximately the same (with turbulent flames).

Table III-E-1 presents the influence of flame gas composition on quantum yields of lithium, sodium, potassium, thallium, and lead. The quantum yields in Table III-E-1 were calculated from quenching cross sections measured by Jenkins (see table for references) for the specified elements in highly laminar, shielded flames.

c. MEASUREMENT OF ATOMIC FLUORESCENCE LIFETIMES

The lifetimes of the excited states of several elements have been measured by a variety of techniques. No measurements of metal atom fluorescence lifetimes in analytical flames seem to have been measured, although existing techniques could presumably be used.

In the "phase-shift" method,[6] the atom being studied is optically excited

**Table III-E-1. Fluorescence Quantum Yields for Several
Atoms in Several Flames**[a]

Flame Composition[b]	Temperature (°K)	Fluorescence Quantum Yields				
		Li	Na	K	Tl	Pb
$2H_2/O_2/10Ar$	1800	0.15	0.75	0.37	0.33	0.22
$6H_2/O_2/4N_2$	1800	0.105	0.049	0.049	0.099	0.10
$2H_2/O_2/4N_2$	2100	0.021	0.066	0.047	0.070	0.079
$H_2/O_2/4N_2$	1600	—	0.044	0.03	0.051	0.069
$0.4C_2H_2/O_2/4N_2$	2200	0.017	0.042	0.028	0.042	0.067

[a] Taken from D. R. Jenkins, *Proc. Roy Soc.* (*London*), **A293**, 493(1966); *Ibid.*, **303**, 453 (1968); *Ibid.*, **303**, 467 (1968); *Ibid.*, **306**, 413 (1968).
[b] Flow rate ratios of gases.

to the desired resonance state by means of an intensity-modulated excitation source. The phase shift ϕ between the exciting light and the fluorescence light is measured electronically and related to the lifetime τ by means of the relation

$$\tau = \frac{\tan \phi}{2\pi f_s} \qquad \text{(III-E-2)}$$

where f_s is the source modulation frequency. Phase-sensitive (synchronous or lock-in) amplifier systems are used. Modulated electron beams have also been used for excitation. The phase shift works well for simple exponential (first-order) decay schemes. However, an error is introduced if radiative cascading from higher states results in a more complicated decay behavior.

In the "method of delayed coincidence" a repetitively pulsed electron beam is used to excite the gaseous atoms.[7] Each electron beam pulse also starts a time-to-pulse-height converter. Each fluorescence light pulse, slightly delayed with respect to the excitation pulse, is detected in the usual way and turns off the time-to-pulse-height converter. The resulting pulse height, proportional to the time delay, is sorted and stored in a multichannel pulse-height analyzer. In this way, a distribution of delay times is built up in the memory of the analyzer, which is read out when a sufficient number of photons has been accumulated.

Direct measurement of the decay of excited states is also possible.[8] Individual decays may be photographed by means of a high speed oscilloscope with camera.

d. MEASUREMENT OF ANALYTE CONCENTRATION[9,10]

Atomic fluorescence spectrometry, like most optical methods, is a *relative technique* requiring standard solutions in order to determine concentrations of unknown samples. Three types of samples or solutions are required: the *analytical sample*, which contains the analyte, along with solvent, matrix species, and interferents; the *blank* or *control*, which strictly speaking should contain everything in the analytical sample *except* the analyte; and one or more *standards*, which are similar in composition to the analytical sample except that the analyte concentration is known.

The preparation of a satisfactory blank solution is seldom very difficult in atomic spectroscopic techniques because the spectral specificity is relatively high. The appearance of specific line fluorescence at the analytical wavelength is usually due to the analyte itself, and so a simple reagent blank solution is often a satisfactory means to correct for interferences, particularly for analytes in solutions containing a matrix of high volatility. In some cases, however, particularly if the analyte is present in a solution containing a matrix of low volatility, chemical interferences can be appreciable.* Spectral interferences can also be appreciable when the analyte is present in a multi-element sample and excitation proceeds via a continuum source*. With line sources, spectral interferences are generally negligible.*

The most common method of calibration in flame fluorescence spectrometry involves the preparation of a series of standard solutions (*external standards*) of known analyte concentration. To minimize reagent matrix errors, all standards must be made up to the same reagent concentration as the sample itself.[10] The fluorescence signals are measured for all the standards and for the blank solution. The blank reading is subtracted from each standard reading, and the *analytical curve* is prepared by plotting the blank-corrected standard signals versus the volumetric concentration of the standards, usually on log-log paper. Linear analytical curves over four decades are not uncommon. The concentration of the analytical sample (unknown) is then interpolated from the analytical curve. For greater precision in a particular analysis, it is possible to prepare a pair of standards bracketing the estimated sample concentration.

The principal disadvantage of the *external standard method* (analytical curve) is that uncontrolled change(s) (measurement errors) in the experimental conditions may cause serious errors until a new analytical curve is prepared. This is true in all types of luminescence spectrometry, because there are so

---

* Refer to Section IV-A for a thorough discussion of factors influencing sensitivity and selectivity of analysis.

many experimental variables which effect the measured radiance of fluorescence (e.g., source radiance, flame parameters, slit widths, phototube gain). This disadvantage can be overcome to some extent by the use of the *internal standard* method,[10] in which a constant concentration of some reference element is already present or is added to all samples and standards. The *ratio* of signal due to the analyte to that due to the internal standard (reference) element is measured. Ideally, any variation in experimental conditions will affect the sample and internal standard signals similarly, so that the ratio will remain unchanged. The selection of the internal standard element and line is critical; obviously the internal standard and analyte should be subject to similar interferences. No research has been published yet involving selection of internal standard elements and lines for particular analyte elements and lines.

Matrix effects may be reduced or eliminated by the use of the *standard addition* method, wherein a small volume of a relatively concentrated standard solution is added to an aliquot of the analytical sample. The resulting solution serves as a standard. If the analytical curves are linear in the concentration range of concern, only a single addition standard need be prepared. The concentration of the analyte C_a is given by

$$C_a = \frac{i_s C_s V_s}{i_s (V_a + V_s) - i_a V_a} \qquad \text{(III-E-3)}$$

where C_a and C_s are the concentrations of the analytical sample and the original standard solution, respectively; V_a and V_s are the volumes of the sample aliquot and standard solution added to the sample aliquot, respectively; and i_a and i_s are the blank-corrected relative instrument responses for the analytical sample and for the addition standard solution, respectively. This assumes that the instrument response is directly proportional to concentration: if the matter is in doubt, a series of addition standards of increasing concentration in analyte may be prepared.

The standard addition method has the significant advantage that the effects of matrix species on the analytical signal will be the same as that on the added standard material. Thus it is not necessary to have detailed knowledge of the sample matrix. However, proper application of the standard addition method does require that some conditions be met:

1. A proper blank solution must be used; the standard addition method will not reduce blank errors.

2. The sample aliquot must not be significantly diluted by the addition of the standard solution.

3. The standard solution must contain the same form of analytical element as present in the sample (e.g., with respect to oxidation state, complexation).

4. For the single-addition procedure, the instrument response must be proportional to concentration.

Most of these conditions are relatively easily met in the flame fluorescence spectrometry as well as other flame methods.

e. SOURCES OF ERROR IN MEASURING CONCENTRATIONS*

Any quantitative analysis by an instrumental method is composed of four steps leading to errors: (a) the measurement of the signal which is a measure of the analyte concentration and the concomitant errors due to noise and drift in the signal, (b) the estimation of the portion of the signal due to the analyte and the concomitant errors associated with estimation of the blank; (c) the estimation of the signal due to the analyte in a real sample and the errors associated with the influence of the sample matrix upon the analyte signal; and (d) a fourth source of error is in the calibration procedure.

(1) *Signal Reading Errors.* Signal reading errors are often the largest and most important source of errors at very low concentrations where the signal-to-noise ratio approaches unity (i.e., near the limiting detectable signal). Methods of increasing the signal-to-noise ratio by optimization of experimental conditions have been discussed in Section III-D-3-c.

(2) *Blank Errors.* In most practical applications of atomic and molecular luminescence spectrometry with real samples, blank errors can be significant sources of error, especially for concentrations above the detection limit. It is *always* desirable to have the smallest blank signal possible to avoid an error-prone subtraction of a large blank signal from an only slightly larger blank plus sample signal. In real analyses, it is seldom possible to prepare a *true blank* (i.e., a blank containing *everything except the analyte.* Generally we use a *solvent blank*—a solution containing the solvent used for the analyte—or a *reagent blank*—a solution containing the solvent and all reagents used in the preparation of the sample solution.

Blank errors primarily arise as a result of spectral interferences† in luminescence spectrometry. These interferences are attributable to the blank, and they produce an erroneous analyte signal and therefore an erroneous estimation of analyte concentration.

---

* These sources of error also apply to molecular luminescence spectrometry. Errors imply systematic (nonrandom) events rather than random events such as noise.

† Spectral interferences are due to overlap of lines or bands of analyte and interferent(s). Nonspectral interferents are ones that influence production of the measured analyte species. Interferences in atomic fluorescence and in solution fluorimetry and phosphorimetry are discussed in detail in Sections IV-A-2 and IV-B-2, respectively. Interferences can lead to either an *enhancement* or a *depression* of analyte signal.

(3) *Matrix Errors.*   Matrix errors are generally the most difficult ones to compensate for or to minimize because these result in an erroneous signal for the analyte in the sample which *cannot* be corrected by a simple reagent blank correction. Such errors are caused by spectral and nonspectral interferences associated with the matrix.† Of course, a *true blank* would allow compensation for the matrix effect. However, even a true blank would not compensate or correct for random variation in matrix content.

(4) *Calibration Errors.*   One of the prime requirements in any analytical procedure is a reliable analytical curve or some means to relate the analyte signal to analyte concentration. If several samples are measured, an analytical curve is generally used; if only one or two samples are measured, it is necessary to choose one of the other comparison procedures. In either approach, it is critical that the standard solution(s) be properly prepared from pure, stable materials and be in the same form with regard to oxidation state, complexation, isomerization, and so on, as the analyte in the unknown sample. Of course, the usual considerations of volumetric and gravimetric precision and accuracy in preparation of solutions and samples must be observed. If the same relative error is made in all standards, then the analytical curve will have the correct shape but the wrong position. Generally errors in preparation of standard solutions do *not result* in the wrong *shape* of analytical curve.

2. MOLECULAR LUMINESCENCE SPECTROMETRY[11–14]

a. MEASUREMENT OF LUMINESCENCE EXCITATION AND EMISSION SPECTRA

Fluorescence of phosphorescence excitation spectra are measured experimentally by adjusting the emission monochromator to the wavelength of maximum luminescence emission and recording the output signal (relative luminescence emission signal) as a function of excitation wavelength λ. On an uncorrected luminescence spectrometer, the excitation spectrum will depend not only on the absorption spectrum of the sample but also on the transmission factor of the excitation monochromator T_λ and the spectral distribution of the excitation source radiancy $B_{C\lambda}$. This generally means that uncorrected excitation spectra are similar in shape to absorption spectra but are skewed to longer wavelength, because $T_\lambda B_{C\lambda}$ increases with wavelength in the visible-ultraviolet region. To correct excitation spectra, all luminescence signals must be multiplied by $(T_\lambda B_{C\lambda})^{-1}$ and replotted versus the excitation wavelength λ. Several commercial instruments accomplish this process automatically (see Section III-C-4).

Emission spectra are measured by adjusting the excitation monochromator to the wavelength of maximal excitation (as determined from the excitation spectrum) and recording the output signal (relative signal of luminescence emission) as a function of emission wavelength λ'. On an uncorrected instrument, the emission spectrum will depend not only on the spectral radiance of emission $B_{L\lambda'}$ of the sample but also on the transmission factor of the emission monochromator $T_{\lambda'}$ and the spectral response of the multiplier phototube $\gamma_{\lambda'}$. As a result, uncorrected spectra may be used as a qualitative fingerprint of organic materials only for a particular instrument or group of similar instruments. To correct emission spectra, all luminescence signals must be multiplied by $(T_{\lambda'}\cdot\gamma_{\lambda'})^{-1}$. Commercial instruments are available to perform this process automatically (see Section III-C-4).

For pure compounds, the shapes of the fluorescence and phosphorescence emission spectra are independent of the wavelength of excitation, although, of course, the radiance of emission is a function of the excitation wavelength. Conversely, the shape of the excitation spectrum is independent of the wavelength of emission. Corrected excitation spectra are identical to or very similar to the absorption spectra. *These facts may be used as criteria of purity for organic compounds*[12].

The effect of the monochromator slit width must be considered when spectra are being recorded. Excitation spectra should be recorded with a sufficiently narrow excitation monochromator slit width to obtain the desired resolution; the emission monochromator slit width may be made considerably wider, however, in order to obtain the intensity required for good signal-to-noise ratio. Conversely, emission spectra may be recorded with a narrow emission monochromator slit width and a wide excitation monochromator slit width. In either case, the slit widths should not be so wide that the monochromators are allowed to pass adjacent interferent bands.

Some types of luminescent compounds exhibit particularly sharp bands. The emission spectra of many fused-ring aromatic hydrocarbons measured in crystalline n-paraffin solid solutions at 77°K show remarkably well-resolved fine structure. Band half widths of the order of 10 Å have been observed. Obviously, an emission monochromator of high resolution is required in these cases.

The effect of the response time of the amplifier-readout system on the recording of spectra must be considered. As discussed in Section III-D-1-d, the scan speed (nm sec^{-1}) must be no larger than $\Delta\lambda'/t_R$, where $\Delta\lambda'$ is the half width (nm) of the most narrow band or spectral feature to be observed and t_R is the system response time (sec). Most commercial luminescence spectrometers feature variable scan speed.

A special effect, which must be considered in recording phosphorescence or slow fluorescence *excitation* spectra, is the effect of the luminescence decay time on the instrument response time. The sample decay time adds quadratically to the response time of the amplifier-readout system. The total system response time t_R, including the effect of the luminescence lifetime τ is given by

$$t_R = \sqrt{t_r^2 + 16\tau^2} \qquad \text{(III-E-4)}$$

where t_r is the response time of the amplifier-readout system alone and is equal to the reciprocal of the electrical noise bandwidth. It is t_R (not t_r alone) which must be used to determine the maximum permissible scan speed when the luminescence *excitation* spectra of luminescent species of long lifetimes are being recorded.

b. MEASUREMENT OF LUMINESCENCE QUANTUM YIELDS

The luminescence quantum yield Y of a molecular species may be determined experimentally by dividing the area under the corrected emission spectrum by the area under the corrected excitation (absorption) spectrum. The power yield Y_L is given by

$$Y_L = \frac{\lambda_o}{\lambda_o'} Y \qquad \text{(III-E-5)}$$

where λ_o is the wavelength of maximum absorption and λ_x' is the wavelength of maximum emission.

The experimentally measured fluorescence quantum yields of a number of organic molecules have been listed by Parker.[12] Reported quantum yield values typically fall between 0.1 and 1.0 for analytically useful determinations.

In the determination of phosphorescence quantum yields with an instrument using a mechanical phosphoroscope, the emission intensity must be divided by the phosphoroscope α factor, given by equation III-B-1. Table III-E-2 presents a number of experimentally measured fluorescence and phosphorescence quantum yields and lifetimes of several aromatic hydrocarbon molecules in several different solvents and at several temperatures. By comparison of the data, the influence of environment on quantum yields and lifetimes can be determined.

c. MEASUREMENT OF LUMINESCENCE LIFETIME

Since the decay time of the prompt (normal) fluorescence of organic molecules is generally of the order of several nanoseconds, techniques for its measurement are similar to those used for the measurement of atomic fluorescence lifetimes (see Section III-E-1-c). These include the phase-shift technique[15] and the stroboscopic technique.[16] Special high-speed flash tubes, capable of generating subnanosecond light pulses, are used in the stroboscopic method.[17] Commercial flash lamp systems of this type are available (TRW Model 31A Nanosecond Spectral Source, TRW Instruments, El Segundo, Calif.; and the Xenon Model 437 Nanopulser and 783A Nanopulse lamp, Xenon Corp., Medford, Mass.).

Phosphorescence lifetimes are almost always much longer than fluorescence lifetimes and are therefore easier to measure. The simplest method is to measure the phosphorescence emission signal as a function of time after complete termination of the excitation radiation. For lifetimes longer than about 1.0 sec, a strip-chart recorder may be used. For faster decay times, a wide band oscilloscope is employed. Obviously, the response time of the readout system must be much smaller than the sample decay time. The lifetime is defined as the time required for the measured luminescence signal to decay from any given value to $1/e$ ($= 0.368$) of that value. The exciting radiation must be cut off quickly; an electrically operated guillotine shutter is sometimes used. Lifetimes between 10^{-4} and 10^{-1} sec may be measured by using the rotating phosphoroscope can or disk as the shutter and observing the exponential emission light pulses on an oscilloscope. Lifetimes much faster than about 10^{-4} sec may be measured by the stroboscopic technique or by the direct observation of the luminescence decay on an oscilloscope, following excitation by a short pulse of light.

Most pure organic compounds exhibit first-order (exponential) decays. In such cases, a plot of the log luminescence signal versus time during the decay yields a straight line. Any nonlinearity in this plot is a highly sensitive indication of the presence of impurities. Although an exponential decay plot may be converted to a log plot point by point, it is much more convenient to use an electronic log converter in the circuit ahead of the recorder.[18] The frequency response of most log converters is somewhat limited, however, and so care must be taken to ensure that the converter itself does not distort the decay curve. Very fast decays may be stored in a transient storage unit (e.g., Biomation 610 Transient Recorder, Palo Alto, Calif.) and then read out slowly through a log converter. The slope of a log decay plot is a convenient measure of the lifetime. The time required for the log plot to cover one log cycle (one decade) of luminescence signal is measured and divided by 2.3 to get the lifetime τ.

A very simple method for characterizing exponential waveforms without the use of a log converter has been described by Huen.[19] An analog integrator and an X-Y oscilloscope or X-Y recorder serve to obtain a plot of the phase-plane trajectory of the waveform. Departures from simple exponentiality are easily discerned, and the lifetime is readily measured.

Refer to Table III-E-2 for a listing of fluorescence and phosphorescence lifetimes of several aromatic hydrocarbons in several solvents at different temperatures. The influence of environments on lifetime of luminescence can be ascertained by comparing the data in Table III-E-2.

Table III-E-2. Fluorescence and Phosphorescence Quantum Yields and Lifetimes of Several Aromatic Hydrocarbons in Solution at Several Temperatures[a]

Substance	Solvent	Temperature (°K)	$Y_F{}^b$	$Y_P{}^b$	τ_F (sec)[c]	τ_P (sec)[c]
Benzene	Ethanol	77	—	—	—	$3.6\text{--}16^d$
	Ethanol	293	0.035	—	13×10^{-9}	—
	n-Hexane	293	0.11	—	5.726×10^{-9d}	—
	EPA	77	0.21	0.19	—	8.0
Coronene	Isopropanol	77	—	—	—	8.6
	n-Hexane	77	—	—	—	9.0
	Chloroform	293	0.3	—	4.7×10^{-7}	—
	Benzene	293	0.3	—	2×10^{-7}	—
	Polymethylmethacrylate	77	—	0.3	—	—
Chrysene	EPA	77	—	—	—	9.4
	EPA	77	—	—	—	2.2
	n-Heptane	77	—	—	—	2.2
	Isopropanol	77	—	—	—	2.65
	Polymethylmethacrylate	77	—	0.13	—	—
Anthracene	Glycerin	293	—	—	—	3.7×10^{-2}
	n-Hexane	298	0.31	—	14×10^{-9}	6.3×10^{-3}
	n-Hexane (O$_2$-free)	293	0.463	—	—	9×10^{-4}
	Benzene	298	0.241	—	$4\text{--}14.3 \times 10^{-9d}$	9.1×10^{-3}
	Toluene	—	0.27	—	$4.4\text{--}14.5 \times 10^{-9d}$	—
	EPA	77	—	0.09	—	9×10^{-2}
	Polystyrene	—	0.26	—	5.2×10^{-9}	—
	Cyclohexane	293	—	—	6.2×10^{-9}	—
9,10-Diphenylanthracene	Ethanol	293	0.74–0.81	—	7.91×10^{-9}	—
	Benzene	298	0.85	—	$7.3\text{--}8.5 \times 10^{-9d}$	—
	Chloroform	293	0.65	—	—	—

1,2-Benzanthracene	Kerosene	293	0.74	—	—	—
	Polymethylmethacrylate	298	0.83	—	—	—
	Cyclohexane	293	0.19	—	4.5×10^{-8}	—
	n-Hexane	77	—	0.001	—	1.6×10^{-9}
	EPA	77	—	0.3	—	—
	Propylene ether/isopentane	83	—	0.4	—	—
Eosin	H_2O (basic)	293	0.23	—	—	—
	H_2O	298	0.12	—	4.7×10^{-9}	—
	Ethanol (disodium salt)	77	0.87	0.024	—	8.9×10^{-3}
	Ethanol (disodium salt)	293	0.74	—	3×10^{-9}	—
	Ethanol (dianion)	293	0.80	—	—	—
	Ethanol (dianion)	93	0.93	—	—	—
	Glycerin	77	0.48	0.064	—	1.07×10^{-2}
	Glycerin	293	0.53	—	3×10^{-9}	—
Fluorene	Ethanol	293	—	—	—	5.7–6.9
	Isopropanol	77	0.54	—	—	—
	n-Hexane	293	—	—	1.1×10^{-8}	—
	Cyclohexane	293	—	—	—	5.1
	n-Heptane	77	0.54	—	—	7.1
	EPA	77	0.85	0.07	—	—
Fluorescein	H_2O (basic)	293	0.85	—	$3.9\text{–}4.4 \times 10^{-9}$	—
	H_2O	293	0.65	—	4.7×10^{-9}	—
	Glycerin	293	1.0	—	4.5×10^{-9}	—
	Ethanol	293	—	—	4.4×10^{-9}	—
	Methanol	293	0.19	—	4.7×10^{-9}	—
	Ethanol (cation)	293	1.00	—	—	—
	Ethanol (cation)	93	1.00	—	—	—
	Ethanol (dianion)	293	1.00	—	—	—
	Ethanol (dianion)	93	—	—	—	—
	Boric acid	293	—	—	3.8×10^{-9}	0.65
	Boric acid	173	—	—	4.6×10^{-9}	2.4
	Boric acid glycerin	293	—	—	4.35×10^{-9}	0.85

Table III-E-2 (*Continued*)

Substance	Solvent	Temperature (°K)	Y_F^b	Y_P^b	τ_F (sec)^c	τ_P (sec)^c
	Boric acid glycerin	93	—	—	—	2.8
	Plexiglass	293	—	—	—	2.7×10^{-3}
Naphthalene	Ethanol	293	0.12	—	2.7×10^{-9}	5.57×10^{-5}
	Isopentane	293	—	—	—	2.5
	Isopentane	77	—	—	—	3.3
	Ethyleneglycol	293	—	—	—	1.0×10^{-3}
	n-Heptane	77	—	—	—	2.21
	n-Hexane	293	0.10	—	$8.3\text{--}103 \times 10^{-9d}$	—
	Paraffin	293	—	—	—	$1.2\text{--}2.8 \times 10^{-9}$
	Cyclohexane	293	—	—	9.5×10^{-8}	—
	EPA	77	0.39	0.008	3.3×10^{-5}	2.33
	Ether/ethanol	77	0.29	0.03	3.3×10^{-6}	2.15
	Isooctane	77	—	—	—	2.45
	Isopropanol	77	—	—	—	2.45
Perylene	Ethanol	298	0.84	—	—	—
	Benzene	293	0.89	—	$4.9\text{--}5.1 \times 10^{-9}$	—
	Petroleum ether	298	0.79	—	—	—
	Polymethylmethacrylate	298	0.87	—	—	—
Phenanthrene	Ethanol	293	0.10	—	19×10^{-9}	$0.9\text{--}7.9 \times 10^{-3d}$
	Hexane	77	—	0.23	—	9.0×10^{-4}
	EPA	77	0.14	0.11	—	3.8
	Ethyleneglycol	293	—	—	—	9.1×10^{-4}
	n-Heptane	77	—	—	—	3.35
	Isooctane	77	—	—	—	3.55
	Isopropanol	77	—	—	—	3.65
	Ethanol/ether	77	—	0.14	—	3.3

Compound	Solvent	Temperature	Y_F	Y_P	τ_F	τ_P
	Isopentane/methyl-cyclohexane	77	—	—	—	3.7
	Propylether/isopentane	83	—	—	—	3.0
	Polymethylmethacrylate	77	—	0.31	—	—
	Boric acid	293	—	—	—	1.5
Pyrene	Ethanol (O_2-free)	296	0.65	—	1×10^{-7}	—
	Paraffin 0.1	293	0.36–0.69[d]	—	—	—
	Cyclohexane	293	0.65–0.75[d]	—	2×10^{-8}	—
	EPA	77	—	0.2	—	—
	p-Xylenol	293	0.61	—	9.1×10^{-8}	—
	Polymethylmethacrylate	298	—	—	—	—
p-Triphenyl	Cyclohexane	293	0.36	—	2.7×10^{-9}	—
	n-Heptane	—	—	—	—	—
	Benzene	—	0.065	—	—	—
	Benzene (O_2-free)	302	—	—	2.2×10^{-9}	—
	Toluene	298	—	—	$1.6–13 \times 10^{-9}$[d]	—
	Polystyrene	293	0.75	—	—	—
Triphenylene	n-Hexane	293	—	—	—	5.5×10^{-5}
	EPA	77	0.06	0.28	—	17.1
	Polymethylmethacrylate	77	—	0.46	—	—
	Benzene (O_2-free)	293	—	—	—	5.6×10^{-5}

[a] Data taken from *Landolt Börnstein*, Vol. 3, by A. Schmillen and R. Legler, R. H. Hellwege and M. Hellwege, Eds., Springer, Berlin, 1967.

[b] Y_F = *quantum* yield of fluorescence. Y_P = *quantum* yield of phosphorescence.

[c] τ_F = lifetime of fluorescence. τ_P = lifetime of phosphorescence.

[d] A range of values designates either different results for the same solution or different results for slight variations of the same solution (e.g., concentrations of analyte or slight solvent changes such as 95% ethanol rather than absolute ethanol).

d. MEASUREMENT OF CONCENTRATION*

(1) *Methods.* The measurement of the concentration of molecular species in luminescence spectrometry may be performed by several distinct experimental methods. The simplest and most straightforward, applicable only to intrinsically luminescent analytes, is the *direct method*, which involves the measurement of the luminescence signal of the analyte itself either with or without prior separation from interfering substances. Alternately, a nonluminescent or weakly luminescent substance may be converted into a form more suitable for luminescence analysis by means of appropriate chemical reactions. Such methods are called *chemical methods*. If the substance to be determined is nonluminescent and yet possesses the ability to quench the luminescence of some luminescent compound, then the substance may be determined indirectly by the measurement of the reduction in luminescence intensity of the luminescent compound. Such a process would constitute a *quenching method*. Finally, there is the possibility of *energy transfer methods*, which involve the absorption of excitation light by a *donor* species and the transfer of the energy from the donor to an *acceptor* species, which may be either another molecule (intermolecular) or another portion of the same molecule (intramolecular).†

(2) *Analytical Curve.* All the foregoing methods are relative methods and, as such, require some sort of calibration procedure. The most common calibration method is the analytical curve method discussed previously, which involves the preparation of a series of standard solutions of the analyte. These standards are then treated as the unknown samples themselves will be treated, according to the particular measurement method used. Thus, in the direct method, the luminescence signal of each standard is measured directly. In the chemical method, appropriate reagents are added to each standard, and the mixture is allowed to react before the luminescence signal is measured. In the quenching method, suitable reagents are added to each standard before the luminescence signal is measured. In the intermolecular energy transfer methods, a suitable acceptor or donor species must be added. The relative luminescence signal of each of the standard solutions prepared in the above-mentioned way is measured, blank-corrected, and plotted versus the analyte concentration, usually on log-log coordinates. Luminescence analytical curves are typically linear over several decades of concentration, have log-log slopes near unity, and exhibit a plateau region at high analyte concentrations

---

* Sources of error in quantitative measurements and methods of measuring atomic concentrations in atomic fluorescence also apply to molecular luminescence spectrometry—see Sections III-E-1-d and III-E-1-e; also refer to Section IV-B-2 for a more thorough discussion of interferences in molecular luminescence spectrometry.

† See Section IV-C for a more thorough discussion of types of luminescence.

(see Section II-C). Unknown sample solutions are treated and measured in exactly the same way as the standards, and their concentrations are interpolated from the analytical curve. Linear analytical curves are highly desirable, because interpolation from a linear plot is relatively easy.

(3) *The Blank.* Only in the most ideal analysis are all the measured light emissions from the sample due to the analyte. The purpose of the blank solution is to correct for absorbing species and luminescent species other than those related to analyte concentration.

All readings of the luminescence signals of samples and standards must be blank corrected by subtracting the signal reading given by the blank solutions. Several types of blank solutions are used in practice. The ideal or *true blank* would in principle contain everything contained in the unknown samples in the same concentrations as in the unknown samples *except* the analyte. Needless to say, a true blank is seldom possible for real analyses in complicated systems.

Thus some approximation to a true blank is usually made. The simplest and least satisfactory is a *solvent blank*, consisting merely of the solvent used to make up the standards and dissolve and dilute the samples. Such a blank would correct only for luminescence impurities in the solvent and would be suitable only in the direct measurement of a luminescent compound uncontaminated by luminescence impurities. More satisfactory is a *reagent blank* that contains, in addition to the solvent, each of the various reagents in the same concentrations used in the treatment of samples and standards. A reagent blank is useful in the chemical, quenching, and energy-transfer methods, as it corrects for absorption by and luminescence of impurities in the added reagents. Neither a solvent blank or a reagent blank, however, corrects for the absorption by and luminescence of contaminants and matrix substances originally in the sample itself. If these interferences cannot be distinguished spectrally, a prior chemical and/or physical separation may be necessary.

If a separation step is impossible or undesirable, it may be feasible to compensate for the presence of interfering substances by the preparation of an *internal blank*, which is very nearly a true blank. An internal blank is produced by measuring the luminescence signal of a sample and then adding a non-luminescent compound that specifically reacts with the analyte to yield products that are not luminescent, at least at the excitation and emission wavelengths used for the analysis. Alternately, a compound that specifically quenches the analyte luminescence may be used. In either case the resulting solution serves as the blank, after correction for dilution if necessary. This technique is sometimes used in clinical analyses. For instance, in the fluorescence analysis of urinary estrogen, an internal blank is provided by the addition of nitrate ion, which specifically destroys the luminescence of the estrogens.[20] The highly specific reactions between enzymes and their substrates may also be used as a basis for the luminescence analysis of either the enzyme or the substrate if either is luminescent.

Other methods of compensating for matrix interferences are the internal standard and the standard addition methods previously described for atomic fluorescence flame spectrometry (see Section III-E-1-e). If the instrumental sensitivity should change between the time that the analytical curve is measured and the time that the sample is measured (e.g., aging of the excitation lamp, fatigue of the multiplier phototube), then it is possible to use a *reference solution* to make the necessary correction. The reference solution should be an appropriately concentrated solution of a very stable, easily purified luminescent species that absorbs and luminesces in the approximate wavelength range of interest, (e.g., a dilute solution of quinine sulfate could be used in normal fluorimetry). The concentration of the reference species should be above the limit of detection and below the region of nonlinearity of the analytical curve, that is, on the linear portion of the curve. Therefore, after an analytical curve is measured or during the measurement of the analytical curve, the reference solution is also measured. Then, when unknown samples are to be measured at some later time, the instrumental sensitivity is adjusted to give the same reading for the reference solution as originally obtained. This adjustment is best accomplished by varying the amplifier gain or phototube voltage rather than the monochromator spectral bandwidth; such variations correct for change in position of the analytical curve but not for change in curvature or slope, nor for uncontrolled variations in instrumental parameters.* Also, an *internal standard* can be used to compensate for changes in instrumental sensitivity.

3. GLOSSARY OF SYMBOLS

A_I Area of image of xenon arc lamp on entrance slit of monochromator, cm^2.

A_L Area of flame from which fluoresence is observed, cm^2.

$B_{C\lambda}$ Spectral radiance of source at wavelength λ, watt cm^{-2} sr^{-1} nm^{-1}.

C_u Concentration of analyte in unknown, concentration units.

C_s Concentration of analyte in standard, concentration units.

f_s Source modulation frequency, Hz.

H Height of entrance (or exit) slit of monochromator for absorption measurements, cm.

H' Height of entrance (or exit) slit of monochromator for fluorescence measurements, cm.

---

*A similar technique can be used in atomic fluorescence spectrometry.

i_a Readout signal due to unknown (sample), A (or volt).

i_L Readout signal due to atomic fluorescence, A (or volt).

i_s Readout signal due to unknown (sample) plus standard addition, A (or volt).

t_r Response time of amplifier-readout system, sec.

t_R Total response time of luminescence spectrometer, sec.

T_λ Transmittance of excitation monochromator and entrance optics at wavelength λ, no units.

$T_{\lambda'}$ Transmittance of emission monochromator and entrance optics at wavelength λ', no units.

V_a Volume of unknown (sample) solution, cm³.

V_s Volume of standard added to unknown solution, cm³.

W Width of entrance (or exit) slit of monochromator for absorption measurement, cm.

W' Width of entrance (or exit) slit of monochromator for fluorescence measurement, cm.

Y Luminescence quantum yield, no units.

Y_L Luminescence power yield of process, no units.

$\gamma_{\lambda'}$ Photoanodic sensitivity of photodetector at wavelength λ', A watt^{-1}.

$\Delta\lambda'$ Half width of most narrow band or spectral feature observed, nm.

Δi Readout signal due to atomic absorption (i.e., difference signal), A (or volt).

$\Delta\lambda$ Half width of narrowest spectral feature, nm.

λ Wavelength of excitation, nm.

λ_o Peak wavelength of excitation, nm.

λ' Wavelength of emission, nm.

λ_o' Peak wavelength of emission, nm.

π 3.1416...

τ Luminescence lifetime, sec.

ϕ Phase shift between luminescence light and exciting light, angular degrees.

Ω Solid angle of radiation incident upon the monochromator slit, sr.

4. REFERENCES

1. R. Smith, "Atomic Flame Fluorescence Spectrometry," in *Spectrochemical Methods of Analysis*, J. D. Winefordner, Ed., Wiley-Interscience, New York, 1971, Chapter 4.

2. J. D. Winefordner and R. Smith, "Atomic Fluorescence Flame Spectrometer," in *Analytical Flame Spectroscopy, Selected Topics*, R. Mavrodineanu, Ed., N. V. Philips Gloeilampfabrieken, Eindhoven, Netherlands, in press, Chapter 12.

3. J. D. Winefordner, V. Svoboda, and L. Cline, *Crit. Rev. Anal. Chem.*, **1**, 233 (1970).

4. P. J. T. Zeegers, R. Smith, and J. D. Winefordner, *Anal. Chem.*, **40**, (13), 26A (1968).

5. S. J. Pearce, L. de Galan, and J. D. Winefordner, *Spectrochim. Acta*, **23B**, 793 (1968).
6. J. E. Hesser, *J. Chem. Phys.*, **48**, 2518 (1968).
7. J. Z. Klose, *Phys. Rev.*, **141**, 181 (1966).
8. T. M. Holzberlein, *Rev. Sci. Instr.*, **35**, 1041 (1964).
9. A. Shatkay, *Anal. Chem.*, **40**, 2097 (1968).
10. R. Herrmann and C. T. J. Alkemade, *Flammenphotometrie*, 2nd ed., Springer, Berlin, 1960; *Chemical Analysis by Flame Photometry*, P. T. Gilbert, transl., Wiley, New York, 1963.
11. D. W. Ellis, "Luminescence Instrumentation and Experimental Details," in *Fluorescence and Phosphorescence Analysis*, D. M. Hercules, Ed., Wiley-Interscience, New York, 1966, Chapter 2.
12. C. A. Parker, *Photoluminescence of Solutions*, Elsevier, Amsterdam, 1968.
13. J. D. Winefordner, P. A. St. John, and W. J. McCarthy, "Phosphorimetry as a Means of Chemical Analysis," in *Fluorescence Assay in Biology and Medicine*, S. Udenfriend, Vol. 2, Academic Press, New York, 1969, Chapter 2.
14. J. D. Winefordner, W. J. McCarthy, and P. A St. John, "Phosphorimetry as an Analytical Approach in Biochemistry," in *Methods of Biochemical Analysis*, Vol. 15, D. Glick, Ed., Pergamon Press, New York, 1967.
15. A. Müller, R. Lumbry, and H. Kokubin, *Rev. Sci. Instr.*, **36**, 1214 (1965).
16. T. E. Sisneros, *Appl. Opt.*, **6**, 417 (1967).
17. J. Yguerabide, *Rev. Sci. Instr.*, **36**, 1734 (1965).
18. P. A. St. John and J. D. Winefordner, *Anal. Chem.*, **39**, 500 (1967).
19. T. Huen, *Rev. Sci. Instr.*, **40**, 1067 (1969).
20. M. Rubin, Ed., *Fluorimetry in Clinical Chemistry*, American Instrument Co., Silver Spring, Md., 1965.

ANALYTICAL USE OF
LUMINESCENCE SPECTROMETRY

A. ATOMIC FLUORESCENCE SPECTROMETRY

1. SENSITIVITY OF ANALYSIS

a. EXPERIMENTAL RESULTS—FLAME CELLS

Table IV-A-1 furnishes an extensive listing of elements studied by atomic fluorescence flame spectrometry and of general experimental conditions.

In Table IV-A-2, a comparison is given of the best currently available experimental limits of detection (μg ml^{-1}) for aqueous solutions of elements by atomic fluorescence flame spectrometry with a line source (AFL), atomic fluorescence flame spectrometry with a continuum source (AFC), atomic absorption flame spectrometry with a line source (AAL), and atomic emission flame spectrometry (AE). It is evident that at the present state-of-the-art of atomic fluorescence flame spectrometry, line sources produce lower limits of detection for nearly all elements than continuum sources. Despite this, continuum sources appear to have considerable use in atomic fluorescence spectrometry because they are useful for many elements and have good stability (better than stability of line sources) and long lifetimes.* Another helpful feature is the predictable analytical curves resulting with continuum sources; predictable response is useful in determining interferences and in measuring fundamental parameters, such as quantum yields and degrees of atomization of atoms in flames.

From the results in Table IV-A-2, it is apparent that atomic fluorescence flame spectrometry—particularly with a line source (AFL)—is truly *competitive* with atomic absorption flame spectrometry and *complementary* to atomic emission flame spectrometry (AE). Because the instrumental requirements are similar for both AFL and AE, it would seem that this is an ideal combination of techniques for trace element (mainly metals) analysis. If high temperature (e.g., 20,000°K), stable continuum sources were ever commercially available, then the ideal combination would be AE–AFC (atomic

---

* Useful operation times.

Table IV-A-1. Elements[a] **Studied**[b] **by Atomic Fluorescence Flame and Nonflame Spectrometry**[c]

		Atomic Fluorescence Flame Spectrometry	
Element	Source(s)[c]	Flame(s)[d]	Electronics[e]
Ag	E, H, or C	H-P, H-T, P-P, or Ac-T	DC or AC
Al	E or C	H-P, or Ac-P	AC
As	E, H, or C	H-P or Ac-P	DC or AC
Au	E, H, or C	H-P, or H-T	DC or AC
Be	E or H	Ac-P	AC
Bi	E, H, or C	H-P	DC or AC
Ca	E, H, or C	H-P, H-T, or Ac-T	DC or AC
Cd	E, H, C, or M	H-P, H-T, Ac-P, Ac-T, P-P, or N-P	DC or AC
Co	E, H, or C	H-P, H-T, Ac-P, or P-P	DC or AC
Cr	E, H, or C	H-P, or H-T	DC or AC
Cu	E, H, or C	H-P, H-T, Ac-T, or P-P	DC or AC
Fe	E, H, or C	H-P, H-T, Ac-P, Ac-T, or P-P	DC or AC
Ga	E, H, C, or M	H-P, or H-T	DC or AC
Ge	E or C	Ac-P	AC
Hg	E, C, or M	H-P, H-T, Ac-T, or P-P	DC or AC
In	E, H, C, or M	H-P or H-T	DC or AC
Mg	E, H, or C	H-P, H-T, Ac-P, Ac-T, or P-P	DC or AC
Mn	E, H, or C	H-P, H-T, Ac-T, or P-P	DC or AC
Ni	E, H, or C	H-P, H-T, or Ac-P	DC or AC
Pb	E, H, or C	H-P, H-T, Ac-T, or P-P	DC or AC
Pd	H, C, or M	H-P or H-T	DC or AC
Sb	E, H, or C	H-P, H-T, or Ac-P	DC or AC
Se	E	H-P, H-T, or P-P	DC or AC
Si	E	Ac-P	AC
Sn	E or H	H-P, H-T, or Ac-P	DC or AC
Te	E	H-P, H-T, or P-P	DC or AC
Tl	E, H, C, or M	H-P, H-T, Ac-T, or P-P	DC or AC
Zn	E, H, C, or M	H-P, H-T, Ac-P, Ac-T, or P-P	DC or AC

Table IV-A-1 (*Continued*)

Atomic Fluorescence Nonflame Spectrometry

Element	Source	Nonflame	Electronics
Ag	H	G	AC
As	H	G	AC
Au	H	G	AC
Be	H	G	AC
Ca	H	G	AC
Cd	E or H	G or P	DC or AC
Cu	H	G	AC
Fe	H	G	AC
Ga	E	P	DC
Hg	E or H	P or G	DC or AC
Mg	H	G	AC
Pb	H	G	AC
Sb	H	G	AC
Tl	H	G	AC
Zn	H	G	AC

[a] Only elements measured by more than one author and reported in the literature are measured.

[b] Studies reported in the literature prior to 1970.

[c] Taken from R. Smith, "Flame Fluorescence Spectrometry," in *Spectrochemical Methods of Analysis*, J. D. Winefordner, Ed., John Wiley, New York, 1971.

[d] E = Electrodeless discharge lamp coupled to microwave power generator via either a *cavity* or an *antenna*.

    H = Hollow cathode discharge tube of commercial *shielded* variety or *high intensity* variety or laboratory constructed *demountable* and coolable by water.

    C = Continuum source—150 to 900-watt xenon arc lamp or metal vapor discharge arc lamp.

    M = Metal vapor discharge arc lamp of Osram or Philips variety.

[e] H-P = Hydrogen premixed flames (oxidant can be air, O_2, or N_2O).

    H-T = Hydrogen turbulent flames (oxidant can be air, O_2, or N_2O).

    Ac-P = Acetylene premixed flames (oxidant can be air, O_2, or N_2O).

    Ac-T = Acetylene turbulent flames (oxidant is O_2 or N_2O).

    P-P = Propane or natural gas (oxidant is air).

[f] DC = Direct current amplification.

    AC = Alternating current amplification.

[g] G = Heated graphite cell or rod atomizer (in argon).

    P = Heated platinum loop atomizer.

Table IV-A-2. Comparison of Experimental Limits of Detection in Atomic Absorption Flame Spectrometry, Atomic Fluorescence Flame Spectrometry, and Atomic Emission Flame Spectrometry[a,b]

Element	Wavelength (flame)[c]				Limits of Detection ($\mu g\ ml^{-1}$)			
	AAL (Å)	AFL (Å)	AFC (Å)	AE (Å)	AAL (19)	AFL	AFC	AE
Ag	3281 (A/A)	3281 (H/A)	3281 (H/A)	3281 (A/A)	0.0005	0.0001(28)[d]	0.001(12)[d]	0.02(23)[d]
Al	3962 (A/N)	3962 (H/N)	3092 (H/A)	3962 (A/N)	0.1	2.0(6)	4.0(16)	0.005(24)
As	1937 (H/A)	1937 (H/A)	1972 (E/A)	2350 (A/O)	0.1	0.1(6)	2.0(16)	50.0(14)
Au	2428 (A/A)	2676 (H/A)	2676 (E/O)	2676 (A/N)	0.02	0.05(20)	4.0(26)	4.0(14)
Be	2349 (A/N)	2349 (A/N)	—	2349 (A/N)	0.003	0.01(2, 17)	—	0.1(17)
Bi	2231 (A/A)	3068 (H/A)	3068 (E/A)	2231 (A/O)	0.05	0.005(9)	0.5(16)	2.0(18)
Ca	4227 (A/A)	4227 (H/A)	4227 (H/A)	4227 (A/N)	0.002	0.02(28)	0.02(16)	0.0001(23)
Cd	2288 (H/A)	2288 (H/O)	2288 (H/A)	3261 (A/N)	0.01	0.000001(28)	0.01(16)	2.0(23)
Co	2407 (A/A)	2407 (H/A)	2407 (H/A)	3454 (A/N)	0.005	0.005(15)	0.02(16)	0.05(23)
Cr	3579 (A/A)	3579 (H/A)	3579 (H/A)	4254 (A/N)	0.005	0.05(6)	10.0(20)	0.005(23)
Cu	3247 (A/A)	3247 (H/A)	3247 (H/A)	3274 (A/N)	0.005	0.001(11, 21)	0.01(16)	0.01(23)
Fe	2483 (A/A)	2483 (H/A)	2483 (H/A)	3720 (A/N)	0.005	0.008(6)	0.04(16)	0.05(23)
Ga	2874 (A/A)	4172 (H/A)	4172 (H/A)	4172 (A/N)	0.07	0.3(16)	0.1(16)	0.01(23)
Ge	2651 (A/N)	2651 (H/A)	2651 (H/A)	2651 (A/N)	1.0	2.0(16)	2.0(16)	0.5(23)
Hg	2537 (A/A)	2537 (H/A)	—	2537 (A/O)	0.5	0.08(3)	—	40.0(14)
In	3039 (A/A)	4511 (H/A)	4105 (H/A)	4511 (A/N)	0.05	0.1(28)	2.0(20)	0.002(24)
Mg	2852 (A/A)	2852 (A/A)	2852 (H/A)	2852 (A/N)	0.0003	0.001(27)	0.004(10)	0.005(23)
Mn	2795 (A/N)	2795 (H/A)	2795 (H/A)	4031 (A/N)	0.002	0.006(28)	0.003(16)	0.005(23)
Na	5890 (A/O)	—	5896 (H/A)	5890 (A/N)	0.002	—	0.008(16)	0.0001(14)
Ni	2320 (A/A)	2320 (H/A)	2320 (H/O/Ar)	3415 (A/N)	0.005	0.003(22)	0.5(25)	0.6(14)
Pb	2833 (A/A)	4058 (H/A)	2833 (H/A)	4058 (A/N)	0.01	0.02(21)	0.6(16)	0.2(23)
Rh	3435 (A/A)	3692 (H/A)	3692 (H/A)	3692 (H/A)	0.03	3.0(20)	10.0(20)	0.3(14)
Sb	2175 (A/A)	2311 (P/A)	2311 (H/A)	2598 (A/O)	0.1	0.05(8)	300.0(20)	20.0(14)
Se	1960 (H/A)	1960 (A/A)	1960 (H/A)	—	0.1	1.0(5)	4.0(16)	—
Si	2516 (A/N)	2040 (P/A)	—	2516 (A/O)	0.1	0.1(7)	—	5.0(14)
Sn	2246 (H/A)	3034 (H/A)	—	2840 (A/N)	0.03	0.05(4)	—	0.3(23)
Sr	4607 (A/A)	4607 (H/A)	—	4607 (A/N)	0.01	0.03(28)	—	0.0002(23)
Te	2143 (A/A)	2143 (A/A)	2143 (H/A)	2383 (A/O)	0.1	0.05(7)	8.0(16)	200.0(14)
Tl	2768 (A/A)	3776 (H/A)	3776 (H/A)	3776 (A/N)	0.03	0.006(16)	0.07(10, 12, 13)	0.02(23)
Zn	2138 (A/A)	2138 (H/A)	2138 (H/A)	2138 (A/O)	0.002	0.00005(28)	0.02(16)	50.0(14)

are given because all limits of detection are poorer than by AAL).

AFL = Atomic fluorescence flame spectrometry with a line source.

AFC = Atomic fluorescence flame spectrometry with a continuum source.

AE = Atomic emission flame spectrometry.

[b] All results are for aqueous solutions. Only results are given for those elements measured by AFL and/or AFC.

[c] Flames Types—A/A = acetylene/air; A/N = acetylene/nitrous oxide; A/O = acetylene/oxygen; H/A = hydrogen/air; H/O = hydrogen/oxygen; P/A = propane/air (the ratio of fuel to oxidant is not specified; also in several flames the oxidant is entrained air).

[d] 1. M. P. Bratzel and J. D. Winefordner, *Anal. Letters*, **1**, 43 (1967).

2. M. P. Bratzel, R. M. Dagnall, and J. D. Winefordner, *Anal. Chem.*, **41**, 1527 (1969).

3. R. F. Browner, R. M. Dagnall, and T. S. West, *Talanta*, **16**, 75 (1969).

4. R. F. Browner, R. M. Dagnall, and T. S. West, *Anal. Chim. Acta*, **46**, 207 (1969).

5. M. S. Cresser and T. S. West, *Spectrosc. Letters*, **2**, 9 (1969).

6. R. M. Dagnall, M. R. G. Taylor, and T. S. West, *Spectrosc. Letters*, **1**, 397 (1968).

7. R. M. Dagnall, K. C. Thompson, and T. S. West, *Talanta* **14**, 557 (1967).

8. R. M. Dagnall, K. C. Thompson, and T. S. West, *Talanta*, **14**, 1151 (1967).

9. R. M. Dagnall, K. C. Thompson, and T. S. West, *Talanta*, **14**, 1467 (1967).

10. D. R. Demers and D. W. Ellis, *Anal. Chem.*, **40**, 860 (1968).

11. J. I. Dinnin, *Anal. Chem.*, **39**, 1491 (1967).

12. D. W. Ellis and D. R. Demers, " Atomic Fluorescence Flame Spectrometry," in *Trace Inorganics in Water*, R. A. Baker, Ed., *Adv. Chem. Ser. No. 73*, Washington, D.C., 1968.

13. D. W. Ellis and D. R. Demers, *Anal. Chem.*, **38**, 1943 (1966).

14. V. D. Fassel and D. W. Golightly, *Anal. Chem.*, **39**, 466 (1967).

15. B. Fleet, K. V. Liberty, and T. S. West, *Anal. Chim. Acta*, **45**, 205 (1969).

16. A. Hell and B. Ricchio, talk given at Pittsburgh Conference in Analytical Chemistry and Applied Spectroscopy, Cleveland, Ohio, 1970.

17. D. N. Hingle, G. F. Kirkbright, and T. S. West, *Analyst*, **93**, 522 (1968).

18. R. S. Hobbs, G. F. Kirkbright, and T. S. West, *Analyst*, **94**, 554 (1969).

19. H. Kahn, " Atomic Absorption Spectroscopy," in *Trace Inorganics in Water*, R. A. Baker, Ed., *Adv. Chem. Ser. No. 73*, Washington, D.C., 1968.

20. D. C. Manning and P. Heneage, *At. Absorpt. Newsl.*, **7**, 80 (1968).

21. D. C. Manning and P. Heneage, *At. Absorpt. Newsl.*, **6**, 124 (1967).

22. J. Matousek and V. Sychra, *Anal. Chem*, **41**, 518 (1969).

23. E. E. Pickett and S. R. Koirtyohann, *Spectrochim. Acta*, **23B**, 235 (1968).

24. E. E. Pickett and S. R. Koirtyohann, *Spectrochim. Acta*, **24B**, 325 (1969).

25. R. Smith, C. M. Stufford, and J. D. Winefordner, *Can. Spectrosc.* **14**, 2 (1969).

26. C. Veillon, J. M. Mansfield, M. P. Parsons, and J. D. Winefordner, *Anal. Chem.*, **38**, 204 (1966).

27. T. S. West and X. K. Williams, *Anal. Chim. Acta*, **42**, 27 (1968).

28. K. E. Zacha, M. P. Bratzel, J. D. Winefordner, and J. M. Mansfield, *Anal. Chem.*, **40**, 1733 (1968).

fluorescence flame spectrometry with a continuum source) because such a system could be used for both quantitative and qualitative (wavelength scanning) analysis of trace elements.

b. PREDICTED RESULTS—FLAME CELLS

The ratio of signals of AFL(AFC) with respect to AAL and with respect to AE can be calculated if we make the following assumptions: (a) there is a square flame with a uniform *low* analyte concentration and temperature at all points, (b) there is complete illumination of the flame cell with exciting light and complete measurement of fluorescence and emission emanating from the flame toward the measurement system, (c) the same resonance line of the same element is measured in all flame methods, and (d) the same instrumental system is used in all flame methods. Denoting the ratio of signals by G, it can be shown* (see Section II-C) that

$$G_{AFL/AAL} = \frac{Y_L \Omega_E}{4\pi} \tag{IV-A-1}$$

$$G_{AFC/AAL} = \frac{c_2 B^o_{C\lambda_0} \Delta\lambda_D Y_L \Omega_E}{4\pi B_N{}^o \delta_0} = \left[\frac{c_2 B^o_{C\lambda_0} \Delta\lambda_D}{B_N{}^o \delta_0}\right]\left[\frac{Y_L \Omega_E}{4\pi}\right] \tag{IV-A-2}$$

$$G_{AFL/AFC} = \frac{B_N{}^o \delta_0}{c_2 B^o_{C\lambda_0} \Delta\lambda_D} \tag{IV-A-3}$$

$$G_{AFL/AE} = \frac{B_N{}^o \delta_0 Y_L \Omega_E}{4\pi c_2 B^o_{B\lambda_0} \Delta\lambda_D} \tag{IV-A-4}$$

$$G_{AFC/AE} = \frac{B^o_{C\lambda_0} Y_L \Omega_E}{4\pi B^o_{B\lambda_0}} \tag{IV-A-5}$$

where all symbols were defined in Section II-C (derivations of the atomic fluorescence radiances also appear in Section II-C; i.e., $G_{AFL/AAL}$ is simply B_{AFL}/B_{AAL}, $G_{AFC/AAL} = B_{AFC}/B_{AAL}$, $G_{AFL/AFC} = B_{AFL}/B_{AFC}$, $G_{AFL/AE} = B_{AFL}/B_{AE}$, and $G_{AFC/AE} = B_{AFL}/B_{AE}$, where the B's are radiances for the processes designated by subscripts).

It is evident from equations IV-A-1 to IV-A-5 that the following conclusions can be made:

1. The *radiance (signal) of AFL is always much less than the radiance (signal) of AAL.* That is, because the power yield $Y_L \sim 0.1$ for atoms in flames and the

---

* The c_2 is $\sqrt{\pi}/2\sqrt{\ln 2}$ and is nearly unity.

fractional solid angle of exciting radiation incident upon the flame $\Omega_E/4\pi \sim$ 0.01 for a good spectrometric system, the AAL radiance is about 10^3 greater than the AFL radiance.

2. The *factor* $c_2 B_{C\lambda_0}^o \Delta\lambda_D/B_N^o\delta_o$ *is generally much less than unity* $(c_2 \sim 1)$; $B_N^o\delta_o/\Delta\lambda_D$ is the spectral radiance of the line source, where B_N^o is the source line radiance, δ_o is a profile factor to account for finite width of the source line and profiles of the source and absorption lines and $\Delta\lambda_D$ is the Doppler half width of atomic absorption line in the flame; thus as long as the effective black body temperature of the line source is greater than the continuum source, the factor $c_2 B_{C\lambda_0}^o \Delta\lambda_D/B_N^o\delta_o < 1$, and so the radiance of AFC is usually less than AFL.

3. *The radiauce (signal) of AFL is greater than the radiance (signal) of AE if the effective temperature of the line source is greater than the flame temperature* (source line spectral radiance $B_N^o\delta_o/\Delta\lambda_D$ is then greater than black body spectral radiance corresponding to the flame temperature). For electrodeless discharge tube sources and most analytical flames (2000–3000°K), elements with lines of wavelength less than about 3500 Å will have higher AFL radiances than AE radiances; and for wavelengths longer than about 3500 Å, the reverse is true.

4. *The radiance (signal) of AFC is greater than for AE if the effective black body temperature of the continuum source is greater than the temperature of the flame.* For a xenon arc lamp and most analytical flames, elements with lines of wavelength less than about 3000 Å will give greater AFC radiances than AE radiances, and for wavelengths longer than about 3000 Å, the reverse is true.*

Therefore, according to the conclusions just drawn, it is apparent that atomic absorption flame spectrometry *always* results in greater radiances (signals) than either AFL (or AFC) or AE. However, it is well known (see Table IV-A-2) that AAL *does not necessarily give lower detection limits for all elements* (in fact, *AFL and AE often give lower detection limits*). The discrepancy is not in the estimation of the radiance (signal) ratios but rather in the correlation of signal ratios with inverse limit of detection ratios (a *higher* signal would mean a *lower* limit of detection). The ratio of signals would only be directly related to the inverse ratio of detection limits if the *noise ratios of* the the methods being compared *are identical*. Of course, this is not true; the noise in AAL is *much greater*—especially if high radiance sources are used—because

---

* Actually, for this statement to be valid, it is necessary to assume the use of analytical flames having temperatures of 3000°K or below and the use of a 450-Watt xenon arc lamp ($\sim$6600°K black body temperature; 0.06 emissivity).

of source flicker noise, which is minimal in AFL or AFC,* and does not exist in AE, where no external light source is used.

The ratios of limits of detection for the same ratios given in equations IV-A-1 to IV-A-5 are

$$L_{\text{AAL/AFL}} = G_{\text{AFL/AAL}} N_{\text{AAL/AFL}} \qquad \text{(IV-A-6)}$$

$$L_{\text{AAL/AFC}} = G_{\text{AFL/AAL}} N_{\text{AAL/AFC}} \qquad \text{(IV-A-7)}$$

$$L_{\text{AFC/AFL}} = G_{\text{AFL/AFC}} N_{\text{AFC/AFL}} \qquad \text{(IV-A-8)}$$

$$L_{\text{AE/AFL}} = G_{\text{AFL/AE}} N_{\text{AE/AFL}} \qquad \text{(IV-A-9)}$$

$$L_{\text{AE/AFC}} = G_{\text{AFC/AE}} N_{\text{AE/AFC}} \qquad \text{(IV-A-10)}$$

where $L_{\text{X/Y}}$ represents the ratio†* of detection of method X to method Y for the same element measured in the same flame with the same instrument and $N_{\text{Y/X}}$ represents the ratio of noise at readout for method Y to noise at readout for method X.

If we assume an instrumental system similar to those used in commercial atomic absorption flame spectrometers, then it is possible to estimate the total noise at the readout for each method. The major noise sources in AF and AE will be flame flicker and shot noises; in AA the major noise sources will be flame flicker, shot, and source flicker noises. For the light sources and flames assumed to estimate the G values, the noise ratios are found to be approximately the following values:

$$N_{\text{AAL/AFL}} \sim 10^3 \text{ to } 10^4 \qquad \text{(IV-A-11)}$$

$$N_{\text{AAL/AFC}} \sim 10^3 \text{ to } 10^5 \qquad \text{(IV-A-12)}$$

$$N_{\text{AFC/AFL}} \sim 1 \qquad \text{(IV-A-13)}$$

$$N_{\text{AFL/AE}} \sim 1 \qquad \text{(IV-A-14)}$$

$$N_{\text{AFC/AE}} \sim 1 \qquad \text{(IV-A-15)}$$

Therefore, if noise as well as signal is considered, the limits of detection for AFL and AFC for high temperature flames (e.g., 3000°K) and, for low temperature flames (e.g., 2000°K) often better‡ than the limits of detection for AAL. In any event, if limits of detection are most important, AE should be used for all analyte atoms with resonance lines at wavelengths *above about* 3000 Å and AF or AA for analyte atoms with resonance lines at wavelengths below about 3000 Å when using high temperature, high background flames.

---

* Right angle illumination measurement; and so source radiation *does not* reach the detector for scatter. Source flicker noise also produces fluorescence flicker noise, but at the limit of detection the latter noise is negligible.

† A value of $L_{X/Y} > 1$, means that the limit of detection by method X is greater (poorer) than by method Y.

‡ Lower.

High temperature flames are of general analytical utility also because of the minimization of solute vaporization interferences. Therefore, based purely on theoretical grounds, it should not matter whether AFL (or AFC) or AAL are used with high temperature flames. However, because essentially the same spectrometric measurement system can be used for both AFL (or AFC) and AE, it would seem that the combination of AFL (or AFC) and AE is ideal for experimental work. Of course, when low temperature flames are employed the choice is more obviously AFL (or AFC) and AE.

C. EXPERIMENTAL RESULTS—NONFLAME CELLS

Table IV-A-3 presents an extensive listing of elements measured by atomic fluorescence spectrometry with nonflame cells, as well as general experimental conditions.

In Table IV-A-4, absolute limits of detection (ng) are given for atomic emission flame spectrometry (AE), for atomic fluorescence flame spectro-

Table IV-A-3. Limits of Detection[a] for Several Elements with Nonflame Cells in Atomic Fluorescence Spectrometry

Element	Line (Å)	Cell	Limit of Detection ($\mu g\ ml^{-1}$)	References[b]
Ag	3281		0.0005	1
As	1890		0.5	1
Au	2428		0.01	1
Be	2349		0.03	1
Ca	4247		0.0001	1
Cd	2288		0.00002	2
Cu	3247		0.005	1
Fe	2483		0.01	1
Ga	4172		20.0	2
Hg	2536		0.02	2
Mg	2852		0.0000001	3
Pb	2833		0.02	1
Sb	2311		0.2	4
Tl	3776		2.0	2, 4
Zn	2139		0.00004	4

[a] Best available in July, 1970.
[b] 1. M. D. Amos, P. A. Bennett, K. G. Brodie, P. W. Y. Lung, and T. P. Matousek, talk given at Society of Applied Spectroscopy Meeting, New Orleans, October, 1970.
  2. M. P. Bratzel, R. M. Dagnall, and J. D. Winefordner, *Appl. Spectrosc.*, **24**, 518 (1970).
  3. T. S. West and X. K. Williams, *Anal. Chim. Acta*, **45**, 27 (1969).
  4. H. Massman, *Spectrochim. Acta*, **23B**, 215 (1968).

Table IV-A-4. Absolute Detection Limits (ng) for Several elements Measured by a Variety of Methods.

Elements	NAᵃ	ESᵃ	SSMSᵃ	AFLᵇ	AALᵇ	AFGLᶜ	AAGLr	AEᵇ	IFᵉ
Ar	—	—	0.03	—	—	—	—	—	—
Ag	0.01	8.0	0.2	0.1	0.5	0.0008	0.0005	20.0	4.0
Al	1.0	30.0	0.02	2000.0	100.0	—	0.03	5.0	0.2
As	0.1	600.0	0.06	100.0	100.0	—	—	50,000.0	7000.0
Au	0.05	100.0	0.2	50.0	20.0	—	0.07	4000.0	500.0
B	—	30.0	0.01	—	6000.0	—	5.0	30,000.0	0.5
Ba	5.0	4.0	0.2	10.0	50.0	—	0.1	2.0	—
Be	—	5.0	0.008	10.0	2.0	—	0.003	100.0	0.04
Bi	50.0	4.0	0.2	5.0	50.0	—	0.03	2000.0	200.0
Br	0.5	—	0.1	—	—	—	—	—	—
C	—	—	0.01	—	—	—	—	—	20.0
Ca	100.0	5.0	0.03	20.0	2.0	—	0.03	0.1	10.0
Cd	5.0	100.0	0.3	0.001	10.0	0.00001	0.00006	2000.0	20.0
Ce	10.0	100.0	0.1	—	—	—	—	10,000.0	50.0
Cl	—	—	0.04	—	—	—	—	—	50.0
Co	0.5	100.0	0.05	5.0	5.0	—	0.008	50.0	0.1
Cr	100.0	30.0	0.05	50.0	5.0	—	0.05	5.0	—
Cs	50.0	10,000.0	0.1	—	50.0	—	0.007	8.0	—
Cu	0.1	20.0	0.08	1.0	5.0	0.01	0.006	10.0	0.2
Dy	0.0001	50.0	0.5	—	200.0	—	—	70.0	10.0
Er	0.1	50.0	0.5	—	100.0	—	—	40.0	10,000.0
Eu	0.0005	30.0	0.2	—	40.0	—	—	0.6	5.0
F	100.0	10.0	0.02	—	—	—	—	—	1.0
Fe	5000.0	30.0	0.05	8.0	5.0	0.003	0.03	50.0	0.08

Ga	0.5	100.0	0.09	7.0	70.0	10.0	—	10.0	7.0
Gd	1.0	50.0	0.5	—	4000.0	—	—	2000.0	10,000.0
Ge	0.5	100.0	0.2	2000.0	1000.0	—	—	500.0	4.0
H	—	—	0.0008	—	—	—	—	—	—
He	100.0	—	0.003	—	—	—	—	—	—
Hf	1.0	50.0	0.4	—	8000.0	—	—	80,000.0	100.0
Hg	1.0	100.0	0.6	80.0	500.0	100.0	0.5	40,000.0	2.0
Ho	0.01	50.0	0.1	—	100.0	—	—	20.0	100,000.0
I	0.5	—	0.1	—	—	—	—	—	600.0
In	0.005	100.0	0.1	100.0	50.0	—	0.008	2.0	40.0
Ir	0.01	300.0	0.3	—	2000.0	—	—	100,000.0	2000.0
K	5.0	100.0	0.03	—	5.0	—	0.6	3.0	—
Kr	—	—	0.1	—	—	—	—	—	—
La	0.1	10.0	0.1	—	2000.0	—	0.05	2000.0	—
Li	—	8.0	0.006	—	5.0	—	—	0.03	200.0
Lu	0.005	100.0	0.1	—	3000.0	—	—	200.0	100,000.0
Mg	50.0	8.0	0.03	1.0	0.3	0.0000001	0.003	5.0	0.01
Mn	0.005	10.0	0.05	6.0	2.0	—	0.003	5.0	2.0
Mo	10.0	100.0	0.3	—	30.0	—	0.05	100.0	100.0
N	—	—	0.01	—	—	—	—	—	300.0
Na	0.5	10.0	0.02	8.0	2.0	—	—	0.1	—
Ne	—	—	0.02	—	—	—	—	—	—
Nb	0.5	300.0	0.08	—	3000.0	—	—	60.0	100.0
Nd	10.0	80.0	0.4	—	2000.0	—	—	200.0	5000.0
Ni	5.0	1.0	0.07	5.0	3.0	—	0.03	600.0	0.06
Np	—	200.0	—	—	—	—	—	—	—
O	—	—	0.01	—	—	—	—	—	0.1
Os	5.0	—	0.4	—	1000.0	—	—	10,000.0	50.0

Table IV-4-A (*Continued*)

Elements	NA[a]	ES[a]	SSMS[a]	AFL[b]	AAL[b]	AFGL[c]	AAGLr	AE[b]	IF[e]
P	50.0	10.0	0.03	—	—	—	—	—	0.0006
Pa	—	200.0	—	—	—	—	—	—	—
Pb	1000.0	10.0	0.3	20.0	10.0	0.04	0.03	200.0	5000.0
Pd	0.05	200.0	0.3	—	30.0	—	0.05	50.0	—
Pr	0.05	30.0	0.1	—	10,000.0	—	—	1000.0	500.0
Pt	5.0	2.0	0.5	—	100.0	—	0.3	2000.0	—
Pu	—	200.0	—	—	—	—	—	—	—
Rb	5.0	30.0	0.1	—	5.0	—	0.008	2.0	—
Re	0.05	300.0	0.2	—	1000.0	—	—	1000.0	—
Rh	0.05	200.0	0.09	3000.0	30.0	—	0.06	300.0	—
Ru	1.0	300.0	0.03	—	300.0	—	—	300.0	1000.0
S	500.0	—	0.03	—	—	—	—	—	0.2
Sb	0.5	100.0	0.2	50.0	100.0	0.2	0.05	20,000.0	50.0
Sc	1.0	0.8	0.04	—	100.0	—	—	30.0	2.0
Se	500.0	10,000.0	0.1	1000.0	100.0	—	0.2	—	5.0
Si	5.0	30.0	0.03	100.0	100.0	—	0.003	5000.0	80.0
Sm	0.05	50.0	0.5	—	2000.0	—	—	200.0	100.0
Sn	50.0	10.0	0.3	50.0	30.0	—	0.01	300.0	100.0
Sr	0.05	100.0	0.09	30.0	10.0	—	0.02	0.2	—
Ta	5.0	3000.0	0.2	—	5000.0	—	—	20,000.0	—
Tb	5.0	300.0	0.1	—	2000.0	—	—	400.0	6.0
Tc	—	—	—	—	—	—	—	—	—
Te	5.0	4000.0	0.3	50.0	100.0	—	0.008	20,000.0	200.0
Th	5.0	5000.0	0.2	—	—	—	—	200,000.0	40.0

Ti	5.0	3.0	0.05	—	100.0	—	0.5	200.0	—
Tl	—	30.0	0.2	6.0	30.0	2.0	0.003	20.0	20.0
Tm	1.0	50.0	0.1	—	200.0	—	—	20.0	10,000.0
U	0.5	4000.0	0.2	—	30,000.0	—	—	10,000.0	10.0
V	0.1	80.0	0.04	—	20.0	—	—	10.0	2000.0
W	0.1	600.0	0.5	—	3000.0	—	—	500.0	40.0
Xe	—	—	0.4	—	—	—	—	—	—
Y	—	3.0	0.07	—	300.0	—	—	40.0	20.0
Yb	0.1	50.0	0.5	—	40.0	—	—	2.0	—
Zn	10.0	300.0	0.1	0.04	2.0	0.0008	0.001	5000.0	2.0
Zr	100.0	3.0	0.1	—	5000.0	—	—	5000.0	20.0

[a] Taken from G. H. Morrison and R. K. Skogerboe, "General Aspects of Trace Analysis," in *Trace Analysis—Physical Methods*, G. H. Morrison, Ed., Wiley-Interscience, New York, 1965.

[b] Values taken from Table IV-A-2 assuming measured solution volume is 1 cm³. By proper techniques, smaller volumes—say 0.1 cm³—could be measured, which would reduce the detection values by ten fold.

[c] Taken from Table IV-A-3 with same assumption as in note b.

[d] Taken from B. V. Lvov, *Spectrochim. Acta*, **24B**, 53 (1968).

[e] Taken from lowest values in Table IV-B-1 with same assumptions as in note b.

metry with line sources (AFL), for atomic absorption flame spectrometry with line sources (AAL), for atomic absorption spectrometry with a graphite furnace and line sources (AAGL), for atomic fluorescence spectrometry with a graphite furnace and line sources (AFGL), for inorganic fluorimetry (IF), for neutron activation (NA), for spark source mass spectrometry (SSMS), and for arc-spark emission spectrometry (ES). The absolute values of AFL, AAL, AAGL, and AFGL compare quite favorably with the more exotic and expensive methods of NA, SSMS, and ES. By using a multichannel spectrometer and a continuum source, it would be rather simple to automate AFGL to enable rapid measurement of several elements simultaneously (e.g., trace wear metals in jet engine lubricating oils).

2. SELECTIVITY OF ANALYSIS

a. SPECTRAL INTERFERENCE

The spectral interferences in AFL* and AAL* should be least among the five flame methods (AE, AAL, AAC, AFL, and AFC)* because interference can *only* occur when there is overlap between the analyte resonance line and the interference line or band. This type of interference is possible in AFL and AAL, but it is not too important because of the narrowness of absorption and fluorescence lines (order of 0.1 Å with wings extending out several ångstroms) and because of the relatively small absorption, unless the lines overlap significantly or the interferent has a very large absorption coefficient even in the line wings, where small overlap may occur. Figure IV-A-1 illustrates spectral interferences in atomic fluorescence spectrometry.

The spectral interferences in AE, AAC, and AFC should be approximately identical because in these cases there will be spectral interference by any interferent with an emission line or band, an absorption line or band, and a fluorescence line or band within the spectral bandwidth of the monochromator.

In AFC (and in AFL), any unabsorbed radiation passing through the flame (or graphite cell in AFG) will *not* produce bending of analytical curves or any form of spectral interference. Of course, in AAL, spectral lines of the source that are not absorbed by the analyte (e.g., arc lines from the inert gas within the source) or are absorbed unequally (e.g., mul-

* Because of the great similarity in the flame methods, it is convenient to compare them with respect to interferences and instrumentation as well as with respect to sensitivity. Recall: AFL = atomic fluorescence flame spectrometry with a line source; AFC = atomic fluorescence flame spectrometry with a continuum source; AAL = atomic absorption flame spectrometry with a line source; AAC = atomic absorption flame spectrometry with a continuum source; AE = atomic emission flame spectrometry.

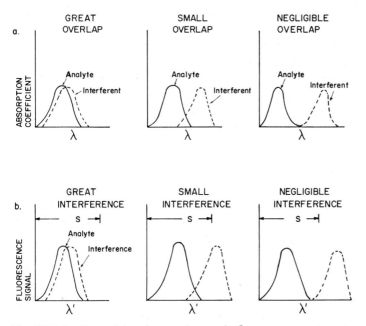

Fig. IV-A-1. Spectral interferences in atomic fluorescence spectrometry: (*a*) overlap of absorption bands, (*b*) overlap of fluorescence bands (*s* = spectral bandpass of monochromator).

tiplets of the analyte) result in curvature of analytical curves, if they are within the spectral bandwidth of the monochromator; that is, they result in reduced slope, but *seldom* in any spectral interferences. It should be stressed that spectral interferences in AE, AAC, and AFC become less significant as the spectral bandwidth *s* of the monochromator is reduced. Assuming a good medium resolution monochromator is used ($s \sim 0.1$ Å); *spectral interferences should only be slightly greater in AE, AAC, and AFC than in AFL and AAL*. Similar conclusions regarding AFG (atomic fluorescence with the graphite furnace) can be made.

b. SCATTERING OF INCIDENT RADIATION

Scattering of incident radiation can cause an increased fluorescence (or absorption) signal in atomic fluorescence (or atomic absorption) methods. The scattering of radiation in flames is a result of incompletely vaporized particles of the sample matrix. Of course, scattering from solvent droplets can be corrected for by use of a blank solution. If line sources are used in atomic fluorescence spectrometry, correction for scattering can be made by using a control sample with the same matrix but no analyte or by measuring the scatter signal from a nonfluorescent line of similar wavelength to the fluorescent line. If continuum sources are used in atomic fluorescence spectrometry, scattering is compensated for simply by scanning the wavelength region of the fluorescent line. Of course, the best means of compensating for such an interference is to eliminate it altogether by complete atomization of the sample; this can be accomplished by use of efficient nebulizers to produce fine aerosols and by use of high temperature flames (or a high temperature graphite cell).

Scattering from incompletely vaporized particles can also cause an interference in atomic absorption methods. This interference, too, may be eliminated by more efficient atomization methods as discussed previously. However, an additional interference can also arise in atomic absorption methods when high matrix concentrations are nebulized into the flame; this absorption interference is a result of molecular species (e.g., NaCl) produced in the flame gases. Molecular absorption can be compensated for when line sources are used by measuring the absorption for a spectral line not absorbed by the analyte or by measuring the absorption at the wavelength of the analyte absorption line with a continuum source. In the latter technique, the spectral bandwidth of the monochromator must be significantly greater than the absorption line width or compensation will be incomplete. (Of course, if the spectral bandwidth is larger, then spectral interferences will be greater.) Some commercial atomic absorption instruments employ the latter method of compensation for scattering and molecular absorption.

C. PHYSICAL INTERFERENCES

Physical interferences are *nonspecific* (independent of analyte type) ones that influence solution transport rate into the flame F and aspirator yield ε when a chamber-type nebulizer is employed. Solution transport rate is dependent on the solution viscosity (F is approximately proportional to the reciprocal of the viscosity). Surface tension and ionic strength of the solution seem to have lesser influence on the transport rate. The aspirator yield depends on surface tension, viscosity, solution density, solution transport rate, and nebulizing gas flow rate, and so all solutions (standards and unknowns) should have nearly identical physical properties. If this can not be accomplished, then it is necessary to correct for the variation of transport rate and aspirator yield with solution properties, which is difficult to do with high accuracy in some cases.

It should be stressed, however, that *all* flame methods are plagued with the physical interferences just named.

d. CHEMICAL INTERFERENCES

Chemical interferences are *specific*—they depend on the analyte type and they influence the production of atoms from particles within the flame gases (or graphite furnace or other cell). *Solute vaporization interference** is the most important type of chemical interference in flames. Solute vaporization interferences can be subdivided into four types according to Alkemade:[5] depression of the signal due to formation of a less volatile compound than the analyte (e.g., $CaCl_2$ in a solution containing phosphate results in a Ca:P compound that is less volatile than $CaCl_2$), enhancement of the signal due to formation of a more volatile compound than the analyte (e.g., $AlCl_3$ in a solution containing HF results in AlF_3), depression of signal due to occlusion of analyte within a less volatile matrix of the interferent than the analyte itself e.g., $SrCl_2$ in a solution containing a large concentration of $CaCl_2$), and enhancement of signal due to dispersion of analyte within a more volatile matrix than the analyte itself (e.g., $CaCl_2$ in a solution containing aliphatic salts). Solute vaporization of high concentrations of the analyte may be incomplete, but this only results in curvature of analytical curves and is not an interference. The first two types of interferences can occur even if the interferent is of lower concentration than the analyte, whereas the latter two interferences occur only if the interferent is of greater concentration than the analyte.

All other types of chemical interferences affect the free-atom fraction β (see Section II-D). The extent of *ionization* of the analyte atoms in flames depends primarily on the flame gas temperature, the analyte type, and the concentration of free electrons in the flame gases. To minimize this interference, it is necessary to add a *normalizer* to the analyte solution (i.e., a high concentration of a compound such as $CsCl_2$ containing an easily ionizable atom such as cesium). The ionization of the normalizer represses the ionization of the analyte atoms; indeed, in very hot flames such as C_2H_2/N_2O it is impossible to completely suppress the ionization of all metals. If solute vaporization interference is negligible, a lower temperature flame should be used to minimize ionization.

The extent of *compound formation* of the analyte atoms with flame gas products, primarily O and OH, can cause the β factor to be significantly less than unity (cf. Section II-D). The β factor due only to compound formation (cf. Section II-D) is dependent on the flame gas temperature, the analyte type, and the concentration of the flame gas product involved in the formation of the compound. *A high temperature, reducing, low burning velocity flame* (e.g., C_2H_2/N_2O with the reducing interconal red feather region) *should be*

---

* This interference affects the value of ε.

used to minimize compound formation. (A graphite furnace at about 3000°K with an inert gas atmosphere certainly seems to be a more ideal system for all atoms.) However, even though *compound formation with flame gas products* results in β being smaller than unity, it *is not strictly an interference, because β should be constant for any given analyte in a given flame* (composition and temperature). *Compound formation interference is only significant in flames if the species introduced into the flame is relatively stable at the flame temperature.* For examples under rather unusual circumstances involving a high concentration of potassium chloride and hydrogen chloride in solution, the hydrogen chloride concentration in the flame gases may influence the value of β. However, *such compound formation interferences are very rare in flame spectrometric methods.*

It should be stressed that *solute vaporization interferences are the most serious in analytical flame spectrometric methods, that ionization interferences can generally be minimized, and that compound formation interference is seldom serious.* However, compound formation of the analyte with flame gas products does result in a β value less than unity and can seriously influence sensitivity, precision, and accuracy of measurement.

e. FLAME TEMPERATURE VARIATION INTERFERENCES

A small variation (less than 100°K) in flame temperature has relatively little effect on most parameters in the radiance expressions and in the relation between atomic concentration n and solution concentration C except for the β factor. Assuming a measurement of analyte concentrations on the linear portion of the analytical curve, then ionization should be negligible and only compound formation need be considered. Suppose that we also assume that only one compound, the monoxide, is important (and this is valid for a number of elements in flames). Then it is possible to estimate the relative change in atomic fluorescence signal S_{AF} to be

$$\frac{dS_{AF}}{S_{AF}} \simeq \left(\frac{V_d}{kT}\right)\frac{dT}{T} \tag{IV-A-16}$$

where V_d is the dissociation energy of the monoxide. For a first approximation, equation IV-A-16 is valid for most analytes in analytical flames. To obtain a more general expression, the variation of the power (quantum) yield Y, the aspiration efficiency ε, and the flame gas expansion factor ε_f must also be considered, as well as the presence of other molecules (e.g., monohydroxides) in the flame gases on β.

We should emphasize that as long as the analyte atoms form a relatively stable ($V_d > 2.5$ eV) compound with flame gas products, then the temperature dependence dS_{AF}/S_{AF} of the signal in atomic fluorescence (or atomic absorption) flame spectrometry is not much less than the sensitivity dS_{AE}/S_{AE}

of the signal in atomic emission* flame spectrometry, which is contrary to many statements in the literature—particularly the literature concerning atomic absorption flame spectrometry. It should further be stressed that there is no such thing as *energy-transfer interferences* in atomic emission flame spectrometry if the flame is indeed in thermal equilibrium, which is approximately the case in analytical flames. Alkemade[6] has shown that this and several other "accepted" misconceptions in flame spectrometry are just fantasies—not facts.

f. QUENCHING INTERFERENCES

Quenching of excited atoms due to change of flame gas composition resulting from nebulization of the sample solution could possibly produce an interference in atomic fluorescence flame spectrometry. On the other hand, no such interferences are possible in atomic emission flame spectrometry because there are as many quenching (deactivating) collisions as activating ones. Furthermore, these interferences cannot occur in atomic absorption flame spectrometry because the mechanism of deactivation is irrelevant as long as the atomic fluorescence radiant flux is negligible compared with the transmitted source flux. However, in atomic fluorescence spectrometry, it is well known that substitution of argon for N_2 as a diluent in a premixed H_2/air flame results in enhanced atomic fluorescence signals owing to the smaller quenching cross sections of argon. Therefore, it might be assumed that such quenching could result in a specific interference in atomic fluorescence spectrometry. According to Smith,[4] the quenching rates for typical flame gas species with the Na–D lines and for a dry air/propane flame (1800°K) are

$$k_{N_2}n_{N_2} = 11.1 \times 10^8 \text{ sec}^{-1}$$
$$k_{CO}n_{CO} = 0.0$$
$$k_{CO_2}n_{CO_2} = 2.9 \times 10^8 \text{ sec}^{-1}$$
$$k_{H_2O}n_{H_2O} = 0.1 \times 10^8 \text{ sec}^{-1}$$
$$k_{O_2}n_{O_2} = 1.4 \times 10^8 \text{ sec}^{-1}$$

$$\sum_i k_{X_i} n_{X_i} = 15.5 \times 10^8 \text{ sec}^{-1}$$

For sodium, the radiative transition probability A_{ul} is $0.63 \times 10^8 \text{ sec}^{-1}$, and so

$$Y_L = \frac{A_{ul}}{\sum\limits_i k_{X_i}n_{X_i}} = \frac{0.63 \times 10^8}{15.5 \times 10^8} = 0.039 \qquad \text{(IV-A-17)}$$

---

* In AE, $dS_{AE}/S_{AE} = [(V_d + E_u)/kT]\, dT/T$, when E_u is the excitation energy of the upper level (eV).

Now to change Y_L by ± 0.001 unit, the total quenching rate would have to be changed by $\pm 0.4 \times 10^8$ sec^{-1}. Assuming that the species introduced with the sample has the same second-order collisional rate constant as N_2, this change would correspond to a proportion of the interferent which is about 3.6% of the nitrogen concentration in the flame gases or about 2.8% of the total flame gas compositions. In this calculation, the introduction of solvent water has been neglected, and it was assumed that the interferent was no more efficient as a quencher than N_2. Certainly it is unlikely that any concomitant at concentrations less than 1000 μg ml^{-1} would result in a quenching interference.

3. ANALYTICAL CURVES

Analytical curves of magnesium (2852-Å resonance line) measured with a continuum source (450-watt xenon arc lamp) and with a line source (electrodeless discharge lamp operated at microwave frequency) are given in Figures IV-A-2 and IV-A-3, respectively. The experimental curves were measured by Zeegers and Winefordner[7]. It is interesting to note the excellent agreement between theory and experiment with respect to shape and even absolute positions of the curves determined using the continuum light source.

It should again be stressed that the cases of *incomplete illumination and incomplete measurement of the fluorescence* (see Figures IV-A-2 and IV-A-3) *result in reduced signals and in deviation from linearity at lower concentrations compared with the case of complete illumination of the flame with exciting radiation and complete measurement of the flourescence over the entire flame volume. Therefore, the former cases are of lesser analytical value, and conditions of incomplete illumination and measurement should be avoided experimentally if possible.*

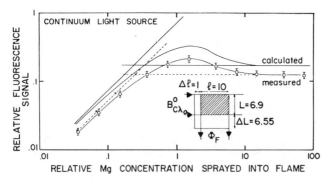

Fig. IV-A-2. Fluorescence analytical curves of magnesium ($\lambda_o = 2852$ Å) in a flame-sheathed C_2H_2/air flame measured with a continuum source: dashed lines indicate asymptotes of measured curve. Geometry of illumination and measurement appear in insert. Curves taken from Ref. 7. (With permission of the authors.)

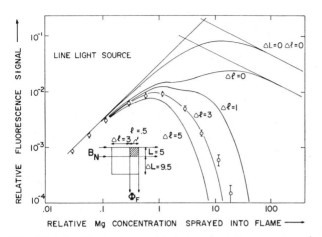

Fig. IC-A-3. Fluorescence analytical curves of magnesium ($\lambda_o = 2852$ Å) in a flame-sheathed C_2H_2/air flame measured with a narrow line source (EDT); for all curves, $\Delta l = 9.5$ mm except one for which $\Delta l = 0$ mm. Curves taken from Ref. 7. (With permission of the authors.)

It should also be noted that even for the cases of complete illumination and complete measurement of fluorescence, the analytical curves are not analytically useful at high concentrations; that is, with a continuum source the slope approaches zero, and with a line source the slope approaches -0.5 at high concentrations (log-log plots). However, this further points out the analytical usefulness of flame methods for trace analysis.

Experimental curves for metals in turbulent diffusion or in unsheathed premixed flames are similar to the curves in Figure IV-A-2 for a continuum source of excitation and in Figure IV-A-3 for a line source of excitation. Unless care is taken to excite the flame with collimated light and to measure collimated fluorescence light as shown in the inserts of Figures IV-A-2 and IV-A-3, much of the detail that appears in those figures is lost. In any event, the limiting slopes of the curves do indeed approach the carefully measured curves in Figures IV-A-2 and IV-A-3. An experimental analytical curve for copper is presented in Figure IV-A-4.

4. COMPENSATION FOR OR MINIMIZATION OF INTERFERENCES

Some methods of minimizing interferences are mentioned in Section IV-A-2. If solute vaporization interference is important, use a more efficient nebulizer and a higher temperature, lower burning velocity flame (e.g., C_2H_2/N_2O). If ionization interference is important, a lower temperature flame such as H_2/air could be used if solute vaporization interference is

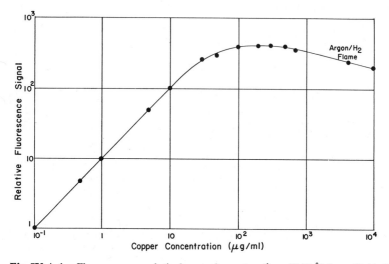

Fig. IV-A-4. Flourescence analytical curve for copper ($\lambda_o = 3247$ Å) in an H_2/Ar/entrained air turbulent diffusion flame; excitation is via a high intensity hollow cathode discharge lamp (limit of detection under experimental conditions, 0.003 μg ml^{-1}).

absent; or a normalizer such as cesium chloride could be added to all solutions, and it would still be possible to use a high temperature flame to minimize solute vaporization problems.

If interferences are still present in the flame method selected, then a compensation technique must be employed. For example, the *simulation method* consists of preparation of an analytical curve with standard solutions containing the same matrix as the unknowns. This method is obviously only useful if all unknowns have about the same known concentration of the interferent(s).

The *parametric method* consists of plotting a family of analytical curves, each curve containing a concentration of interferent different from that of the other curves. In an actual analysis, therefore, the unknown is first analyzed for the interferent and then for the analyte, using the proper analytical curve.

The *method of standard additions* involves addition of a small volume of a standard analyte solution to the unknown analyte solution of known volume. By measuring the signal of the unknown solution and of the unknown plus standard and by knowing the volumes of solutions, it is a simple matter to calculate the concentration of analyte in the unknown solution. This method compensates for most types of interference and is particularly useful if only a few of any given type of sample are measured (cf. Section III-E-1-d).

The addition of a *releasing agent* (a buffer) is not recommended; strontium, for example, should not be added to minimize effect of phosphate on calcium. Sometimes more serious spectral or solute vaporization interferences or addition of analyte impurity in the releasing agent may result, and so this type of interference is best minimized by using a higher temperature, low burning flame and a more efficient nebulizer (e.g., C_2H_2/N_2O) flame produced with a chamber-type nebulizer burner.

A variety of more elaborate techniques to minimize interferences, such as the *dilution method*, the *doubling method*, the *self-standardization method*, and the *substitution method* have been described by Herrmann and Alkemade[8] or flame emission spectrometry. Several of these methods, as well as a more general method of compensating for interferences in spectrochemical methods, were more recently described by Shatkay.[9] These more elaborate and more general methods could be used for compensation of interferences in atomic fluorescence flame spectrometry if necessary.

5. ATOMIC FLUORESCENCE FLAME SPECTROMETRY OF THE ELEMENTS

To avoid unnecessary duplication of phrases and sentences and to maximize the information available on the various articles published on atomic fluorescence flame spectrometry, some of the characteristics of experimental systems used for elements studied by various authors with atomic fluorescence flame spectrometry are listed in Table IV-A-1.

6. ATOMIC FLUORESCENCE NONFLAME SPECTROMETRY OF THE ELEMENTS

Some experimental characteristics used for elements studied by various authors with atomic fluorescence spectrometry of elements vaporized and atomized by nonflame cells also appear in Table IV-A-1.

7. APPLICATIONS OF ATOMIC FLUORESCENCE SPECTROMETRY

All applications using atomic fluorescence spectrometry as the measurement step will be given in a subsequent volume in this series.

8. GLOSSARY OF SYMBOLS

AAC   Atomic absorption flame spectrometry with a continuum source

AAGL   Atomic absorption spectrometry with a nonflame cell and a line source.

AAL   Atomic absorption flame spectrometry with a line source.

AE	Atomic emission flame spectrometry.
AFC	Atomic fluorescence flame spectrometry with a continuum source.
AFGL	Atomic fluorescence spectrometry with a nonflame cell and a line source.
AFL	Atomic fluorescence flame spectrometry with a line source.
A_{ul}	Einstein coefficient of spontaneous emission, $\sec^{-1}$.
$B_{B\lambda_0}^o$	Spectral radiance of black body source, watt $cm^{-2}\,sr^{-1}\,nm^{-1}$.
$B_{C\lambda_0}^o$	Spectral radiance of continuum source, watt $cm^{-2}\,sr^{-1}\,nm^{-1}$.
B_N^o	Radiance of line source, watt $cm^{-2}\,sr$.
c_2	$\sqrt{\pi}/2\sqrt{\ln 2}$, no units.
ES	Emission spectrometry.
E_u	Excitation energy of resonance line, eV.
k	Boltzmann constant, eV $°K^{-1}\,mol^{-1}$.
k_X	Second-order rate constant for collisional deactivation, $cm^3\,\sec^{-1}$.
$G_{AFC/AAL}$	Ratio of AFC radiance to AAL radiance, no units.
$G_{AFC/AE}$	Ratio of AFC radiance to AE radiance, no units.
$G_{AFL/AAL}$	Ratio of AFL radiance to AAL radiance, no units.
$G_{AFL/AE}$	Ratio of AFL radiance to AE radiance, no units.
$L_{AAL/AFC}$	Ratio of AAL limit of detection to AFC limit of detection, no units.
$L_{AE/AFC}$	Ratio of AE limit of detection to AFC limit of detection, no units.
$L_{AE/AFL}$	Ratio of AE limit of detection to AFL limit of detection, no units.
$L_{AFC/AFL}$	Ratio of AFC limit of detection to AFL limit of detection, no units
n_X	Concentration of species X.
NA	Neutron activation analysis.
$N_{AAL/AFC}$	Ratio of AAL noise to AFC noise, no units.
$N_{AAL/AFL}$	Ratio of AAL noise to AFL noise, no units.
$N_{AE/AFC}$	Ratio of AE noise to AFC noise, no units.
$N_{AE/AFL}$	Ratio of AE noise to AFL noise, no units.
$N_{AFC/AFL}$	Ratio of AFC noise to AFL noise, no units.
S_{AE}	Signal due to atomic emission, A.
S_{AF}	Signal due to atomic fluorescence, A.
SSMS	Spark source mass spectrometry.
T	Temperature of flame, °K.
V_d	Dissociation energy of metal monoxide, eV.
Y_L	Quantum (power) yield of resonance line, no units.
δ_o	Profile factor to account for profiles of source line and absorption line, no units.
$\Delta\lambda_D$	Doppler half width, nm (or cm).
π	$3.1418\ldots$
Ω_E	Solid angle of exciting radiation collected by entrance optics and impinging upon flame (or nonflame) cell, sr.

6. REFERENCES

1. J. D. Winefordner, V. Svoboda, and L. J. Cline, *Crit. Rev. Anal. Chem.*, **1**, 233 (1970).
2. P. J. T. Zeegers, R. Smith, and J. D. Winefordner, *Anal. Chem.*, **40**, *(13)*, 26A, (1968).
3. J. D. Winefordner, "Non-Flame Cells In Atomic Fluorescence Spectrometry," in *Proceedings of Atomic Absorption Conference, Sheffield, England, 1969*, Butterworths, London, 1971.
4. R. Smith, "Flame Fluorescence Spectrometry," in *Spectrochemical Methods of Analysis*, J. D. Winefordner, Ed., Wiley, New York, 1971.
5. C. T. J. Alkemade, *Anal. Chem.*, **38**, 1252 (1966).
6. C. T. J. Alkemade, *Appl. Opt.*, **7**, 1261 (1968).
7. P. J. T. Zeegers, and J. D. Winefordner, *Spectrochim. Acta*, **26B**, 161 (1971).
8. R. Herrmann and C. T. J. Alkemade, *Flammenphotometrie*, Springer, Berlin, 1960; *Chemical Analysis by Flame Photometry*, P. T. Gilbert, Transl., Wiley, New York, 1963.
9. A. Shatkay, *Anal. Chem.*, **40**, 2097 (1968).

B. SOLUTION FLUORIMETRY AND PHOSPHORIMETRY[1,2]

1. SENSITIVITY OF ANALYSIS

a. METHODS OF EXPRESSING SENSITIVITY IN MOLECULAR LUMINESCENCE SPECTROMETRY

Sensitivity according to the International Union of Pure and Applied Chemistry (IUPAC) should refer to only the slope of the analytical curve dS/dC (this is true for both atomic and molecular spectrometric methods). However, it also has served to refer to the *minimum detectable analyte concentration* producing a minimum detectable signal or minimum acceptable signal-to-noise ratio at the readout. The factors influencing the minimum detectable signal-to-noise ratio and the *limit of detection* (minimum detectable analyte concentration) are discussed in Section III-D-3-c. Each limit of detection expressed in this manner is a function of the instrumental system (the light source—type, power, window, and age; the type of monochromators—the transmittance, the light gathering power, the resolution, and the wavelength settings; the type of chopper used, if any; the photodetector; and the electronic measurement system—particularly the frequency response bandwidth and the measurement frequency). Actually it should be stressed that the limit of detection may never be reached because of nonrandom variations in the blank luminescence and drift in the source radiance. The blank signal is due to luminescence from solvent impurities and reagent impurities, scattered light, luminescence from the cuvette and surrounding, luminescence of concommitants, and Raman emission from the solvent. If the variation in the luminescence of a series of blanks is greater than the instrumental noise level, then the limit of detection as defined by the signal-to-noise ratio will *not* be reached. However, in dilute solutions of the sample, the limiting detectable signal-to-noise ratio is often attained.

Some workers prefer to express sensitivity in a more fundamental manner. The *absolute sensitivity* has been defined by Parker[1] and is the factor $Y_L D_\lambda / \Delta\lambda'$, where Y_L is the quantum yield, D_λ is the absorbance* per centimeter for a concentration of 1 μg ml^{-1}, and $\Delta\lambda'$ is the effective half bandwidth of the luminescence emission spectrum. The quantum yield† is independent of the excitation wavelength λ for most molecules, and the factor D_λ is independent of emission wavelength λ' The factor $D_\lambda Y_L / \Delta\lambda'$ is *dependent* only on the molecule in question and is *independent* of the instrumental factors. Of course, the absolute sensitivity (i.e., $D_\lambda Y_L / \Delta\lambda'$) may not give a good estimation of the relative limits of detection as determined by the signal-to-noise ratio because instrumental factors have been neglected in the absolute sensitivity and because noise is not considered. However, absolute sensitivity is a very useful means of tabulating molecular characteristics of importance in luminescence spectrometry. To convert the absolute sensitivity to an instrumental signal, it is only necessary to multiply by an instrumental factor (see Sections II-C and III-D-3).

b. EXPERIMENTAL RESULTS

Table IV-B-1 presents some of the best reported limits of detection (μg ml^{-1}) of ions measured by inorganic fluorimetry. In most instances, the limits of detection resulting in inorganic fluorimetry are much lower than by inorganic absorption spectrophotometry and are comparable with other sensitive methods of trace element analysis (cf. Table IV-A-3 for an extensive comparison of the major methods of trace element analysis).

In Table IV-B-2, a tabulation is given of a number of reagents useful in inorganic fluorimetry. There is so much literature in organic fluorimetry and phosphorimetry, that space does not permit a general listing of limits of detection. Winefordner, McCarthy, and St. John[2] and Winefordner, St. John and McCarthy[3] (in 1969) made an extensive listing of limits of detection of organic molecules in phosphorimetry. Also Zander[4] has given a number of limits of detection in phosphorimetry. White and Weissler[5] (in 1963) published an extensive listing of limits of detection of organic molecules in

---

* The D refers to older term, optical density.

† Parker[1] uses *quantum* yield in his definition of absolute sensitivity. However, it would also seem quite reasonable to use power yield—at least in view of the approach given in Section III-C.

Table IV-B-1. Comparison of Experimental Limits of Detection in Inorganic Fluorimetry[a]

Element	Reagent	Limit of Detection (μg ml^{-1})	References[b]
Ag	2,3-Naphthotriazole	0.02	27
	Eosin + 1,10-phenanthroline	0.004	23
	8-Hydroxyquinoline-5-sulfonic acid	0.01	55
Al	3-Hydroxy-2-naphthoic acid	0.0002	37
	Morin	0.0002	68
	Salicylidene-o-aminophenol	0.0003	20
	N-Salicylidene-2-amino-3-hydroxyfluorene	0.0008	69
	Mordanted blue 9	0.0005	22
As	Ce(IV)-As(III) reaction OS catalyst	7.0	
As	Uranyl nitrate	50.0	40
Au	Rhodamine B	0.5	64
B	Benzoin	0.01	70
	Quinizarin	0.01	31
	Dibenzoylmethane	0.0005	46
	Acetylsalicylic acid	0.01	53
Be	3-hydroxy-2-naphthoic acid	0.0002	37
	Morin	0.00004	70
	1-Amino-4-hydroxyanthraquinone	0.2	68
	3-Hydroxyquinaldine	0.001	68
Bi	Rhodamine B	0.5	60
C(CN$^-$)	p-Benzoquinone	0.2	29
	o-(p-Nitrobenzene sulfonyl) quinone monoxime	0.2	30
	Pd complex of sulfo-8-hydroxyquinoline	0.02	67
C(C$_2$O$_4$$^{2-}$)	Ce(IV)-As(III) reaction Os catalyst	9.0	38
Ca	8-Quinolyhydrozaone of 8-hydroxyquinalde-hyde	0.02	11
	Dicarboxymethylaminomethyl-2,6-dihydroxynaphthalene	0.01	15
Cd	p-Tosylaminoquinoline	0.02	10
Ce	4-[(2,4-Dihydroxyphenyl)azo]-3-hydroxy-naphthalene sulfonic acid	0.05	33
Cl	Luminol + H$_2$O$_2$	0.05	6
Co	Salicylfluorene + H$_2$O$_2$	0.0001	12
	1-(2-Pyridylazo)-2-naphthol	0.06	57
Cu	Benzamido (p-Dimethyibenzylidene) acetic acid	0.0002	2
	2-Amino-1-propene	0.01	54
	Tetrachlorotetraiodofluorescein + o-phenanthroline	0.001	68
	Thiamine	3.0	72
	2,9-Dimethyl-4,7,diphenyl-1,10-phenanthroline	0.01	13
	Salicylalazine	0.05	40

Table IV-B-1 (*Continued*)

Element	Reagent	Limit of Detection (μg ml^{-1})	References[b]
Dy	Sodium tungstate	0.01	1, 17
	Calcium tungstate	5.0	40
Er	Calcium tungstate	10.0	40
Eu	Thenoyltrifluoroacetone	0.01	61, 63
	Benzoyltrifluoroacetone + trioctylphosphine oxide	0.07	69
	Sodium tungstate	0.005	1
	Calcium tungstate	0.5	40
F	Al complex of Alizarin Garnet R	0.001	68
	Al complex of Eriochrome B	0.001	68
Fe	Stilbexone + H_2O_2	0.001	43
	Luminol + H_2O_2	0.0008	5
Ga	2′,2′,4′-Trihydroxy-5′-chlorobenzene-3′-sulfonic acid	0.001	10
	Lumogallion	0.002	48
	2,2′-Pyridylbenzimidazole	0.07	7
	8-Hydroxyquinoline	0.05	68
	8-Hydroxyquinaldine	0.02	68
	Salicylidene-*o*-aminophenol	0.007	68
Gd	Thenolyltrifluoroacetone	100.0	63
	Calcium tungstate	10.0	40
Ge	2,2′,4′-Trihydroxy-3-arseno-5-chlorobenzene	0.004	45
	Benzoin	2.0	68
H(H_2O_2)	Luminol + hemin	0.0003	42
Hf	Flavanol	0.1	67
	Quercetin	1.0	14
Hg	Rhodamine B	0.002	34
Ho	Thenoyltrifluoroacetone	~ 100.0	63
I(I$^-$)	Ce(IV)-As(III) reaction Os catalyst	0.6	44
In	2,2′ Pyridylbenzimidazole	0.1	7
	Rhodamine S	0,5	49
	Dimethyl-[6-(dimethylamino)xanthen-3-Xylidene]ammonium chloride	5.0	9
	8-Hydroxyquinoline	0.04	68
	8-Hydroxyquinaldine	0.2	68
Ir	2,2′,2″-Terpyridine	2.0	26
Li	8-Hydroxyquinoline	0.2	47, 67
	Dibenzothiazolylmethane	1.0	52
Lu	Thenolytrifluoroacetone	~ 100.0	63
Mg	*N*,*N*′-Bissalicylidene ethylenediamine	0.00001	66
	N,*N*′-Bissalicylidene-2,3-diaminebenzofuran	0.002	19
	2-Hydroxy-3-sulfo-5-chlorophenylazo barbituric acid	0.004	10
Mn	8-Hydroxyquinoline	0.002	51
Mo	Carminic acid	0.1	68

Table IV-B-1 (*Continued*)

Element	Reagent	Limit of Detection (μg ml^{-1})	References[b]
N	Resorcinol + H_2SO_4	0.3	40
Nb	Lumogallion	0.1	39
Nd	Calcium tungstate	5.0	40
Ni	Al-1-(2-Pyridylazo)-2-naphthol	0.00006	57
$O(O_2)$	Trypaflavine	0.0001	70
Os	4,6-Bis(methylthio-3-amino-pyrimidine)	0.05	16
P	Glycogen, phosphatase, TPN$^+$	0.0000006	58
	Al-morin complex	0.05	44
Pb	Morin	5.0	40
Pr	Thenolyltrifluoroacetone	$\sim$100.0	63
	Calcium tungstate	0.5	40
Ru	5-Methyl-1,10-phenanthroline	1.0	67
$S(H_2S)$	Fluorescein	0.0002	3
$S(S^{2-})$	Fluorescein	0.001	28
$S(SO_3^{2-})$	Quinine + H_2O_2 + acid	10.0	40
Sb	Luminol	0.05	39
	Rhodamine 6 Zh	0.1	35
Sc	Morin + antipyrine	0.01	50
Se	2,3-Diaminonaphthalene	0.005	25
	2,3-Diaminobenzidine	0.01	18
Si	Benzoin	0.08	24
Sm	Thenoyltrifluoroacetone	2.0	63
	Sodium tungstate	0.5	1
	Calcium tungstate	0.1	40
Sn	Flavanol	0.1	68, 70
Tb	HCl	3.0	67
	Sodium tungstate	0.1	1
	Calcium tungstate	5.0	40
	Thenoyltrifluoroacetone	$\sim$100.0	63
	Bis[1,2)pyridyl-3-methyl-5-pyrazalonyl)-4,4'-methane	0.1	17
	EDTA + sulfosalicylic acid	0.006	21
Te	Butylrhodamine B	0.2	70
Th	Quercetin	0.5	4
	Morin	0.02	67
	1-Amino-4-hydroxyanthraquinone	0.04	70
Tl	HCl (77°K)	0.02	8, 68
	HBr (77°K)	0.02	62, 68
	Rhodamine B	0.1	68
Tm	Calcium tungstate	10.0	40
U	Morin	0.5	67
	Rhodamine B	0.01	70
V	Resorcinol	2.0	67
W	Carminic acid	0.04	68
Y	8-Hydroxyquinoline	0.02	67

Table IV-B-1 (*Continued*)

Element	Reagent	Limit of Detection (μg ml^{-1})	References[b]
	5,7-Dibromohydroxyquinoline	0.1	41
Zn	2,2'-Pyridylbenzimidazole	0.01	7
	p-Tosylaminoquinoline	0.02	10
	2,2'-Methylenebibenzothiazol	0.002	65
	Picolinealdehyde-2-quinoylhydrazone	0.03	36
	1,1,3-Tricyano-2-amino-1-propene	0.02	54
	Benzothiazolmethane	0.002	56
Zr	Morin	0.02	67

[a] A comprehensive listing of reagents or detection limits is not intended. These limits of detection are in most cases indicative of values obtained under ideal (interference-free) conditions. In most cases, the analyte results in a fluorescent species (in a few cases, the analyte quences a fluorescent species or changes the rate of a chemical reaction). Limits of detection are given only for reagents commonly used for analytical determinations, and not all metals forming fluorescent complexes with a given reagent are listed (e.g., 8-hydroxyquinoline, morin, and quercetin, form many other fluorescent complexes besides those listed). Finally, better limits of detection for some elements by inorganic fluorimetry have undoubtedly appeared in the literature prior to January 1, 1970 (the latest date for references to the table) which are not in the table; the reasons for these omissions are oversight, unavailable literature, or lack of suitable information in the original article.

[b]1. G. Alberti and M. A. Massucci, *Anal. Chem.*, **38**, 214 (1966).

2. J. B. Allred and D. G. Guy, *Anal. Biochem*, **29**, 293 (1969).

3. H. D. Axelrod, J. H. Cury, J. E. Bonelli, and J. P. Lodge, *Anal. Chem.*, **41**, 1856 (1969).

4. A. K. Babko, T. H. Chan, A. I. Volkova, and T. E. Gef'man, *Ukr. Khim. Zh.*, **35**, 642 (1969).

5. A. K. Babko and I. E. Kalinichenko, *Ukr. Khim. Zh.*, **31**, 1316 (1965).

6. A. K. Babko, A. V. Terletskaya, and L. I. Dubovenko, *Ukr. Khim. Zh.*, **32**, 728 (1966).

7. L. S. Bark and A. Rixon, *Anal. Chim. Acta*, **45**, 425 (1969).

8. M. U. Belyi and I. Ya. Kushnirenko, *Zh. Prikl. Spektrosk.*, **9**, 272 (1968).

9. A. Bordea, *Bul. Inst. Politeh. Iasi.*, **13**, 209 (1967).

10. E. A. Bozhevol'nov, *Oesterr. Chemik.-Z.* **66**, 74 (1965).

11. E. A. Bozhevol'nev, L. F. Fedorova, I. A. Krasavin, and V. M. Dziomko, *J. Anal. Chem. (USSR)*, **24**, 399 (1969).

12. E. A. Bozhevol'nov and S. U. Kreingold, *Tr. Vses. Nauchn. Issled. Inst. Khim. Reakt. Osob. Chist. Khim. Veshchestv.*, **26**, 204 (1966).

13. D. A. Britton and J. C. Guyon, *Anal. Chim. Acta*, **44**, 397 (1967).

14. A. Brookes and A. Townsend, *Chem. Commun.*, **24**, 1660 (1968).

15. B. Budesinsky and T. S. West, *Talanta*, **16**, 399 (1969).

16. A. S. Burchett, *Dissertation Abstr.*, **B27**, 1384 (1966).

17. E. Butter, U. Kolowos, and H. Holzapfel. *Talanta*, **15**, 901 (1968).

18. M. Costa, *Rev. Port. Quim.*, **8**, (3), 136 (1966).

19. G. A. Crosby, *Mol. Cryst.*, **1**, 37 (1966).

20. R. M. Dagnall, R. Smith, and T. S. West, *Talanta*, **13**, 609 (1966).

21. R. M. Dagnall, R. Smith, and T. S. West, *Analyst*, **92**, 358 (1967).

Table IV-B-1 (*Continued*)

22. J. P. F. DeAlbinati, *An. Asoc. Quim. Argent.* **53**, 61 (1965).
23. M. T. El-Ghamry, W. Frei, and G. W. Higgs, *Anal. Chim. Acta*, **47**, 41 (1969).
24. G. Elliot and J. A. Radley, *Analyst*, **33**, 1623 (1961).
25. R. C. Ewan, C. A. Baumann, and A. L. Pope, *J. Agr. Food. Chem.*, **16** (2), 212 (1968).
26. D. W. Fink and W. E. Ohnesorge, *Anal. Chem.*, **41**, 39 (1969).
27. M. P. Grigor'eva, E. N. Stepanova, and G. A. Sapozhnikova, *Vop. Pitan*, **28**, (3), 65 (1969).
28. A. Grünert, K. Ballschmiter, and G. Tölg, *Talanta*, **15**, 45 (1968).
29. G. G. Guilbault and D. N. Kramer, *Anal. Chem.*, **37**, 918 (1965).
30. G. G. Guilbaut and D. N. Kramer, *Anal. Chem.*, **37**, 1395 (1965).
31. A. Holme, *Acta Chem. Scand.*, **21**, 1679 (1967).
32. Z. Holzbecher, *Microchem. J.*, **9**, 288 (1965).
33. C. Huu, A. I. Volkova, and T. E. Get'man, *Zh. Anal. Chim.*, **24**, 688 (1969).
34. H. Imai, *Nippon Kagaka Zasshi*, **90**, 275 (1969).
35. A. I. Ivankova and D. P. Shcherbov, *Issled. Razrab. Fotometrich. Metod Opred. Mikrokolichestv. Elem. Miner. Syr'e 1967*, 138.
36. E. R. Jensen and R. T. Pflaum, *Anal. Chem.*, **38**, 1268 (1966).
37. G. F. Kirkbright and W. I. Stephen, *Anal. Chim. Acta*, **32**, 544 (1965).
38. G. F. Kirkbright, T. S. West, and C. Woodward, *Anal. Chim. Acta*, **36**, 298 (1966).
39. O. I. Komlev and V. K. Zinchuk, *Visn. l'viv. Univ. Ser. Khim. 1967*, (9), 50.
40. M. A. Konstantinova-Shlezinger, *Fluorimetric Analysis*, N. Kaner, Transl. Israel Program for Scientific Translations, Jerusalem, 1965. This book has many references prior to 1964 to reagents and detection limits for inorganic and organic species. General references in Table IV-B-1 are to the book by Konstantinova-Shlezinger, but specific references can be found in the references in the book itself.
41. A. I. Krillov, R. S. Lauer, and N. J. Poluektov, *J. Anal. Chem. USSR*, **22**, 1123 (1967).
42. J. Kubal, *Chem. Listy*, **62**, 1478 (1968).
43. M. Laamaa, M. L. Allaalu, and H. Kokk, *Tartu Riikliku Ulti Kaali Toim.* **219**, 199 (1968).
44. D. B. Land and S. M. Edmonds. *Mikrochim. Acta. 1966*, 1013.
45. A. M. Lukin, O. A. Efremenko, and G. S. Petrova, *J. Anal. Chem. USSR*, **22**, 1040 (1967).
46. M. Marcantonatos, G. Gamba, and D. Monnier, *Helv. Chim. Acta*, **52**, 538 (1969).
47. A. L. Markman and S. A. Strel'tsova, *Tr. Tashkent Politekh. Inst.*, **42**, 50 (1968).
48. M. A. Matveets and D. P. Shcherbov, *Issled. Razrab. Fotometrich. Metod Opred. Mikrokolichestv. Elem. Miner. Syr'e 1967*, 122.
49. E. P. Mulikovskaya, *Nov. Metody Anal. Khim. Sustava Podzemn. Vod. 1967*, 78.
50. V. A. Nazarenko and V. P. Antonvich, *J. Anal. Chem. USSR*, **24**, 254 (1969).
51. B. K. Pal and D. E. Ryan, *Anal. Chim. Acta*, **47**, 35 (1969).
52. A. E. Pitts and D. E. Ryan, *Anal. Chim. Acta*, **37**, 460 (1967).
53. V. N. Podchainova, L. V. Skornyakova, and B. L. Dvinyaninov, *Izv. Vyssh. Ucheb. Zaved. Khim. Khim. Tekhnol.*, **11**, 241 (1968).
54. K. Ritchie and J. Harris, *Anal. Chem.*, **41**, 163 (1969).
55. D. E. Ryan, and B. K. Pal. *Anal. Chim. Acta*, **44**, 385 (1969).
56. E. Sawicki and R. A. Carnes, *Anal. Chim. Acta*, **41**, 178 (1968).
57. G. H. Schenk, K. P. Dilloway, and J. S. Coulter, *Anal. Chem.*, **41**, 510 (1969).
58. D. W. Schultz, J. V. Passoneau, and O. H. Lowry, *Anal. Biochem.*, **19**, 300 (1967).
59. T. Shigematsu, M. Matsui, and T. Sumida, *Bull. Inst. Chem. Res. Kyoto Univ. 46*, 249 (1968).
60. R. Sivori and A. H. Guerro, *An. Asoc. Quim. Argent. 55*, (**3-4**), 157 (1967).

Table IV-B-1 (Continued)

61. V. Skramovsky and P. Heaberle, *Clin. Chim. Acta*, **22**, 161 (1968).
62. E. A. Solov'ev, A. P. Golovina, E. A. Bozhevol'nov, and I. M. Plotnikova, *Vestn. Mosk. Gos. Univ., Ser. Khim. 1966*, (5), 89.
63. E. C. Stanley, B. I. Kinneberg, and L. P. Varga, *Anal. Chem.*, **38**, 1362 (1966).
64. B. T. Taskarin and D. P. Shcherbov, *Sb. Stat. Aspir. Soiskatel., Min. Vyssh. Sredneaziat., Obrazov. Kaz. SSR, Khim. Khim. Tekhnol.*, **3-4**, 208 (1965).
65. R. R. Trenholm and D. E. Ryan, *Anal. Chim. Acta*, **32**, 317 (1965).
66. S. J. Weisman, *J. Chem. Phys.*, **10**, 214 (1942).
67. A. Weissler and C. E. White, "Fluorescence Assay," in *Handbook of Analytical Chemistry*, L. Meites, Ed. McGraw-Hill, New York, 1963. This handbook has many references prior to 1963 to detection limits of inorganic and organic species. References in Table IV-B-1 to A. Weissler and C. E. White are not the original references but only refer to a table in this general reference. Specific references can be found only in the handbook, edited by L. Meites, just cited.
68. T. S. West, "Chemical Spectrophotometry in Trace Characterization," in *Trace Characterization Chemical and Physical*, W. W. Meinke and B. F. Scribner, Ed. *Nat. Bur. Std. Monogr. 100* (Superintendent of Documents, Government Printing Office, Washing-ington, D.C., 1967).
69. C. E. White, H. C. E. McFarlane, J. Fugt, and R. Fuchs, *Anal. Chem.*, **39**, 367 (1967).
70. C. E. White and A. Weissler, *Fluorescence Analysis—A Practical Approach*, Dekker, New York, 1970. This book contains numerous references prior to 1969 to detection limits, reagents, and procedures for inorganic and organic species. This book is recommended by the authors of the present book.
71. F. Will, *Anal. Chem.*, **33**, 1360 (1961).
72. Y. Yamane, Y. Yamada, and S. Kunihiro, *Bunseki Kagaku*, **17**, 973 (1968).

Table IV-B-2. Fluorimetric Reagents and Elements Detected with Reagents[a] in Inorganic Fluorimetry

Reagent[b]	Metals	Nonmetals
Acetylsalicylic acid	Be	
Acid Chrome Blue-Black [(2-2'-dihydroxy-(1-azo-1')-4-(naphthalene sulfonic acid)]	Al, Ga	
Acridine	Cr, Se, Te	(O_3)
Alizarin Garnet R	Al	(F)
Alizarin Red S	B	
1-Amino-4-hydroxyanthraqinone	B, Be, Li, Th	
7-Amino-3-nitronaphthalene sulfonic acid	Sn	
1-(0-Arsenophenylazo)-2-hydroxy-3,6-napthalene disulfonic acid	B	
Benzamido acetic acid	Cu	
8-(Benzene sulfoamino)-quinoline	Cd, Zn	

276

Table IV-B-2 (*Continued*)

Reagent[b]	Metals	Nonmetals
Benzoin	B, Be, Cu, Ge, Sb, Zn	
Benzothiazoylmethane	Li, Zn	
Benzoyltrifluoroacetone	Eu, Sm, Tb	
Benzoyltrifluoroacetone	Eu, Sm, Tb	
2,2′-Bipyridine	Ir, rare earths	
Bissaliccylidene diaminofuran	Mg	
Butyl Rhodamine B	B, Ta, Te	
Cacotheline	Mo, Nb, W	
Calcein (calcein type)	Ca, Cu	(P)
3,3′-Bis(bis(carboxymethyl)aminomethyl)-4, 4′-dihydroxy-*trans*-stilbene	Cd	
Carminic acid	B, Co, Cu, Fe, Ga, Mn, Mo, Ni Tl, W, Zr	(F)
Chloramine T		(CN⁻)
4′-Chloro-2 hydroxy-4-methoxybenzophenone	B	
6-Chloroanthraquinone	B	
Chromogen Black E.T.-00 (Mordant Black II)	Al	
Curcumin	Ba, Mg	
Diethyldithiocarbamate	Cr	
Datiscetin [3,5,7,2′-tetrahykroxyflavone]	Al, Ga, In, Zr and others detected with morin	
Diaminobenzidine	Se, Sc	
4,4′-Diamino-2,2′-disulfostilbene-$N,N,N',$ N'-tetraacetic acid	Fe	
2,3-Diaminonaphthalene (DAN)	Se	
Dibenzoylmethane	B, Eu, Sm, Tb	
Dibenzothiazolylmethane	Li	
5,7-Dibromo-8-hydroxyquinoline	Ga, La, Lu, Sc, Y	
Dicarboxymethylaminomethyl-2,6 dihydroxynaphthalene	Al, Ba, Be, La, Mg, Sr	
5.7-Dichloro-8-hydroxyquinoline	La, Sc, Y	
1,4-Dihydroxyanthraquinone	Be	
1,8-Dihydroxyanthraquinone	Be, Th	
O,O'-Dihydroxyazobenzene	Al, Ga, Mg	
Dihydroxy-2,4-benzophenone	B	
1-(2,4-Dihydroxyazo)-2-naphthol-4-sulfonic acid	Ga	
4-[(2,4-Dihydroxyphenyl)azo]-3-hydroxy-1- naphthalenesulfonic acid	Ce	
Eosin	Ag, Hg	(P)
Eosin + 1,10 phenanthroline	Ag	
Eriochrome Red B		(F)
Erythrosin		(P)
Flavanol	Al, Hf, Sn, W, Zr	(F)
Fluorescein (disodium salt)	Ag, Cr	(Br, F, O_2 , S)

Table IV-B-2 (*Continued*)

Reagent[b]	Metals	Nonmetals
Fluorexone (bis-di(carboxymethyl)amino-methyl-fluorescein)	Ba, Ca, Mg, Sr	
HBr (77–298°K) [also HCl]	Bi, Ce, Cu, Pb, Sb, Sn, Te, Tl	
2-Hydroxy-4-methoxy-4′-chlorobenzophenone	B	
2-Hydroxy-3-naphthoic acid	Al, Be, Hf, In, Mg, Se, Th, Y, Zr	
2-(*o*-Hydroxyphenyl)-benzothiazole	Be	
2-(*o*-Hydroxyphenyl)-benzoxazole	Cd, Cu, Ga	
8-Hydroxyquinaldine	Ba, Ga, In	
8-Hydroxyquinoline	Al, Be, Ca, Cd, Cs, Ga, In, K, Li, Mg, Mn, Na, Rb, Sn, Zn, Zr	(F)
8-Hydroxyquinoline-5-sulfonic acid	Ag, Cd, Mg, Zn	(S)
1-(2-hydroxy-3-sulfo-5-chloro-phenylazo)-2′-hydroxynaphthalene	Mg	
Leucofluorescein		(O$_2$)
Luminol	Many metals	(H$_2$O$_2$)
Lumocupferron	Cu	
Lumogallion [2,2′,4′-trihydroxy-5-chloro-1,1′-azobenzene-3-sulfonic acid]	Al, Ga, Nb	
Lumomagneson	Mg	
8-(Mesitylsulfonamido)-quinoline	Cd, Zn	
4,6′ bis(methylthio-3-amino-pyridine)	Os	
β-Methylambelliferone	Ag, Al	(P)
2-Methyl-8 hydroxyquinoline	Al, Zn	
Mordant Blue 9—C.I. 14855	Al	
Morin	Al, Be, Ca, Cd, Ga, Ge, In, Mo, Pb, Sb, Sc, Sn, Sr, Th, W, Zn, Zr	(F)
NaF	U	
α-Naphthoflavone [7,8-benzoflavone]	Au, Co, Cu, Fe	
2,3-Naphthotriazole	Ag	
4-Nitro-2-naphthol	Sn	
6-Nitro-2-naphthylamine-8-sulfonic acid	Sn	
1,10-Phenanthroline	Dy, Eu, Ir, Sm, Tb	
Phenylfluorone	B	
Phenylsalicylic acid	Tb, Y	
8-(Phenylsulfonamido)-quinoline	Cd, Zn	
Picolinealdehyde-2-quinolylhydrazone	Cd, Zn	
Pontachrome BBR	Al	
Pontachrome VSW	Al	
Primulin	Ag, Cr	
Pyridine	Cd, Pb	

Table IV-B-2 (*Continued*)

Reagent[b]	Metals	Nonmetals
1-(2-Pyridylazo)-2-naphthol (PAN)	Al	(F)
((2)-Pyridyl-3-methyl-5-pyrazolonyl)-4,		
4'-methane	Tb, Dy	
2,2'-Pyridylbenzimidazole	Ga, In, Zn	
Pyronine G	In	
Quercetin	Al, B, Ce, Eu, Ge, Hg,	
	Li, Sc, Th, Zr	(F)
Quinine sulfate	Cr, Hg, Se, Te	(S)
Quinizarin		(P)
Resacetophenone	B, Ge	
Resorcinol		(N)
Rezarson	Ge	
Rhodamine B	Au, Co, Fe, Ga, Hg, In,	
	Mn, Sb, Sn, Te, Tl,	
	U, W	
Rhodamine 6G	Ag, Ga, In, Re, Sb, Ta	
Rhodamine 6G + *o*-hydroxybenzoic acid	B	
Rhodamine S	Au, Ga, Hg, In, Sb, Te,	
	Tl	
Rose Bengal + 1, 10-phenanthroline	Cu	
Salicylaldehyde formylhydrazone	Al	
Salicylaldehydeacethydrazone	Ga, Sc, Y	
Salicylalazine	Cu	
Salicyclic acid + 1,10-phenanthroline	Eu, Tb	
Salicylidene-*o*-aminophenol	Al, Ga	
N,N'-Bis(salicylidene)-2,3-diaminofuran	Mg	
N-Salicylidene-2-amino-3 hydroxyfluorene	Al	
Solochrome Black	Ga	
Solochrome Red	Ga	
Stilbexone	Fe	
Sulfonaphtholazoresorcinol	Ga	
5-Sulfo-8-hydroxyquinoline		(S, CN$^-$)
4-Sulfophenyl-3-methyl-pyrazolone	Dy, Tb	
Sulfosalicyclic acid	Ni, Tb	
2,2',2"-Terpyridine	Ir	
1,2,5,8-Tetrahydroxyanthraquinone	Ge	
2-Thenoyltrifluoroacetone	Eu, Sm, Tb, other rare	
	earths	
8-(*p*-Toluene sulfonilylamino)-quinoline	Cd, Zn	
1,1,3-Tricyano-2-amino-1-propene	Cu	
1,2,4-Trihydroxyanthraquinone	Ge	
2,2',4',Trihydroxy-5-chloro-1,1'-azobenzene-		
3-sulfoninic acid	Ga	
8-(*p*-Tosylsulfonamido)quinoline	Cd, Zn	

Table IV-B-2 (*Continued*)

Reagent[b]	Metals	Nonmetals
Tropeolein 000	Mg	
Trypaflavine		(O$_2$)
Trypan Red	V	
Uranyl acetate	Ag	(Br, Cl, F, S, CN, CNS)
Uranyl nitrate	Ag, Tl	
Uranyl sulfate	Li	

[a] An excellent coverage of spectrofluorimetric procedures for inorganic species has been given in the biannual fundamental reviews by C. E. White (and more recently A. Weissler) for many years in *Analytical Chemistry*. Because these reviews are comprehensive and readily available, only a tabular, unreferenced summary of selected reagents is given here. Besides the above-mentioned reviews, the reader is referred to M. A. Konstantinova-Shlezinger, *Fluorimetric Analysis*, Israel Program for Scientific Translations, Jerusalem, 1965; A. Weissler and C. E. White, "Fluorescence Analysis," in *Handbook of Analytical Chemistry*, L. Meites, Ed., McGraw-Hill, New York, 1963; and T. S. West, "Chemical Spectrophotometry in Trace Characterization," W. W. Meinke and R. F. Scribner, Ed. *Nat. Bur. Std. Monogr. 100* (Government Printing Office, Washington, D.C., 1967).

[b] The common (accepted or trade) name and/or chemical name is given for all reagents.

[c] Species in parentheses are nonmetals that are generally measured by some indirect method. Because of the great number of reagents for boron, it is not placed in parentheses. Of course, some of the metals are also measured by indirect methods (e.g., quenching or catalytic effects in chemical reaction).

fluorimetry. For the years following 1963, the reader is forced to go to the literature to obtain information concerning limits of detection of organic molecules.*

C. PREDICTED RESULTS

Assuming a square sample cell having a uniform concentration of absorbers and fluorescers at all points and assuming the same instrumental system to be used for fluorimetry (or phosphorimetry) and absorption spectrometry of a

* White and Weissler,[6] in their biannual reviews in the "Fundamental Reviews" issue of *Analytical Chemistry* (1970, 1968, 1966, 1964, 1962, ...) often give limits of detection. Even if limits of detection are not supplied, it is a rather simple matter to find a number of articles on the desired molecular species and then to go directly to the source article(s). Also if one is concerned primarily with organic (and inorganic) molecules which are of interest in toxicology,[7] he should consult the extensive listing of molecules and relative detection limits (good, fair, and poor) given for these molecules. For a comparison of limits of detection for several molecules by fluorimetry, phosphorimetry, and absorptiometry refer to Table IV-B-3.

Table IV-B-3. Comparison of Experimental Limits of Detection of Several Organic Molecules Measured by Absorptiometry, Fluorimetry, and Phosphorimetry

Item	Molecule	Limits of Detection (μg ml^{-1})[a]		
		Absorptiometry[b]	Fluorimetry[c]	Phosphorimetry[c]
1	1,2-Benzanthracene	0.03	0.03	0.05
2	2,3-Benzfluorene	—	0.01	0.4
3	1,2-Benzpyrene	0.08	0.03(0.5)	0.02(0.1)
4	3.4-Benzpyrene	0.08	0.003	2.0
5	1,12-Benzperylene	0.08	0.005	0.09
6	Chrysene	0.03	—(0.1)	—(0.1)
7	Coronene	—	0.004(0.1)	0.004(0.02)
8	1,2,3,4-Dibenzanthracene	0.01	0.007	0.09
9	1,2,5,6-Dibenzanthracene	0.01	0.008	0.02
10	1,2,3,4-Dibenzpyrene	0.08	0.07	N
11	1,2,4,5-Dibenzpyrene	0.08	0.007	0.6
12	3,4,8,9-Dibenzpyrene	0.08	0.003	N
13	3,4,9,10-Dibenzpyrene	0.08	0.05	0.5
14	20-Methylcholanthrene	0.3	0.008	N
15	Perylene	0.1	0.002	N
16	Pyrene	0.04	0.002	0.4
17	Triphenylene	0.05	0.3(0.1)	0.003(0.002)
18	Dicumarol	—	0.01	0.001
19	Diphenadione	—	N	1.0
20	Phenindione	—	N	1.0
21	Tromexan	—	10.0	0.01
22	Warfarin	—	0.001	0.01

[a] All fluorimetry and phosphorimetry values for items 1 through 17 except those in parentheses were taken from the article by L. V. S. Hood and J. D. Winefordner, *Anal. Chim. Acta*, **42**, 199 (1968). All fluorimetry and phosphorimetry values in parentheses are due to H. Sauerland and M. Zander, *Erdoel Kohle*, **19**, 502 (1966). All fluorimetry and phosphorimetry values for items 18–22 are due to H. H. Hollifield and J. D. Winefordner, *Talante*, **14**, 103 (1967).

[b] Apparently no tabulation of limits of detection in absorptiometry has been published. Therefore all absorption limits of detection have been estimated using the theoretical expression of $C_{lim} = 2M/\varepsilon b$, where 0.002 is assumed to be the limiting detectable absorbance, M is the molecular weight of the molecule, and ε is the molar extinction coefficient at the peak absorption wavelength. All values of ε are taken from *Spectra Data 1956–57*, O. H. Wheeler and L. A. Kaplan, Ed., Wiley-Interscience, New York, 1966. The cell length is assumed to be 1 cm. The calculated limit of detection is assumed to be limited only by the error in reading the absorbance scale. If experimental conditions are such that the signal-to-noise ratio is less than about unity, then the values in this table will be too low. On the other hand, if the signal-to-noise ratio is greater than about unity, it should be possible to scale expand or reduce the limits of detection below those listed in this table [cf. J. J. Cetorelli and J. D. Winefordner, *Talanta*, **14**, 705 (1967)].

[c] N means not analytically useful.

given species, it is possible to estimate the ratio G of radiances (signals) obtained for a given species measured by fluorimetry (or phosphorimetry) to absorption spectrometry. Equations for the radiances in molecular luminescence spectrometry are given in Section II-C. Because a narrow band of radiation produced with a continuum source-excitation monochromator (or a continuum source-excitation interference filter) or a narrow-line source with a continuum background and an interference filter (e.g., a mercury arc lamp and an interference filter) is generally used in fluorimetry or phosphorimetry, equation IV-A-1 for $G_{\mathrm{AFL/AAL}}$ can be used to estimate the ratio of molecular luminescence ML, radiance (signal) to molecular absorption MA, radiance (signal); that is,

$$G_{\mathrm{ML/MA}} = Y_L \left(\frac{\Omega_E}{4\pi} \right) \qquad \text{(IV-B-1)}$$

where Y_L is the power yield of the luminescence process and $\Omega_E/4\pi$ is the fraction solid angle of exciting radiation collected by the source optics and impinging on the sample cell. Typical values of Y_L and $\Omega_E/4\pi$ are 0.5 and 0.002, and so the value of $G_{\mathrm{ML/MA}}$ will be about 10^{-3} for most organic fluorophors or phosphors. Therefore, the luminescence signals are *always* smaller than the absorption signals—just as for atomic fluorescence signals versus atomic absorption signals. Of course, it is reasonable to suppose that the luminescence signals must be less than the absorption signals because absorption must occur prior to luminescence.

However, it is well known that the limits of detection of many organic molecules in fluorimetry and phosphorimetry are often *much lower* than limits of detection in absorption spectrophotometry (see Table IV-B-3 for a comparison of limits of detection in fluorimetry, phosphorimetry, and absorptiometry for several molecules). Of course, the reason for this is simply that signal size alone does not determine detection limits but rather signal-to-noise ratios.

If noise sources are considered as well as signal sources, then the ratio $L_{\mathrm{MA/ML}}$ of limits of detection in molecular absorption to molecular luminescence spectrometry is

$$L_{\mathrm{MA/ML}} = G_{\mathrm{ML/MA}} \, N_{\mathrm{MA/ML}} \qquad \text{(IV-B-2)}$$

where $N_{\mathrm{MA/ML}}$ is the noise ratio of molecular absorption to molecular luminescence spectrometry for a given molecular species. If $L_{\mathrm{MA/ML}}$ is greater than unity, then the detection limit of absorption spectrometry is greater (poorer) than by luminescence spectrometry; that is, luminescence spectrometry is a more sensitive method.

If an instrumental system, similar to commercial spectrophotofluorometers, is assumed, then it is possible to estimate the total noise for a species measured by a luminescence method and an absorption method assuming the same source. A spectrometric system is used for each method; that is, a spectro-photofluorometer can serve quite nicely for absorption spectrometry if the radiation from the excitation monochromator passes through the sample and then is reflected by a mirror into the emission monochromator (such a double-monochromator arrangement minimizes fluorescence interference in absorption spectrometry with the sample cell after the monochromator and minimizes photodecomposition interference in absorption spectrometry with the sample cell between the source and entrance slit of the monochromator) Assuming the use of a 150-watt xenon arc lamp source and right angle illumination measurement (complete illumination measurement of all luminescence emitted from the sample cell toward the detector), the noise ratio $N_{MA/ML}$ will be about 10^{-4}, which more than compensates for the lower signal ratio.

Therefore, if *noise as well as* signal is accounted for, *lower limits* of detection should result by *inorganic fluorimetry* than by *inorganic absorption spectrometry* and by *organic fluorimetry* (prompt) and *phosphorimetry* than by *organic absorption spectrometry*. This prediction is evident from Table IV-B-3 and by a comparison of the detection limits in Tables IV-B-3, IV-B-2, and IV-B-1 with absorption values listed in the literature. Therefore, the reasons for the popularity of luminescence spectrometric methods for trace analysis of inorganic and organic materials are clear. On the other hand, absorption spectrometry is not as plagued with curvature of analytical curves toward the abscissa at high concentrations, and so absorption spectrometry is generally used for high and intermediate concentrations of absorbers.

2. SELECTIVITY OF ANALYSIS AND METHODS OF COMPENSATION OF INTERFERENCES

a. SPECTRAL INTERFERENCES

(1) *Luminescence Blank Background.* In principle, luminescence background from the blank can be subtracted from the sample luminescence. However, if the background depends directly on the sample luminescence, an interference can arise from the blank.

Raman spectra from many solvents, especially those with carbon–hydrogen and oxygen–hydrogen groups, can cause sharp bands (with narrow emission monochromator slits) or broad bands (with wide emission monochromator slits). Such bands can be mistaken for part of the analyte spectrum. Because most analytically useful solvents contain groups producing Raman bands, the analyst must find means of minimizing their effects. Generally the Raman bands can be distinguished by their sharpness (with narrow emission monochromator slits) or by their variation in wavelength with a change in excitation wavelength.

Raman bands, if troublesome, can be minimized by use of an emission cutoff filter to remove all wavelengths below and including the Raman band; for example, a cutoff filter of about 430 nm will remove the 418 nm Raman band resulting in water excited with 366 nm. Of course, in phosphorimetry with a phosphoroscope, Raman bands do not appear (except as scattered light and then they are weak) because Raman scattering occurs within the lifetime of a vibration. Refer to Figure IV-B-1 for an example of Raman scattering in fluorimetry.

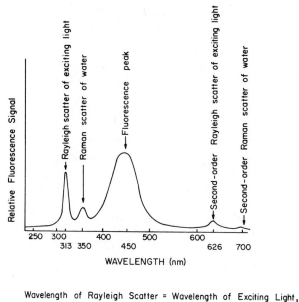

Wavelength of Rayleigh Scatter = Wavelength of Exciting Light, λ_i (in nm)

Wavelength of Raman Scatter = $\lambda_i \left[\dfrac{1}{1 - \lambda_i \widetilde{\Delta\lambda}} \right]$

$\widetilde{\Delta\lambda}$ = Raman shift for solvent, in nm^{-1}

Fig. IV-B-1. Hypothetical fluorescence spectrum of a species in water excited by the 313-nm mercury line. The hypothetical molecule is assumed to have a fluorescence maximum at 450 nm (cf. quinine sulfate in acidic solution).

Fused quartz cells and optics produce a fluorescence and phosphorescence background when short wavelength radiation passes through them. Fused quartz cells used in the front surface excitation mode are particularly prone to this troublesome effect. For high sensitivity, it is necessary to use synthetic quartz and right angle excitation-emission. If the sample cell is placed in a

fused quartz Dewar flask as in low temperature fluorimetry or phosphorimetry, then there is a large blank in the former and a smaller but still troublesome blank in the latter. (A phosphoroscope removes most of the fluorescence
problem.)

Thin-layer-, *paper-*, and *column*-chromatographic substrates containing a
separated sample also produce a troublesome fluorescence and phosphorescence background because of binders used in preparation of the substrates.

Scattered radiation can be either a constant luminescence background upon
which analyte luminescence is superimposed, or it can be a variable luminescence that constitutes a real interference. In the former case, the constant
luminescence is due to nonanalyte (matrix) luminescence, whereas in the
latter case, the variable luminescence is due to the analyte matrix. To minimize scattered radiation, filters can be inserted at the exit slits of the excitation and/or emission monochromators to assure the passage of only the band
of interest. The best means of minimization of scatter signals (of wavelength
different from that of the band of interest) passing through the monochromator exit slits is to employ double monochromators. However, the latter
case is not possible with existing single monochromators (i.e., monochromators having *one* dispersing element) because such a change must generally
be made at the factory. Some commercial spectrophotofluorimeters employ
double-excitation and emission monochromators; these monochromators
have much lower scattered light levels, slightly better resolution than similar
single monochromators, and lower optical speed.

(2) *Luminescence Concomitant Background.* Impurities in solvents, reagents,
and the samples are often quite fluorescent or phosphorescent. This is most
troublesome at low analyte concentrations. If such impurities are the
same in all standards and blanks, then it is possible to *subtract out* the luminescence contribution. However, even if they can be subtracted out, noise contributions due to concomitant luminescence (both directly measured luminescence and scattered luminescence) will still affect the signal-to-noise ratio.
Of course, if the impurities vary in concentration from sample to sample or
from blank to sample, then these nonrandom errors will restrict the limiting
detectable concentration and will often also lead to misinterpretation of the
luminescence signal.

Some solvents* can be easily purified. For example, it is possible to purify

* A tabulation of solvents that form clear, rigid glasses at 77°K appears in Table IV-B-4.
These solvents, as well as many others, are also useful at room temperature where no restrictions are present except those pertinent to the chemistry of the analyte (e.g., pH, ionic
strength, viscosity). It should be stressed that solvents forming clear, rigid solids at 77°K (or
other low temperature) are *not* needed if the new rotating sample cell assembly described in
Section III-C-1 is used. In fact, with the capillary quartz cell, aqueous solutions can now be
studied.

absolute ethanol as received from the supplier by refluxing in a 5-ft, vacuum-jacketed distillation column (1 in. i.d.) packed with 3/32-in. helices and collecting only the middle 70% of the distillate (at reflux ratio of 20:1) after an 8-hour reflux. The product, which is purer than any commercially purchased ethanol, should be stored in glass bottles with aluminum-lined caps; this ethanol has a lower background than the residual quartz fluorescence. Tap water should be doubly distilled to remove fluorescence impurities. Deionizers are also effective purifiers of tap water as long as the first portion of the deionized water is rejected.

Hydrocarbons, such as hexane, isopentane, and heptane, can be purified simply by distillation as described for ethanol or by drying over anhydrous sodium sulfate or sodium-ribbon and then passed through a 2-ft column containing 200 mesh activated silica gel. Ethyl ether can also be purified by the distillation method.

A good means of checking for blank luminescence is to excite the blank at a short wavelength ($\sim$250 nm) and to observe all background above the Raman band ($\sim$280 nm). The absence of any luminescence would imply either no luminescence interference or complete absorption of the exciting radiation by the blank (the latter effect can be easily checked by an absorption measurement).

It should be stressed that contamination of glassware is probably the most serious form of interference in all types of solution fluorimetry and phosphorimetry. In phosphorimetry (and probably also in fluorimetry), contamination can be minimized by rinsing the sample cells at least three times with the new sample and placing the sample cell with solution each time in an ultrasonic cleaner for about 30 sec. Cells should be wiped with Kim Wipe or similar tissue each time the cell is used. All glassware used in luminescence studies should be soaked in a concentrated nitric acid bath for about 24 hours, then rinsed with tap water and redistilled water, and oven dried. An ultrasonic cleaner is an effective means of cleaning glassware. Such procedures as those just described remove tightly adsorbed organic films, and the resulting background can be lowered by as much as 100-fold by the procedures named. One word of caution concerning cleanliness—if the nitric acid bath becomes seriously contaminated, it should be replaced with nitric acid.

If the samples (e.g., tobacco tars, cigarette smoke, blood, urine, petroleum fractions) contain *complex mixtures* of aromatic and heterocyclic hydrocarbon molecules—especially polynuclear systems—then an *instrumental separation method, a chemical separation method, or a physical separation method* must be used. These methods are described in the next section. Of course, *if instrumental separation techniques can be used, then they should be, since they are much faster, simpler, and less prone to additional contamination than the physical separation methods.*

b. PHOTODECOMPOSITION

Photodecomposition[1] may result in two basic types of errors. If the analyte itself undergoes photochemical reaction, it may be consumed appreciably during the time of measurement, yielding low results (case I). If the reagent used to react with the analyte to form a luminescent product undergoes photochemical reaction to produce a different fluorescent product, then the blank luminescence will change during the measurement, causing high results (case II).

According to Parker,[1] the percentage rate of decomposition of the analyte (case I) is given by

$$\% \text{ decomp} = \left(\frac{100}{C_A}\right)\frac{dC_A}{dt} = 2300\varepsilon_A E_o Y_{Ph} \qquad \text{(IV-B-3)}$$

where E_o is the incident light irradiance, (photon sec^{-1} cm^{-2} of cross section of beam), Y_{Ph} is the photochemical quantum yield, ε_A is the extinction coefficient of the analyte at the exciting wavelength, and $2300\varepsilon_A C_A E_o$ is the rate of light absorption* (einstein sec^{-1} liter^{-1}). Substituting typical values ($\varepsilon_A \sim 10^4$, $E_o \sim 8.6 \times 10^{-8}$ einstein^{-1} cm^{-2}, $Y_{Ph} \sim 0.1$ for an unfavorable case), then the percentage decomposition rate would be about 20% per second for an unstirred solution. If the solution were stirred and the total volume of the cuvette were ten times the illuminated volume, then the rate of decomposition would be about 2% per second. To minimize the photodecomposition rate, it is necessary to decrease the intensity of the source and increase the sensitivity of the detector, which would also probably necessitate use of a longer instrumental time constant.

For the second case of photodecomposition interference, the analyte A reacts with reagent R to produce a fluorescent reaction product B; also, the excess reagent B undergoes a photodecomposition to produce a fluorescent product P. The fractional change in fluorescence with time of the fluorescent decomposition product dS_P/dt per fluorescence of the analyte normalized to unit concentration S_A/C_A is given by

$$\frac{dS_P/dt}{S_A/C_A} = 1000 E_o Y_{Ph} k_R \left(\frac{\varepsilon_P Y_P}{\varepsilon_B Y_B}\right) \qquad \text{(IV-B-4)}$$

where ε_B and ε_P are the extinction coefficients (liter mole^{-1} cm^{-1}) for B and P, Y_B and Y_P are the fluorescence quantum yields of B and P, E_o is the incident irradiance (einstein sec^{-1} cm^{-2}), and k_R is the fraction of radiation absorbed per centimeter by the excess reagent R. Again according to Parker,[1] if reasonable values ($E_o \sim 8.6 \times 10^{-8}$ einstein sec^{-1} cm^{-2}, $Y_{Ph} \approx 1.0$,

---

* 1 einstein = $Nh\nu$ ergs, where N = Avogadro's number and $h\nu$ is the energy of the photon.

Table IV-B-4. Solvents forming Clear Glasses at 77°K[a]

Type	Composition (Volume Ratio)
Hydrocarbons	Isopentane
	3-Methyl pentane
	Pentane [tech grade][n-pentane : isopentane (1:1)]
	Petroleum ether (58–60°C fraction)
	n-Pentane : n-heptane (1:1)
	Methyl cyclohexane : n-pentane (4:1 to 3:2)
	Methyl cyclohexane : isopentane (4:1 to 1:5)
	Methyl cyclohexane : methyl cyclopentane (1:1)
	Pentene [mixed isomers]
	Cyclohexane : decalin (1:3)
Basic	Ethanol : NH$_3$ [28% aq] (< 20:1)
	Ethanol : NaOH [0.5% aq] (< 20:1)
	Triethylamine : diethyl ether : isopentane)3:1:3)
	Triethylamine : diethyl ether : n-pentane (2:5:5)
	Methyl hydrazine : methylamine (2:4)
	Diethyl ether : ethanol : NH$_3$ [28% aq] (10:9:1)
Acidic	Trifluoroacetic acid : methylamine : trimethyl amine (4:11:5)
	Ethanol : HCl concn (19:1)
Alcoholic	Ethanol [< 5% H$_2$O]
	n-Propanol
	Ethanol : methanol (4:1 to 5:2)
	Isopropanol : isopentane (3:7)
	Ethanol : isopentane : diethyl ether [EPA] (2:5:5)
	Isopropanol : isopentane : diethyl ether (2:5:5)
	n-Propanol : isopentane (3:7)
	n-Propanol : isopentane (2:8)
	Isopropanol : isopentane (2:8)
	Ethanol (96%) : diethyl ether (2:1)
	Ethanol : glycerol (11:1)
	n-Propanol : diethyl ether (2:5)

$\varepsilon_P Y_P \approx \varepsilon_B Y_P$, and $k_R \sim 0.01$) are assumed, then $(dS_P/dt)/(S_A/C_A)$ 8.6 × 10^{-7} mole liter^{-1} sec^{-1}, and so in a still solution, the fluorescence blank would increase by an amount equivalent to a concentration of about 10^{-6} mole liter^{-1} for every second of exposure to the exciting beam; this is certainly an appreciable change for dilute solutions. Even if the solution were mixed and the cuvette volume were larger, this type of photodecomposition would seriously limit the lowest detectable concentration of analyte.

Table IV-B-4 (*Continued*)

Type	Composition (Volume Ratio)
	n-Butanol : Diethyl ether (2 : 5)
	Ethanol : methanol : diethyl ether (8 : 2 : 1)
	Diethyl ether : isooctane : isopropanol (3 : 3 : 1)
	Diethyl ether : isooctane : ethanol (3 : 3 : 1)
	Diethyl ether : isopropanol (3 : 1)
	Isopropanol : methyl cyclohexane : isooctane (1 : 3 : 3)
	Ethyl cellusolve : *n*-butanol : *n*-pentane (1 : 2 : 10)
Halides	2-Bromobutane
	EPA : CHCl$_3$ (12 : 1)
	Ethyl iodide : isopentane : diethyl ether (1 : 2 : 2)
	Ethyl bromide : methyl cyclohexane : isopentane : methyl cyclopentane (1 : 4 : 7 : 7)
	Ethanol : methanol : ethyl iodide (16 : 4 : 1)
	Ethanol : methanol : propyl chloride (16 : 4 : 1)
	Ethanol : methanol : propyl bromide (16 : 4 : 1)
	Ethyl iodide : ethanol : methanol (5 : 16 : 4)
	Ethanol : isopentane : diethyl ether : dry HCl (1 : 1 : 1 : 0.01 M)
Ethers	Diethyl ether
	2-Methyl tetrahydrafuran
	Di-*n*-propyl ether : isopentane (3 : 1)
	Di-*n*-propyl ether : methyl cyclohexane (3 : 1)
	Di-*n*-propyl ether : *n*-pentane (2 : 1)
	Diethyl ether : isopentane (1 : 1 to 1 : 2)
	n-Butyl ether : isopropyl ether : diethyl ether (3 : 5 : 12)
	Diethyl ether : isopentane : DMF : ethanol (12 : 10 : 16 : 1)

[a] This table is taken in part from a chapter by J. D. Winefordner, P. A. St. John, and W. J. McCarthy, "Phosphorimetry as a means of Chemical Analysis," in *Fluorescence Assay in Biology and Medicine*, S. Udenfriend, Academic Press, New York, 1969.

C. INSTRUMENTAL SEPARATION METHODS

Four major instrumental means exist for minimizing spectral interferences.

(1) *Excitation Resolution.* *Excitation resolution* can be achieved if the analyte can be excited by a wavelength band that will not excite the interferences (see Figure IV-B-2).

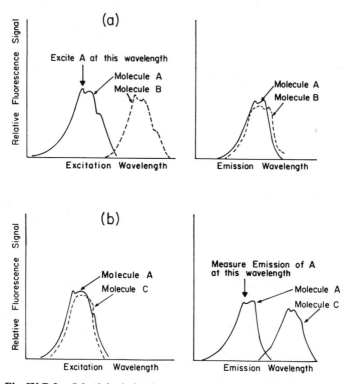

Fig. IV-B-2. Selectivity in luminescence spectrometry: (*a*) excitation resolution of molecule A in a mixture with a similar luminescent molecule B, (*b*) emission resolution of molecule A in a mixture with a similar absorbing molecule C.

(2) *Emission Resolution.* *Emission resolution* can be achieved if the analyte luminescence can be measured with minimal interference from the concomitants; that is, minimal overlap of the analyte luminescence and the concomitant luminescence (see Figure IV-B-2). The foregoing methods of resolution can be carried out if it is possible to know the excitation and emission luminescence spectra of the analyte and concomitants.

(3) *Time Resolution.* *Time resolution luminescence spectrometry* is a rather new method but has even greater potential than the above-mentioned resolution methods because the luminescence lifetimes of molecules are more variable than the peaks of luminescence and absorption spectra. For example, if two molecules have different phosphorescence lifetimes, we can separate their luminescence by either a *phosphoroscopic resolution method* or a *graphical method.* In the latter (see Figure IV-B-3), two phosphors of different lifetimes and

similar steady state radiances result in a decaying signal once the exciting radiation is completely terminated. If the logarithm of the decaying signal is observed as a function of time after termination of the exciting radiation, then it is a simple process to extrapolate the linear portion of the slow decaying species to the termination time to determine the steady state signal of the low decaying species. The fast decayer signal (assuming *only* a two-component mixture) can be determined readily by subtraction. This technique is analogous to the one used to separate radioactive isotope lifetimes. This method of time resolution utilizes the basic phosphorimetric system with a means of rapidly terminating the light source and a means (recorder, oscilloscope, oscillograph, etc.) of following the phosphorescence decay. This method could serve in fluorimetry, but with far greater experimental difficulties; namely, the instantaneous termination of the light source is much more difficult and must be done electronically (e.g., some flash lamps produce sufficiently fast pulses) and the rapidly decaying signal must be followed with rapid responding electronics. The latter problem is simpler to overcome than the former (a pulsed laser could be used). Therefore, time resolution by the graphical method has only been applied to long-lived phosphorescent species, but it could also be applied to rare earths excited by means of energy transfer (luminescence lifetimes of order of 1 msec).

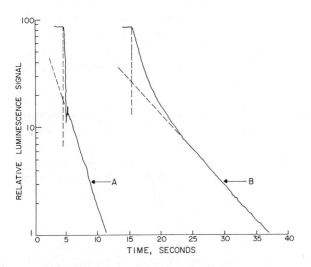

Fig. IV-B-3.   Graphical method of time resolution: logarithmic decay curve *A*, benzoic acid $(3.00 \times 10^{-4} \ M, \ \tau = 2.3 \ \text{sec})$ and benzaldehyde $(6.45 \times 10^{-5} \ M, \ \tau < 0.1 \ \text{sec})$; logarithmic decay curve *B*), tryptophan $(6.04 \times 10^{-6} \ M, \ \tau = 6.4 \ \text{sec})$ and tyrosine $(5.11 \times 10^{-6} \ M, \ \tau = 1.4 \ \text{sec})$. Curves are identical to those in P. A. St. John and J. D. Winefordner, *Anal. Chem.* **39**, 500 (1967). (With permission of the authors.)

Time resolution phosphorimetry can also be achieved by means of phosphoroscopic resolution. By variation of the speed of the phosphoroscope can,* which varies the time between the end of excitation and the start of measurement of luminescence, it is possible to measure a spectrally similar two-component mixture. By using a slow can speed, the slow decayer can be measured; with a fast can speed, both components can be measured. Because phosphors emit independently, it should be possible to resolve a multicomponent mixture by making signal measurements at as many phosphoroscope can speeds as components and then to solve simultaneous equations. For each can speed, analytical curves must be obtained of pure standards of each component of the mixture. This method has only been applied analytically in phosphorimetry. Obviously, the mechanical-type phosphoroscope will not be usable for fluorimetry; that is, the can will not spin rapidly enough (see Section III-C-3). A repetitively modulated flash lamp, however, should serve for time resolution of fast decaying fluorophors.

(4) *Thermal Resolution.* It is often possible to resolve two molecules by simply lowering the temperature of the system. This is especially effective if one molecule is appreciably vibrationally excited, for then lowering the temperature makes excitation resolution feasible.

d. CHEMICAL SEPARATION METHODS

Simple chemical reactions of the analyte with a reagent may lead to a luminescent species possessing unique luminescence characteristics that allow the analyst to use one of the above instrumental means of separation just described. However, this method depends on the analyte itself, the chemical reactions possible with the analyte, and the resulting chemical and spectral properties of the reaction product.

Selective quenching of fluorescence has been used by Zander[4] to determine certain polynuclear aromatic molecules. In this method, a heavy atom solvent is added which *greatly* reduces the fluorescence of most polynuclear aromatic molecules. One of the exceptions is perylene, which can be determined in a complex mixture of polynuclear hydrocarbons by use of a solvent containing iodomethane and EPA; all species other than perylene have greatly reduced fluorescence† signals.

e. PHYSICAL SEPARATION METHODS

In some instances, the only possible means of minimizing interferences is to physically separate the analyte from the interferences. Both single-stage (extraction) and multistage (chromatography) methods can be used. A

---

* With a flash lamp system, the flashing rate could be varied.
† But the phosphorescence signals are greatly increased.

discussion of these methods is outside the scope of this volume, but the literature on them is extensive, as is the literature on the combination of chromatographic or extraction separation step preceding the luminescence measurement step.

3. ANALYTICAL CURVES[2,3]

Typical experimental analytical curves for the fluorescence of quinine sulfate and anthracene appear in Figure IV-B-4; Figure IV-B-5 displays the phosphorescence values of indole butyric acid and benzaldehyde. Also illustrated in Figure IV-B-4 is the influence of incomplete illumination of the cell and incomplete measurement of the luminescence. As in atomic fluorescence spectrometry, incomplete illumination of the sample cell and incomplete measurement of the luminescence results in smaller signals at low concentrations and in deviation from linearity at lower concentrations than in

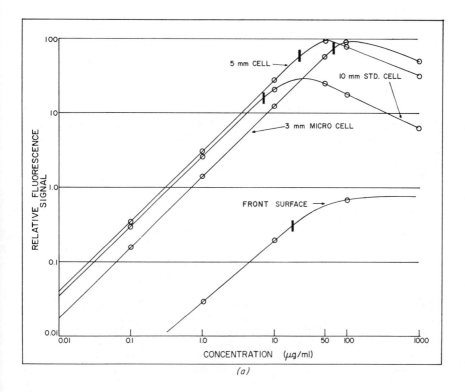

(a)

Fig. IV-B-4. Fluorescence analytical curves: (a) quinine sulfate.

the case of complete illumination measurement. Although front surface il-lumination minimizes these problems, other difficulties associated with light scattering and background cuvette luminescence arise. Therefore, *right angle illumination with complete illumination of the cell and complete measurement of the luminescence in solution luminescence spectrometry is recommended for all analytical studies*.

4. APPLICATIONS OF INORGANIC FLUORIMETRY AND ORGANIC FLUORIMETRY AND PHOSPHORIMETRY

Two recent reviews of fluorimetry covering the time periods of 1966 to 1968 and 1968 to 1970 list nearly 1500 references on the analysis of various inorganic and organic compounds utilizing fluorimetry and phosphorimetry

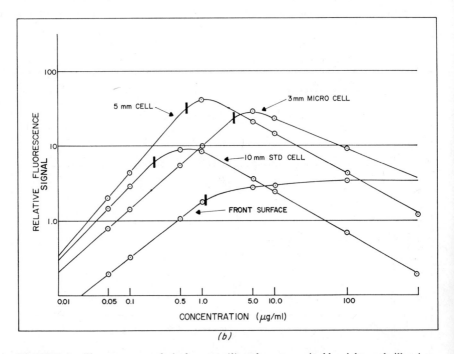

(b)

Fig. IV-B-4. Fluorescence analytical curves: (*b*) anthracene excited by right-angle illumina-tion measurement in square 10-mm, square 5-mm and round micro cells (only center 5 millimeters of 10-mm cell is illuminated and measured) and by front surface illumination measurement. These curves were measured with an Aminco SPF and the corresponding cells (the front surface cell holds less sample than the 5-mm cell, resulting in lower signals at low concentrations). These curves are courtesy of American Instrument Company, Silver Spring, Md.

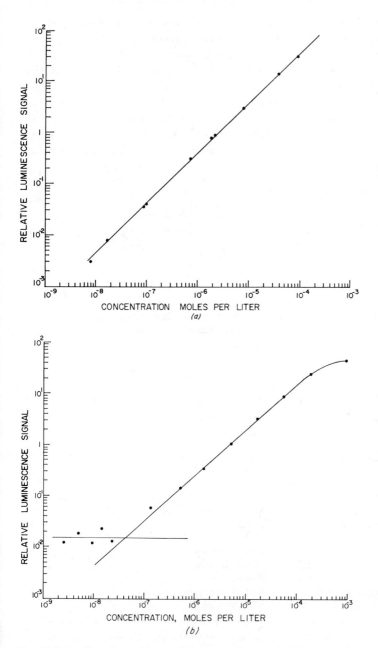

Fig. IV-B-5. Phosphorescence analytical curves: (*a*) indole butyric acid, (*b*) benzaldehyde excited by right angle illumination measurement in 2-mm round cells. These curves were measured with an Aminco SPF with phosphoroscope attachment. Samples were at 77°K.

295

as the measurement step. In addition, several textbooks on fluorimetry and one on phosphorimetry may be consulted for previous uses of these luminescence techniques. Also, in a subsequent volume in this series, P. A. St. John will give many applications of fluorimetry and phosphorimetry. Therefore, no detailed procedures for actual analyses are given here, and only a tabular summary of applications is provided.

Table IV-B-5 lists the general types of applications of quantitative inorganic fluorimetry. Table IV-B-6 supplies an extensive listing of the general types of applications of quantitative organic fluorimetry and phosphorimetry. Table IV-B-7 consists of a partial listing of organic molecules studied previously by quantitative organic fluorimetry and phosphorimetry. By combinations of Tables IV-B-1 and IV-B-5, the specific types of analyses possible by inorganic fluorimetry can be deduced. Similarly, by combination of Tables IV-B-6 and IV-B-7, the specific types of analyses possible by organic fluorimetry and phosphorimetry can also be determined. These tables are not meant to be exhaustive but rather illustrative of the types of studies possible using luminescence methods.

Table IV-B-5. Application[a] of Quantitative Inorganic Fluorimetry and Atomic Fluorescence Flame Spectrometry

Type of Analysis	Type of Species	Method[b]
Agricultural	Water	IF, AF
Agricultural	Plant materials	IF
Agricultural	Food, beverages	IF
Agricultural	Soils	IF
Biological	Blood, spinal fluid	IF, AF
Biological	Saliva	IF
Biological	Urine, feces	IF
Biological	Tissue, bone	IF
Mineral	Rocks, ores, minerals, etc.	IF
Mineral	Petroleum (and products)	IF, AF
Metallurgical	Metals	IF, AF
Industrial products	Cement	IF
Industrial products	Industrial solutions	IF
Industrial products	Chemical products	IF
Industrial products	Plastics	IF
Industrial products	Refractory materials	IF
Industrial products	Glass, ceramics	IF

[a] Application means IF or AF have been used for the type of analysis designated and results have been published.
[b] IF = Inorganic fluorimetry.
   AF = Atomic fluorescence spectrometry.

Table IV-B-6. Applications of Quantitative Organic Fluorimetry and Phosphorimetry[a]

Classification of Analysis	Type of Species[b]	Type of Study	Method[c]
Organic	Aromatic hydrocarbons, heterocycles (native)	In smoke condensate	F, P
Organic	Aromatic hydrocarbons, heterocycles (native)	In air pollution condensate	F, P
Organic	Aromatic hydrocarbons, heterocycles (native)	In automobile exhaust	F
Organic	Aromatic hydrocarbons, heterocycles (native)	In coal tar	F, P
Organic	Aldehydes (Rx)	In air pollution condensate	F
Biochemical	Amino acids (dyes)	In biological materials	F
Biochemical	Amino acids (aromatic, native)	In biological materials	F
Biochemical	Proteins (native)	In biological materials	F, P
Biochemical	Proteins (native)	In biological materials	F
Biochemical	Proteins (dye)	In biological materials	F
Biochemical	Peptides (dye)	In biological materials	F
Biochemical	Proteins (native)	Structure, confirmation studies	F, P
Biochemical	Proteins (dye)	Structure, confirmation studies	F
Biochemical	Antigens, antibodies (dyes)	Identification studies	F
Biochemical	Antigens, antibodies (dyes or Rx)	Strichiometry, kinetics of interaction	F
Biochemical	Vitamins (native or Rx)	In biological materials	F
Biochemical	Coenzymes (Rx)	In biological materials	F
Biochemical	Coenzymes (Rx)	Structures, mechanism of Enzyme–Coenzyme interaction, and photochemical reaction	F
Biochemical	Carbohydrate (dehydration + Rx)	In biological materials	F
Biochemical	Keto acids (Rx)	In biological materials	F
Biochemical	Aldehydes and ketones (Rx)	In biological materials	F
Biochemical	Lipids (Rx)	In biological materials	F
Biochemical	Nucleic acids (native or Rx)	In biological materials	F
Biochemical	Nucleic acids (native)	Energy-transfer studies	F, P
Biochemical	Nucleic acids (dyes)	Structure, confirmation	F

Table IV-B-6 (*Continued*)

Classification of Analysis	Type of Species[b]	Type of Study	Method[c]
Biochemical	Porphyrins (native)	In biological materials	F
Biochemical	Porphyrins (native)	Structure	F
Biochemical	Enzymes (Rx)	In biological materials	F
Biochemical	Enzymes (Rx)	Mechanisms of interaction of enzymes and substrates	F
Pharmacological	Antimalarial drugs (native or Rx)	In biological systems or drugs	F, P
Pharmacological	Antibiotics (Rx)	In biological systems or drugs	F, P
Pharmacological	Sulfonamides (native)	In biological systems or drugs	F, P
Pharmacological	Antitubercular (Rx)	In biological systems or drugs	F
Pharmacological	Antiparasitic (native or Rx)	In biological systems or drugs	F
Pharmacological	Analgesics (native or Rx)	In biological systems or drugs	F, P
Pharmacological	Sedatives and tranquilizers (native or Rx)	In biological systems or drugs	F, P
Pharmacological	Antidepressants (native or Rx)	In biological systems or drugs	F
Pharmacological	Hallucinogenic agents (native or Rx)	In biological systems or drugs	F
Pharmacological	Parasympathetic agents (native or Rx)	In biological systems or drugs	F, P
Pharmacological	Antihistamines (Rx)	In biological materials or drugs	F, P
Pharmacological	Sympathomimetic agents (Rx)	In biological materials or drugs	F, P
Histological	Cell membranes (dyes)	Membranes, transport mechanism studies	F
Histological	Cell membranes (dyes)	Amino acid transport	F
Histological	Intact cells (native)	Oxidation-reduction mechanisms	F
Histological	Intact cells (native)	Enzyme assay	F
Histological	Intact cells (native)	Protein study–energy transfer	F
Histological	Intact cells (native)	Interaction of dyes and nucleic acids	F

298

Histological	Intact cells (native)	Identification of foreign substances	F
Agricultural	Chlorophyll (native)	In plants	F
Agricultural	Chlorophyll (native)	Mechanism of photosynthesis	F
Agricultural	Alkaloids (native or Rx)	In plants	F
Agricultural	Protein (native)	In foods	F
Agricultural	Pesticides (native or Rx)	In foods	F
Agricultural	Aflatoxins (native)	In foods	F
Agricultural	Foods (native)	Grading of foods	F
Agricultural	Seeds (native)	Identification	F
Industrial	Metals (native)	Detection of flaws	F
Industiral	Rubber (native)	Study of aging of vulcanization and grading	F
Industrial	Oils (native)	Purity check of oils	F, P
Industrial	Oils (native)	Oils in water, industrial solutions, etc.	F
Industrial	Fibers (native)	Identification	F
Geological	Oils (native)	Detection of oil deposits	F
Geological	Coals (native)	Quality grading	F
Geological	Minerals (native)	Identification and assay	F

[a] Listing is not meant to be exhaustive but indicative of the many possible areas of quantitative organic fluorimetry and phosphorimetry for real analyses (application means techniques have been used for real analyses and results published).

[b] Only the general classes of species are listed.

The type of measured species appears in parentheses; "native" means that the native organic molecule is measured, "dye" mean that a dye is formed with the species of concern, "Rx" means a chemical reaction is necessary prior to measurement. Only the common measurement forms are listed. If, for example, only "native" is listed, this means that generally—not necessarily always—the most common measurement form is the native organic molecule.

[c] F = Fluorimetry.

P = Phosphorimetry.

Table IV-B-7. Molecular Species Studied by Quantitative Fluorimetry and Phosphorimetry[a]

Species[b]	Measurement Method[c]	Species[b]	Measurement Method[c]
Acenaphthene	F/P	Amino methylhydroxy-	
Acenaphtho quinoxalines	F	benzyl alcohol	F
Acetoacetic acids	F/P	Amino methyl	
Acetol	F	phthalimides	F
Acetophenone	P	Amino methyl	
Acetylcholine	F	pyrimidine	P
Acetyl coenzyme A	F	Aminopeptidase	F
Acetyl fluphenazine	F	Aminophenol	F
Acetyl β-gluco saminidase	F	Aminopterin	F
Acetyl pyridine	F/P	Aminopyrene	F
Acetylsalicylic acid	F/P	Aminoquinolines	F
Acid anhydrides	F	Aminosalicylic acid	F
Acid chlorides	F	Aminotyrosine	P
Acridanone	F	Amobarbital	F
Acridines	F/P	Amoprolium	F
Acrolein	F	AMP	F
Acrylaldehyde	F	Amphetamine	F
ACTH	F	Amylase	F
Actinomycin	F	Anabasine	P
Adenine	F	Analine	F
Adenochrome	F/P	Antazocine	F
Adenosine	F	Anthracene	F/P
Adenylic acid	F	Anthranilic acid	F/P
ADP	F	Anthrone	F
Aflatoxins	F	Antimycin	F
Ajmalicine	F	Antrycide	F
Albumin	F	APAD	F
Alcohol dehydrogenase	F	Apomorphine	P
Aldosterone	F	Aramite	P
Alkyl hydantoin	F	Arginine	F
Allethrin	F	Arterenol bitartrate	P
Alloxan	F	Ascorbic acid	F
Allylamine	F	Asparagine	F
Allyl glycine	F	Aspartic acid	F
Allyl morphine	F	Atabrine	F
Amidases (trypsinlike)	F	ATP	F
Amidinohydrazones	F	Atropine	F/P
Aminacrine	F	Ayapin	F
Amino acid oxidases	F	Azafluoranthene	P
Amino acridine	F	Azafluorene	P
Aminobenzoic acid	F/P	Azaguanine	F/P
Aminoethyl phenol	F	Azaindoles	F
Aminofluorene	P	Azapyrene	P

300

Table IV-B-7 *(Continued)*

Species[b]	Measurement Method[c]	Species[b]	Measurement Method[c]
Barbiturates	F	Carphenazine	F
Barbituric acid	F	Catalase	F
Barthrin	F	Catechol *o*-methyl-	
Bayer pesticides	F/P	transferase	F
Benhepazone	F	Cellulase	F
Benomyl	F	Chinaldin	P
Benzacridines	F	Chinolin	P
Benzaldehyde	P	Chlorcyclizine	F
Benzanthracenes	F/P	Chlordiazepoxide	F
Benzanthrone	F	Chloridazine	F
Benzil	F	Chloroaminobenzoic	
Benzimidazole	P	acid	P
Benzocaine	P	Chlorobenzilate	P
Benzocarbazoles	F/P	Chlorophenol	P
Benzochinolin	P	Chlorophenothiazine	F
Benzochrysenes	F	Chlorophenoxyacetic	
Benzoflavines	F	acid	P
Benzofluoranthenes	F/P	Chlorophyll	F
Benzofluorenes	F/P	Chloroquine	F
Benzoic acid	F/P	Chlorothen	F
Benzoperylenes	F/P	Chlorpheniramine	F
Benzophenanthridines	F	Chlorphenirazine	F
Benzopyrenes	F/P	Chlorpromazine	F/P
Benzoquinolines	F	Chlorprothixene	F
Benzylamine	F	Chlortetracycline	F/P
Bile acids	F	Chymotrypsinogin	F
Biphenyl	P	Cholesterol	F
Biphenylene	F	Cholinesterase	F
Brazan	P	Chrysene	F/P
Brazanquinone	F	Chymotrypsin	P
Bromoacetophenone	P	Cinchona alkaloids	F
Bromo-LSD	F	Cinchonine	F/P
Bromridazine	F	Cinchonipine	F
Brucine	F/P	Cinchophen	P
Buquinolate	F	Citric acid	F
Bufotenine	F	Cocaine	P
Butacaine	P	Codeine	F/P
Caffeine	P	Colchicine	P
Camptothecin	F	Coproporphyrin	F
Carbaryl	F	Co-Ral	P
Carbazoles	F/P	Coronene	F/P
Carbinoxamine	F	Corticosterone	F
Carbohydrates	F	Cortisol	F
Carotene	F	Cortisone	F

301

Table IV-B-7 *(Continued)*

Species[b]	Measurement Method[c]	Species[b]	Measurement Method[c]
Coumaphos	F	Digitalis	F
Coumarin	F	Digitoxigenin	F
Creatine	F	Digitoxin	F
Creatine phosphokinase	F	Dihydroxyfolate	
Cyclaine	P	reductase	F
Cyclizine	F	Dihydroxymandelic	
Cyclohexane 1.4 dione	F	acid	F
Cyclopentaphenanthrene	F	Dihydroxynaphthacene	F
Cysteine	F	Dihydroxyphenylacetic	
Cystine	F	acid	F
Cytochrome C	F	Dihydroxyphenylalanine	
Dantrolene	F	(DOPA)	F
Dapsone	F	Dihydroxyphenyl-	
DDD	P	ethylamine	
DDE	P	(DOPA-amine)	F/P
DDT	P	Dihydroxyphenylserine	F
DEF	F	Dimethoxy-	
Dehydroacetic acid	E	phenylenediamine	F
Dehydrocorticosterone	F	Dimethyl-	
Dehydrogenases	F	aminobenzaldehyde	F
Demethyl-		Dimethylaminoguanine	F
chlortetracycline	F	Dimethylbenzacridines	F
Deoxicholic acid	F	Dimethyl-	
Deoxyribose	F	benzanthracene	F/P
Deserpamine	F	Dimethylguanidine	F
Deserpidine	F	Dimethylphthalate	F
Desmethoxyreserpine	F	Dimetilin	F
Desoxypyridine	P	Diphenyl	F/P
Dexpanthenol	F	Diphenylamine	P
Diacetyl	F	Diphenylbutadiane	F
Diacetyl sulfanilamide	P	Diphenylhydrazine	F
Diaminopimelic acid	F	Diphenylketene	F
Diaminopurine	P	Dipyridazole	F
Diazinon	P	DNA	F
Dibenzacridines	F	Dodecapentanoic acid	F
Dibenzanthracenes	F/P	Doxylamine	F
Dibenzocarbazoles	F/P	Dromoran	F
Dibenzochrysenes	F	Dulcin	F
Dibenzoxanthenes	F	Elastase	F
Dibromoacetophenone	P	Emetine	F
Dichlorophenoxyacetic		Ephedrine	P
acid	P	Epinephrine	F/P
Dicumarol	P	Equilin	F
Diethylstilbesterol	F	Equilinine	F

Table IV-B-7 (*Continued*)

Species[b]	Measurement Method[c]	Species[b]	Measurement Method[c]
Ergocalciferol	F	dehydrogenase	F
Ergot alkaloids	F	Glutamicpyruvic	
Ergotoxin	F	transaminase	F
Esculetin	F	Glutathione	F
Eserine	F	Glutathione reductase	F
Estradiol	F/P	Glycerol	F
Estriol	F	Glycerol dehydrogenase	F
Estrone	F	Glycine	F
Ethanol	F	Glycogen	F
Ethacridine	F	Glyoxylic acid	F
Ethinylestradiol	F	Glyochoclic acid	F
Ethoxydihydro-		Gossypol	F
trimethylquinoline	F	Griseofulvin	F
Ethoxyquin	F	Guanidine	F
Ethylindole acetate	P	Guanidoacetic acid	F
FAD	F	Guanine	F
Flavonoids	F	Guanisoquin	F
Flexin	F	Guanosine	F
Fluoranthenes	F/P	Guthion	F/P
Fluorenes	F	Harmine	F
Fluorescein	F	Heme proteins	F
Fluphenazines	F	Heroin	F
Folic acid	F	Hippuric acid	F/P
Folinic acid	F	Histamine	F
Formaldehyde	F	Histidine	F
Gallic acid	F	Histone	F
Galactose-1-phosphate		Homogentisic acid	F
uridylyl transferase	F	Homovanillic acid	F
Galactosidase	F	Human bone	P
Galanthamine	F	Human dentine	P
Gentisic acid	F	Human enamel	P
Gibberellic acid	F	Hyaluronidase	F
Gitoxigenin	F	Hydrastine	F
Gitoxin	F	Hydrocortisone	F
Glucobrasinine	F	Hydroxyamphetamine	F
Glucose	F	Hydroxyanthranilic	
Glucose dehydrogenase	F	acids	F/P
Glucose phosphate	F	Hydroxybenzaldehydes	F
Glucose-6-phosphate		Hydroxybenzoic acids	F
dehydrogenase	F	Hydroxybutyrate	
Glucosidase	F	dehydrogenase	F
Glucoronidase	F	Hydroxybutyric acid	F
Glutamic acid	F	Hydroxycinnamic	
Glutamic acid		acid (D)	F

Table IV-B-7 (*Continued*)

Species[b]	Measurement Method[c]	Species[b]	Measurement Method[c]
Hydroxycorticosteroids	F	Kelthane	P
Hydroxycorticosterones	F	Kynurenine	F/P
Hydroxycoumarin		Kynuric acid	F
(umbelliferone)	F	Lactic acid	F
Hydroxyethythiamine	F	Lactic acid	
Hydroxyindole	F	dehydrogenase	F
Hydroxyindoleacetic		Lapachol	F
acid	F	Lecithin	F
Hydroxyisopropyl-		Leucine	F
aminopropoxyindole	F	Leucine	
Hydroxykynurenine	F	aminopeptidase	F
Hydroxymandelic acid	F	Leucocyte alkaline	
Hydroxymethylphenyl		phosphatose	F
lactic acid	F	Lidocaine	P
Hydroxyphenylacetic		Lignin sulfones	F
acid	F	Lipases	F
Hydroxyphenylpyruvic		Lithocholic acid	F
acid	F	Lutein	F
Hydroxyphenylserine	F	Lutein epoxide	F
Hydroxytryptamine	F	LSD	F
Hydroxytryptaphane	F/P	Malic acid	F/P
Imidan	P	Malic acid	
Imidazoles	F	dehydrogenase	F
Imipramine	F	Malonaldehyde	F
Inanone	F	Mannosidase	F
Indeno-acenaphthol		Maretin	F
fluoranthenes	F	Matacil	F
Indeno-isoquinoline	F	Mebaral	P
Indeno-pyrene	F	Mechlorethamine	F
Indican	P	Meclizine	F
Indole	F/P	Medazepam	F
Indole acetic acid	F/P	Medroxyprogesterone	
Indole acetyl nitrile	P	acetate	F
Indole butyic acid	P	Menadione	F
Indole carboxylic acid	P	Mephenasin	F
Indole propionic acid	P	Mercaptopurine	F
Indole pyruvic acid	P	Mescaline	F
Indolmethacine	F	Metaaminol	F
Indoxyl	F	Metanephizine	F
Indoxylsulfate	P	Metapyrone	F
Inulin	F	Methandrosterone	F
Isochinolin	P	Methaperylene	F
Isolan	P	Methaqualone	F
Isoniazid	F	Methiomeprazine	F

Table IV-B-7 (*Continued*)

Species[b]	Measurement Method[c]	Species[b]	Measurement Method[c]
Methiophenothiazine	F	Nalidixic acid	F
Methotrexate	F	Naphthacene	P
Methoxybenzoxazolinone	F	Naphthalene	F/P
Methoxychlor	P	Naphthalene acetamide	F/P
Methoxypromazine	F	Naphthalene acetic acid	F/P
Methylacridine	F	Naphthol	F/P
Methylaminocarbostyril	F	Naphthol pyrene	F
Methylanthracene	F	Naphthoxy acetic acid	F
Methylbenzacridines	F	Narceine	P
Methylbenzoquinoline	F	Narcotine	P
Methylchinolin	P	Neocinchophen	F
Methylcholanthrene	F	Nerve gas	F
Methyldibenzopyrene	F	NIA 10242	F/P
Methyldiethyl- aminocoumarin	F	Nicotinamide	F
Methylfluoroindole acetate	P	Nicotine	P
		Nicotinic acid	P
		Nitronaphthalene	F
Methylfluoroindole butyrate	P	Nitrophenols	F/P
Methylfluoroindole propionate	P	Norchlorpromazines	F
		Norepinephrine	F/P
Methylguanidine	F	Normetnephrine	P
Methylhydroxy- carbostyril	F	Nornicotine	P
		Ochratoxin	F
Methylhydroxy- coumarin	F	Orphenaprine	F
		Orthotran	P
Methylindole	F	Oxalic acid	F
Methylisochinolin	P	Oxidases	F
Methylnicotinamide	F	Oxychloroquin	F
Methylphenanthrenes	F	Oxytetracycline	F
Methylphenanthridine	F	Oxythiamine	P
Methylpyrenes	F	PAM	F
Methyltestosterone	F	Pamaquine	F
Methyltrifluoperazine	F	Papaverine	P
Metycaine	F	Parathione	P
MKG 264	P	Penicillin	F
Mobil MC-A-600	F	Pentobarbital	F
Mofin	F	Pentothal	F
Morphine	F/P	Perphenazine	F
Myoglobin	F	Perylene	F/P
NAD	F	Perylenetetracarboxylic dianhydride	F
NADH	F	Phenacetin	P
NADP	F	Phenalen-1-one	F
NADPH	F	Phenanthrenes	F/P

Table IV-B-7 (*Continued*)

Species[b]	Measurement Method[c]	Species[b]	Measurement Method[c]
Phenanthridine	F/P	Propylisome	F
Phenthridone	P	Psilocin	F
Phenazines	P	Pteroic acid	F
Phenindione	P	Pyrazines	F
Pheniramine	F	Pyrenes	F/P
Phenobarbital	F/P	Pyrenoline	F
Phenol	F	Pyrethrum	F
Phenothiazines	F	Pyribenzamide	F
Phenyl acetic acid	P	Pyridines	F
Phenyl alanine	F/P	Pyridoxals	F
Phenyl dibenzacridines	F	Pyridoxamine	F
Phenyl ephrine	P	Pyridoxic acid	F
Phenyl ethylamine	F	Pyridoxine	F
Phenylenediamene	F	Pyridylchlor-	
Phenylenepyrene	F	phenothiazine	F
Phenyl phenol	F	Pyrilamine	F
Phosphamidon	P	Pyrothyldione	F
Phosphatases	F	Pyrolan	F
Phosphoglucoisomerase	F	Pyruvaldehyde	F
Phosphogluconate		Pyruvic acid	F
dehydrogenase	F	Quercetin	F/P
Phthalic acid	F	Quinaldic acid	P
Phthalsulfacetamide	P	Quinidine	F/P
Phthalysulfathiazole	P	Quinine	F/P
Phthocyanines	F	Quinilic acid	P
Physostigmine	F	Rauwolfia alkaloids	F
Picene	F	Rescinnamine	F
Picolinic acid	P	Reserpine	F
Piperonyl butoxide	F	Retene	P
Piperonyl cyclonene	F	Riboflavine	F
Piperoxan	F	RNA	F
Podophyllotoxin	F	Ronnel	P
Porpyrins (including		Rotenone	F
metal complexes)	F	Rutin	F
Potasan	F	Rutonal	F
Prednisolone	F	Salicylaldehyde	F
Presidon	F	Salicylic acid	F/P
Primethamine	F	Salicyloylhydrazide	F
Procainamide	F/P	Salicyluric acid	F
Procaine	F	Sarcosine	F
Prochlorperazine	F	Sarin	F
Progesterone	F	Scoparone	F
Promazine	F	Scopoletin	F
Promethazine	F	Sebacic acid	F

306

Table IV-B-7 (*Continued*)

Species[b]	Measurement Method[c]	Species[b]	Measurement Method[c]
Serotonin	F	Thiamlal	F
Sesamex	F	Thiethylperazine	F
Sesamolin	F	Thiochrome	F
Sevin	P	Thioglycolic acid	F
Skatole	F	Thioridazines	F
Sodium sulfathiazole	P	Thioridothixin	F
Spermine	F	Thiopental	F
Spermidine	F	Thiouracil	P
Stilbamidine	F	Thymidine	F
Streptomycin	F	Thymidylic acid	F
Strychnine	P	Thyine	F
Succinate semialdehyde		Thymol	F
dehydrogenase	F	Tocopherols	F
Succinic acid	F	Toluidine	P
Sulfabenzamide	P	Toluene	P
Sulfacetamide	P	Toxaphene	P
Sulfadiazine	P	TPN	F
Sulfadimethoxine	F	TPNH	F
Sulfafurazole	F	Trichlorophenol	P
Sulfaguanidine	P	Trichlorophenoxyacetic	
Sulfamerizine	P	acid	P
Sulfamethazine	P	Trifluoromeprazines	F
Sulfanilamide	F/P	Trifluoperazine	F
Sulfatases	F	Trifluorophenothiazine	F
Sulfathiazole	P	Triosephosphate	
Sulfenone	P	isomerase	F
Sulfoanthranilic acid	F	Triphenylene	F/P
Surfen	F	Trithion	P
Synephrin	F	Tromexan	P
Tedion	P	Tronothane	P
Telodrin	P	Trypsinogen	F
Terephthalic acid	F	Tubucurarine	F
Terephthalanilide	F	Tyramine	F
Terephthalic acid	F	Tyrosine	F/P
Terphenyl	F	Tyrosine hydroxylase	F
Terramycin	F	Tryptophan	F/P
Testosterone	F	UC 10854	F/P
Tetracycline	F	UC 21149	P
Tetrahydrocortisone	F	Umbelliferone	F
Tetrahydrofolate	F	Urea	F
Tetramethylbenzene	P	Uric acid	F
Thebaine	P	Urobilin	F
Thenyldiamine	F	Valine	F
Thiamine	F	Violaxanthin	F

307

Table IV-B-7 (*Continued*)

Species[b]	Measurement Method[c]	Species[b]	Measurement Method[c]
Violuric acid	F	Xanthine	F
Vitamin A	F	Xanthine oxidase	F
Vitamin B_6	F	Xanthopterin	F
Vitamin B_{12}	F	Xanthurenic acid	F/P
Vitamin C	F	Xylosidase	F
Vitamin D	F	Yohimbine	F/P
Vitamin D_2	F	Zearalenone	F
Vitamin D_3	F	Zechtran	F/P
Warfarin	F/P	Zoxazolamine	F
Xanthen-9-one	F		

[a] An excellent coverage of fluorimetric and phosphorimetric procedures for *quantitative analysis* of organic species has been given by C. E. White (and more recently A. Weissler) for many years in the biennial reviews in *Analytical Chemistry*. Because these reviews are comprehensive and readily available, only a tabular, unreferenced summary of selected organic molecules is given here. Besides the previous reviews, the reader is also referred to M. A. Konstantinova-Shlezinger, *Fluorimetric Analysis*, Israel Program for Scientific Translations, Jerusalem, 1965; A. Weissler and C. E. White, "Fluorescence Analysis," in *Handbook of Analytical Chemistry*, L. Meites, Ed., McGraw-Hill, New York, 1963; S. Udenfriend, *Fluorescence Assay in Biology and Medicine*, Vols. I and II, Academic Press, New York, 1969; J. D. Winefordner, P. A. St. John, and W. J. McCarthy, "Phosphorimetry as a Means of Chemical Analysis," in *Fluorescence Assay in Biology and Medicine*, S. Udenfriend, Academic Press, New York, 1969; M. Zander, *Phosphorimetry*, Academic Press, New York, 1968.

[b] The list of species is *not* meant to be exhaustive. To economize on length, many species appear in the plural form to indicate that certain derivatives may also fluoresce or phosphoresce. In some cases, some of the derivatives are listed separately in the table if the derivative is of considerable importance to medicine, biology, pharmacology, or agriculture. For most molecules, the position of groups within the molecule is not indicated, both to save space and for simplicity. In many cases the accepted (common or trade) name of the species is used rather than the chemical name (in a few cases, both the chemical and accepted names are used in the appropriate position in the table).

[c] An F means that quantitative fluorimetry has been carried out by one or more workers. AP means that quantitative phosphorimetry has been carried out by one or more workers.

5. GLOSSARY OF SYMBOLS

C_A Concentration of analyte, mole liter^{-1}.

D_λ Absorbance per centimeter for a concentration of 1 μg ml^{-1}, cm^{-1}.

E_o Incident light irradiance upon analyte, photon sec^{-1} cm^{-2}.

$G_{ML/MA}$ Ratio of signal (radiance) in molecular luminescence to signal (radiance) in molecular absorption spectrometry for same molecule, no units.

h	Planck constant, erg sec.
k_R	Fraction of radiation absorbed by R per centimeter, cm^{-1}.
$L_{MA/ML}$	Ratio of molecular absorption limit of detection to molecular luminescence limit of detection for same species, no units.
MA	Molecular absorption spectrometry.
ML	Molecular luminescence spectrometry.
$N_{MA/ML}$	Ratio of noise in molecular absorption spectrometry to noise in molecular luminescence spectrometry, no units.
S_A	Luminescence signal due to analyte A, A.
S_B	Luminescence signal due to reaction product B, A.
S_P	Luminescence signal due to photodecomposition product P, A.
t	Time, sec.
Y	Power yield for analyte, no units.
Y_B	Quantum yield for species B, no units.
Y_P	Quantum yield for species P, no units.
Y_{Ph}	Photochemical quantum yield, no units.
$\Delta\lambda'$	Half-intensity width of luminescence band, nm.
ε_A	Molar extinction coefficient of analyte A, liter $mole^{-1}$ cm^{-1}.
ε_B	Molar extinction coefficient of reaction product B, liter $mole^{-1}$ cm^{-1}.
ε_P	Molar extinction coefficient of photodecomposition product P, liter $mole^{-1}$ cm^{-1}.
π	$3.1418\ldots$
Ω_E	Solid angle of exciting radiation collected and incident upon sample cell, sr.

6. REFERENCES

1. C. A. Parker, *Photoluminescence of Solutions*, Elsevier, Amsterdam, 1968.
2. J. D. Winefordner, W. J. McCarthy, and P. A. St. John, "Phosphorimetry as an Analytical Approach in Biochemistry," in *Methods of Biochemical Analysis*, D. Glick, Ed., Vol. 15, Wiley-Interscience, New York, 1967.
3. J. D. Winefordner, P. A. St. John, and W. J. McCarthy, "Phosphorimetry as a Means of Chemical Analysis," in S. Udenfriend, *Fluorescence Assay in Biology and Medicine*, Vol. 2, Academic Press, New York, 1969.
4. M. Zander, *The Application of Phosphorescence to the Analysis of Organic Compounds*, Academic Press, New York, 1968.
5. A. Weissler and C. E. White, "Fluorescence Analysis," in *Handbook of Analytica Chemistry*, L. Meites, Ed., McGraw-Hill, New York, 1963.
6. C. E. White and A. Weissler, "Fluorometric Analysis," in *Anal. Chem.*, **42** (5), 57R (1970); **40**, (5), 116R (1968); **38**, (5), 155R (1966); etc.
7. J. W. Bridges, "Fluorescence Assay," in *Handbook of Analytical Toxicology*, I. Sunshine, Ed., Chemical Rubber Co., Cleveland, Ohio, 1969.

C. COMPARISON OF METHODS OF
LUMINESCENCE SPECTROMETRY

1. GENERAL DISCUSSION OF ATOMIC FLUORESCENCE SPECTROMETRIC METHOD

Atomic fluorescence, including resonance fluorescence, direct-line fluorescence (normal and thermally assisted), and stepwise-line fluorescence (normal and thermally assisted) has already been discussed in detail (Section II-A) because of the great analytical utility of these methods for trace elemental analysis. On the other hand, sensitized fluorescence (also discussed briefly in Section II-A) has been of no analytical use and is unlikely to be analytically useful in flames or other high temperature plasmas.

2. GENERAL DISCUSSION OF SOLUTION LUMINESCENCE SPECTROMETRIC METHODS[1,2]

An excellent treatment of a variety of solution luminescence methods other than fluorimetry and phosphorimetry has been presented by Parker.[1] We have considered normal (prompt) fluorimetry and phosphorimetry in detail (see Section II-B) because these methods can be of significant analytical service for trace analysis of inorganic and organic molecules. Since most solution techniques other than normal fluorimetry and phosphorimetry have been of limited analytical use, only a brief qualitative description of those methods is given here.

a. E-TYPE DELAYED FLUORESCENCE

E-Type delayed fluorescence (named after eosin) is produced via the following mechanism:*

$$S_o \xrightarrow[\text{activation}]{\text{radiative}} S_1 \xrightarrow[\substack{\text{intersystem}\\\text{crossing}}]{\text{radiationless}} T_1 \xrightarrow[\text{activation}]{\text{thermal}} S_1 \xrightarrow[\text{deactivation}]{\text{radiative}} S_o$$

where S_o and S_1 are ground and excited singlet states and T_1 is the lowest triplet state. The radiative deactivation due to $S_1 \rightarrow S_o$ is called E-type delayed fluorescence because it has a long lifetime (two forbidden transitions) and the spectrum of prompt (normal) fluorescence. This type of delayed fluorescence is unlikely to be of great analytical utility.

* Also refer to Section II-B.

b. P-TYPE DELAYED FLUORESCENCE*

According to Parker,[1] P-type delayed fluorescence (named after pyrene) is produced via a triplet, T_1—triplet, T_1 annihilation, namely,

$$2S_0 \xrightarrow[\text{activation}]{\text{radiative}} 2S_1 \xrightarrow[\substack{\text{intersystem} \\ \text{crossing}}]{\text{radiationless}} 2T_1 \xrightarrow[\text{annihilation}]{\text{radiationless}}$$

$$S_{1,2} \xrightarrow[\text{dissociation}]{\text{radiationless}} S_1 + S_0 \xrightarrow[\text{deactivation}]{\text{radiative}} S_0$$

where $S_{1,2}$ is an excited singlet dimer produced by transfer of energy between the triplet molecules. The excited dimer can also emit its own characteristic fluorescence, but the P-type delayed fluorescence is the final radiative deactivation process. This type of delayed fluorescence is also possible without the excited dimer mechanism (i.e., triplet-triplet energy transfer can occur over long distances to produce S_1 directly without need to resort to the excited dimer mechanism. Whatever mechanism is present, P-type delayed fluorescence has considerable intensity at room temperatures for many molecules. It has also been observed at low temperatures (rigid media), but analytical uses of this process (at low temperatures) seem unlikely.

c. SENSITIZED FLUORESCENCE

Energy transfer between an excited singlet donor D_{S1} and a ground singlet acceptor A_{S_0} to produce an excited acceptor A_{S1} which undergoes radiative deactivation is called sensitized fluorescence.

$$D_{S1} + A_{S_0} \xrightarrow[\text{transfer}]{\text{energy}} A_{S1} + D_{S_0}$$
$$\xrightarrow[\text{deactivation}]{\text{radiative}} A_{S_0}$$

According to Förster, the transfer of energy occurs over long distances (e.g., 100 Å) and results most efficiently when there is large overlap between the first absorption band of A and the fluorescence emission band of D and when the donor D has a high fluorescence quantum yield. Also, the transfer of energy rate must exceed (or at least not be appreciable smaller than) the diffusion-controlled quenching rate if the process of sensitized fluorescence is to be of great analytical use. Few analytical applications of sensitized fluorescence have been reported, and the susceptibility of quenching processes and the lack of selectivity make it appear unlikely that many will come to light.

---

* Also refer to Section II-D.

d. SENSITIZED PHOSPHORESCENCE

Energy transfer between an excited (triplet) donor D_{T_1} and a ground singlet acceptor A_{S_0} to produce an excited (triplet) acceptor A_{T_1} which undergoes radiative deactivation is called sensitized phosphorescence.

$$D_{T_1} + A_{S_0} \xrightarrow[\text{transfer}]{\text{energy}} A_{T_1} + D_{S_0}$$

$$\xrightarrow[\text{deactivation}]{\text{radiative}} A_{S_0}$$

This process is spin allowed and has been observed in a number of pairs of organic systems at 77°K in rigid media. For efficient energy transfer via this mechanism, the orbitals of the two molecules must overlap and the triplet vibrational levels of the donor and acceptor must also overlap. Sensitized phosphorescence is quite intense as long as solutions are highly purified to remove oxygen, which is an efficient quencher. Because of the latter necessity, it is unlikely that sensitized phosphorescence will serve in many analytical procedures.

e. SENSITIZED DELAYED FLUORESCENCE

The mechanism of sensitized delayed fluorescence is quite similar to the one for sensitized phosphorescence. However, the entire system is maintained at room temperature where triplet-triplet annihilation of triplet acceptors or triplet acceptor-triplet donor encounters result in production of first excited singlets, which can undergo radiative deactivation.

$$D_{T_1} + A_{S_0} \xrightarrow[\text{transfer}]{\text{energy}} A_{T_1} + D_{S_0}$$

triplet-triplet annihilation $\longrightarrow A_{S_1} \xrightarrow[\text{deactivation}]{\text{radiative}}$

or

$$A_{T_1} + D_{T_1} \longrightarrow A_{S_1} \xrightarrow[\text{deactivation}]{\text{radiative}} A_{S_0}$$

Whatever the mechanism of sensitized delayed fluorescence, Parker[1] has found it to be very sensitive for some binary systems. However, since it seems that the process would be plagued with many sites for quenching, there would be a stringent requirement for the purity of all chemicals.

f. ANTI-STOKES SENSITIZED DELAYED FLUORESCENCE

The excited singlet acceptor A_{S_1} is of greater energy than the excited singlet donor in the anti-Stokes sensitized delayed fluorescence process; therefore it seems likely that energy transfer is via A_{T_1} and D_{T_1} or via two A_{T_1}'s. In any event, the resulting fluorescence has a shorter wavelength than the exciting radiation. Also anti-Stokes delayed fluorescence has not yet been of analytical use, and its of poor sensitivity and selectivity compared with normal fluorimetry and phosphorimetry would seem to make the process a poor candidate for analytical utility in the future.

g. INTRAMOLECULAR ENERGY TRANSFER—DONOR CHELATE-TO-ACCEPTOR METAL ION

Intramolecular energy transfer is a useful means of increasing the sensitivity and selectivity of analysis of some metal ions. If a metal chelate is formed between a metal ion with unfilled d (transition metals) or f (rare earths) levels, transitions of electrons between d or f levels in the metal ion can occur in addition to transitions involving π^*-π and π^*-n singlet and triplet states within the organic portion of the molecule. If the d or f levels are lower than the singlet or triplet π^*-π or π^*-n levels, then intramolecular energy transfer can occur and result in selective excitation of the metal ion.

If the metal ion has energy levels *below* the lowest excited singlet-triplet manifolds, then intersystem crossing can occur to excite the metal and produce narrow line luminescence from the metal ions. Although transition metals are capable of this process, in reality such luminescence seldom occurs because the incompletely shielded d levels are quenched. With rare earth ions, however, the f electrons are well shielded and luminescence appears. More than *one* luminescence line will be produced if more than *one* excited level is between the triplet state and the ground singlet states of the organic.

The energy levels of gadolinium, lanthanum, and lutetium are too high for most donors to excite. The energy levels of praseodymium, neodymium, holmium, erbium, thulium, and ytterbium are low enough for many donors to excite, but quenching processes generally deactivate these ions by nonradiational processes. The energy levels of samarium, dysprosium, europium, and terbium are low enough for many donors to excite, and luminescence is quite intense from these species. Intramolecular energy transfer is a very sensitive means of exciting the latter four rare earths and has been used for actual analytical studies. Of course, as in all types of energy transfer, the possibility of interference is increased because of the increased number of excited species and steps involved in the process

h. INTERMOLECULAR ENERGY TRANSFER—DONOR ORGANIC MOLECULE-TO-ACCEPTOR METAL ION

Intermolecular energy transfer from an organic donor (e.g., benzophenone, has been used to excite low concentrations of samarium, terbium, europium, and dysprosium ions in solution. This process is diffusion controlled, and so measurements should be made at room temperature.

i. ADDITIONAL TYPES OF LUMINESCENCE

Recombination luminescence has been observed for molecules excited in rigid media at low temperature and then warmed up.

Excited dimers can be produced by three basic processes. If A is the monomer, then

$$\text{ordinary dimer} \quad A + A \longrightarrow AA \xrightarrow[\text{excitation}]{\text{radiative}} (AA)^* \qquad (\text{IV-C-1})$$

$$\text{excimer} \quad A + A \xrightarrow[\text{excitation}]{\text{radiative}} A^* + A \longrightarrow (AA)^* \qquad (\text{IV-C-2})$$

$$\text{photodimer} \quad A + A \xrightarrow[\text{excitation}]{\text{radiative}} A^* + A \longrightarrow A_2 \qquad (\text{IV-C-3})$$

Because dimers are significant only at high concentrations of the monomer, they have relatively little analytical use. However *dimers can influence the shape of spectra of species at high concentrations and the shape of analytical curves at high concentrations.* Therefore, *the analyst should be careful to avoid the use of high analyte concentrations* (greater than about 10^{-4} M) *in qualitative and quantitative studies.*

3. COMPARISON OF METHODS OF LUMINESCENCE SPECTROMETRY

Table IV-C-1 presents the general types of luminescence spectrometry and some of the characteristics and analytical capabilities of each type. The method of tabulation is similar to the one used in Parker's[1] fine treatise. This comparison should assist the analyst in his choice of a method for a given analysis.

4. REFERENCES

1. C. A. Parker, *Photoluminescence of Solutions*, Elsevier, Amsterdam, 1968.
2. D. M. Hercules, *Fluorescence and Phosphorescence Analysis*, Wiley-Interscience, New York, 1966.

Table IV-C-1. Comparison of Methods of Luminescence Spectrometry[a]

Method[b]	Species[c]	Type[d]	Substance[e]	Conditions[f]	Desired Temperature (°K)	Radiance of Luminescence[g] Increased Temperature	Increased Radiance Source	Instrument[h]	Sensitivity[i]	Selectivity[i]	Precision[i]
F	Atom	N	IMO	Gf	2000–3000	Complex	B_A^1	Fl. (flame)	VH	VH	H
F	Atom	N	IMO	Gf	2000–3000	Complex	B_A^1	Fl. (flame)	H	VH	H
F	Mol.	N	$\pi^* - \pi 0$.	Fluid	~298	Mod. decrease	B_A^1	Fl.	H	M	VH
				Glass	~298	Mod. decrease	B_A^1	Fl.	L	H	M
				Glass	~77	Mod. decrease	B_A^1	Fl.	H	H	M
				Adsorption	~77–298	Mod. decrease	B_A^1	Fl.	L	M	L
F	In. ion	C	$\pi^* - \pi O.$ (possible $\pi^* - n O$)	Fluid	~298	Mod. decrease	B_A^1	Fl.	VH	M	VH
F	Mol.	C	$\pi^* - \pi 0$	Fluid	~298	Mod. decrease	B_A^1	Fl.	M	M	VH
				Glass	~298	Mod. decrease	B_A^1	Fl.	M	M	M
				Glass	~77	Mod. decrease	B_A^1	Fl.	M	M	M
				Adsorption	~77–298	Mod. decrease	B_A^1	Fl.	M	M	L
F	Mol.	Q	Mainly O_2 FO	Fluid	~298	Mod. decrease	B_A^1	Fl.	L	L	M
DF (E-type)	Mol.	N	FO	Adsorption	~77–298	Mod. decrease	B_A^1	Fl.	M	L	L
				Fluid	~298	Increase	B_A^1	Ph.	(No analytical studies yet, but would estimate L, L, L.)		
DF (P-type)	Mol.	N	FO	Fluid	~298	Complex	B_A^2	Ph.	(Few analytical studies yet, but would estimate L, M, L for both fluid and rigid.)		
SF	Mol.	ET	FO	Rigid	~77	Complex	B_A^2	Ph.	(Few analytical studies yet, but would estimate VL, L, L.)		
				Fluid	~298	Complex	B_A^2	Fl.			

Table IV-C-1 (*Continued*)

Method[b]	Species[c]	Type[d]	Substance[e]	Conditions[f]	Desired Temperature (°K)	Radiance of Luminescence[g]			Sensitivity[i]	Selectivity[i]	Precision[i]
						Increased Temperature	Increased Source Radiance	Instrument[h]			
SF	Atom	ET	Few volatile atoms	Vapor	~1000	Complex	B_A^1	Fl.	M	VH	M
SDF	Mol.	ET (Triplet sensitized)	FO	Fluid-Glass	~77–298	Decrease	B_A^2 (LV) B_A^1 (HV)	Ph.	H	H	L
P	Mol.	N	$\pi^* - \pi$ O.	Glass	~77	Great decrease	B_A^1 (LV)	Ph.	H	H	M
			$\pi^* - n$ O.	Solid	~77	Great decrease	$B_A^{1/2}$	Ph.	M	H	L
			FI	Adsorption	~77	Great decrease		Ph.	L	M	L
P	Mol.	C	$\pi^* - \pi$ O	Glass	~77	Great decrease	B_A^1 (LV)	Ph.	(Few analytical studies yet, but would estimate M, L, L for all three media.)		
			$\pi^* - n$ O	Solid	~77	Great decrease	$B_A^{1/2}$ (HV)	Ph.			
			FI	Adsorption	~77	Great decrease		Ph.			
P	Mol.	Q	FO	Glass	~77	Great decrease	B_A^1 (LV)	Ph.	VH	L	M
			Mainly O$_2$	Adsorption	~77	Great decrease	$B_A^{1/2}$	Ph.	VH	L	L
SP	Mol.	ET	FO	Glass	~77	Great decrease	B_A^1 (LV) $B_A^{1/2}$ (HV)	Ph.	(Few analytical studies yet, but would expect M, L, L.)		
ASDF	M	ET	FO	Fluid-Glass	~77–298	Complex	B_A^2	Ph.	(Few analytical studies yet, but would expect M, L, L.)		
RL	M	N or ET	FO	Glass	~77	Increase	Complex	Fl./Ph.	(Few analytical studies yet, and would expect no analytical use.)		
SF	In. ion	IET	RE	Fluid-glass	~77–298	Decrease	B_A^1	Fl.	VH	H	M
SF	In. ion	IET	RE	Fluid	~298	Increase	B_A^1	Fl.	VH	H	M

[a] Taken in part from C. A. Parker, *Photoluminescence of Solutions*, Elsevier, Amsterdam, 1968.
[b] F = fluorescence.
  DF = Delayed fluorescence.

316

SF = Sensitized fluorescence.

SDF = Sensitized delayed fluorescence.

P = Phosphorescence.

SP = Sensitized phosphorescence.

ASDF = Anti-Stokes sensitized delayed fluorescence,

RL = Recombination luminescence.

Mol. = Molecule.

[c] In. ion = Inorganic ion.

[d] N = *Native*—luminescence is a result of the analyte atom or molecule itself.

C = *Chemical*—molecular luminescence is a result of a compound produced by chemical reaction of the analyte with a chemical reagent.

Q = *Quenched*—molecular luminescence of some species is quenched by analyte. These methods are of limited use because they lack selectivity.

Triplet sensitized—Delayed fluorescence is a result of either triplet-to-singlet energy transfer from donor to acceptor followed by triplet self-quenching or a result of mixed triplet quenching.

ET × Energy transfer (intermolecular).

IET = Intramolecular energy transfer.

[e] IMO = Inorganic or metallo-organic.

O = Organic.

FI = Few inorganics.

FO = few organics.

RE = rare earths.

[f] Conditions denote media in which luminescence is observed. Gf = hot gas (flame or furnace).

[g] Radiance of photoluminescence for each type of photoluminescence as function of exciting radiation absorbed B_A (photons absorbed sec^{-1}).

LV = Low values.

HV = High values.

[h] Fl. = Fluorimeter.

Ph = Phosphorimeter.

[i] Only relative magnitudes of sensitivity, selectivity, and precision are given for the average type atom or molecule. It should be stressed that there are atoms or molecules for which much higher and much lower sensitivity, selectivity, and precision result than the relative magnitude given. Also the relative values can be changed if instrumental factors are varied (e.g., solution temperature, monochromator characteristics, signal processing system).

VH = very high.

H = High.

M = Moderate.

L = Low.

317

I

EQUILIBRIUM CONSTANTS OF DISSOCIATIVE AND IONIZATION EQUILIBRIA*

Consider the dissociative equilibrium

$$MX \rightleftharpoons M + X \qquad K_p = \frac{p_M p_X}{p_{MX}} \qquad (A1\text{-}1)$$

where K_p is the equilibrium constant that depends only on temperature and the species MX and the p's are partial pressures of the designated species. Assuming that the ideal gas law is valid (true for analyte species in flames),

$$p = nkT \qquad (A1\text{-}2)$$

then

$$K_p = \left(\frac{n_M n_X}{n_{MX}}\right) kT \qquad (A1\text{-}3)$$

and substituting for the concentration term in parentheses in terms of the appropriate partition functions (f) gives

$$K_p = \left(\frac{f_M f_X}{f_{MX}}\right) kT \exp -\frac{D_o}{kT} \qquad (A1\text{-}4)$$

where D_o is the dissociation energy of the molecule MX. (The final exponential term is needed because the ground state is taken for the molecule MX dissociating into two atoms M and X and so the energy of the lowest state of the molecule is $-D_o$ with respect to the atoms. The complete atomic partition function (electronic and translational) is given by

$$f_A = \left(\frac{2\pi m_A kT}{h^2}\right)^{3/2} \sum_j g_j \exp -\frac{E_j}{kT} \qquad (A1\text{-}5)$$

where m_A is the mass of species A, h is Planck's constant, k is Boltzmann's constant, E_j is the electronic energy of state j (the summation is over all electronic states). The subscript A refers to species M or X. If there are no other electronic levels within about 1 eV of the ground electronic state, then the final term in brackets is simple g_o the statistical weight of the ground state.

* Treatment taken from R. Mavrodineanu and H. Boiteux, *Flame Spectroscopy*, Wiley, New York, 1965.

For a diatomic molecule MX, the complete partition function is given by

$$f_{\mathrm{MX}} = \left(\frac{2\pi m_{\mathrm{MX}} kT}{h^2}\right)^{3/2} \sum_j R_j(T) V_j(T) g_j \exp -\frac{E_j}{kT} \qquad \text{(A1-6)}$$

where $R_j(T)$ is the rotational partition function, $V_j(T)$ is the vibrational partition function, and $g_j \exp(-E_j/kT)$ is the electronic partition function, and $g_j \exp(-E_j/kT)$ is the electronic partition function. The first term in parentheses is again the translational partition function. The rotational partition function for MX is given by

$$R_{\mathrm{MX}}(T) = \frac{8\pi^2 I_{\mathrm{MX}} kT}{h^2 \sigma} \qquad \text{(A1-7)}$$

where I_{MX} is the principal moment of inertia of MX and σ is a symmetry number— the number of indistinguishable orientations of the molecule: $\sigma = 1$ for molecules *without* symmetry (e.g., HCl), $\sigma = 2$ for molecules with a center of symmetry or an axis of symmetry (e.g., H_2 or H_2O), $\sigma = 3$ for molecules with a ternary axis of symmetry (e.g., NH_3), $\sigma = 4$ for molecules with rectangular symmetry (e.g., C_2H_4), and so on. Substituting for I_{MX} in terms of the rotational constant B_{MX} gives

$$R_{\mathrm{MX}}(T) = \frac{kT}{hc B_{\mathrm{MX}} \sigma} \qquad \text{(A1-8)}$$

For polyatomic molecules (nonlinear), more than one moment of inertia must be considered.

The vibrational partition function for a diatomic molecule is given by

$$V_{\mathrm{MX}}(T) = \left(1 - \exp\frac{-hc\tilde{\nu}_o}{kT}\right)^{-1} \qquad \text{(A1-9)}$$

where $\tilde{\nu}_o$ is the fundamental vibration frequency (cm^{-1}) and c is the speed of light. For polyatomic molecules with more than one fundamental vibrational frequency, a product of terms such as in equation A1-9 result.

By combining the foregoing expressions and converting from natural to common logarithms, we find that $\log K_p$ is given by

$$\log K_p = -\frac{5040 D_o}{T} + \frac{3}{2} \log T + \log\left(1 - \exp\frac{-hc\tilde{\nu}_o}{kT}\right) + i_{\mathrm{M}} + i_{\mathrm{X}} - i_{\mathrm{MX}} \qquad \text{(A1-10)}$$

where the i's are chemical constants* defined by

$$i_{\mathrm{poly}} = -1.578 + \frac{3}{2} \log M_{\mathrm{poly}} + \log\left(\sum_{\substack{\text{ground} \\ \text{multiplet, poly}}} g_j \exp E_j/kT\right) - \log \sigma_{\mathrm{poly}} - \frac{1}{2} \log B_A B_B B_C$$

where $B_A B_B B_C$ is the product of the three rotational constants of the polyatomic molecule.

* For a polytomic molecule (nonlinear)

$$i_M \quad (\text{or} \quad i_X) = -1.592 + \frac{3}{2} \log M_M + \log \left(\sum_{\substack{\text{ground} \\ \text{multiplet, M}}} g_j \exp \frac{-E_j}{kT} \right) \quad (A1\text{-}11)$$

$$i_{MX} = -1.750 + \frac{3}{2} \log M_{MX} + \log \left(\sum_{\substack{\text{ground} \\ \text{multiplet, MX}}} g_j \exp \frac{-E_j}{kT} \right) - \log \sigma_{MX} - \log B_{MX}$$

$$(A1\text{-}12)$$

where M_M (or M_X) and M_{MX} are the molecular weights of M (or X) and MX, respectively (amu).

Thus from equation A1-10 it is possible to estimate the value of K_p at a given temperature and to predict the change of K_p with temperature.

Consider the ionization equilibrium

$$M = M^+ + e^- \qquad K_i = \frac{(p_{M^+})(p_e)}{p_M} \qquad (A1\text{-}13)$$

Saha showed that this equilibrium could be treated as any other and therefore, by the same process used previously,

$$K_i = \left(\frac{(f_{M^+})(f_e)}{f_M} \right) kT \exp \frac{-X}{kT} \qquad (A1\text{-}14)$$

where X is the ionization energy of atom M and f_e is the partition function of free electrons, which is given by

$$f_e = 2 \left(\frac{2\pi m_e kT}{h^2} \right)^{3/2} \qquad (A1\text{-}15)$$

where the factor 2 is the statistical weight factor (i.e., two possible orientations of spin in an external field). The chemical constant for free electrons is given by

$$i_e = \log \left[2 \left(\frac{2\pi m_e}{h^2} \right)^{3/2} k^{5/2} \right] = -6.1816 \qquad (A1\text{-}16)$$

By taking logarithms of both sides of equation A1-14 and substituting in terms of chemical constants for M, e, and M^+, we have

$$\log K_i = -\frac{5040X}{T} + \frac{5}{2} \log T + i_{M^+} + i_e - i_M \qquad (A1\text{-}17)$$

where i_M and i_{M^+} are given by equations A1-11 and i_e by equation A1-16.

We should also mention that in this book all equilibrium constants are expressed in terms of concentrations n rather than partial pressures p. To convert from K_p to K_n, it is only necessary to use the ideal gas law. For the dissociation of a diatomic species as MX, K_n is related to K_p by

$$K_n = \frac{n_M n_X}{n_{MX}} = \frac{K_p}{kT} = \left(\frac{p_M p_X}{p_{MX}} \right) \frac{1}{kT} \qquad (A1\text{-}18)$$

APPENDIX

II

RELATION BETWEEN EINSTEIN COEFFICIENT OF SPONTANEOUS EMISSION AND VARIOUS SPECTRAL PARAMETERS

The relation between the Einstein coefficient of spontaneous emission A_{ul} and the emission moment M_{ul} [refer to Section II-B-3-(2)] is given by

$$A_{ul} = \left(\frac{64\pi^4}{3h\lambda_0{}^3}\right)M_{ul}{}^2 \tag{A2-1}$$

where all terms have been defined previously. The magnitude of $M_{ul}{}^2$ for most transitions depends on the interaction of the electromagnetic radiation vector with the oscillating electric dipole moment of the atom or molecule undergoing interaction with light, and the radiation in such cases is called dipole radiation. The transition moment can be separated into three components directed in the three Cartesian coordinates, x, y, and z; and if two of the components are zero, then the radiation is polarized in the same direction as the nonzero component.

An atom or molecule can also create an oscillating electromagnetic field in its environment by means other than electric dipole interaction (e.g., by a quadrupole moment or by a magnetic dipole moment). However, interactions of the electromagnetic field with an atomic or molecular species by nonelectric dipole interaction are usually extremely weak and produce very low absorption, emission, or luminescence signals.

The Einstein coefficient of spontaneous emission A_{ul} is related to the Einstein coefficient of induced emission B_{ul} by

$$A_{ul} = \frac{8\pi hc}{\lambda_0{}^3} B_{ul} \tag{A2-2}$$

The induced emission coefficient has units of seconds per gram; the product of B_{ul} and the radiant energy per unit wavenumber interval per unit volume is the total number of photons undergoing induced emission per second. The relation between the Einstein coefficient of induced emission B_{ul} and the Einstein coefficient of induced absorption B_{ul}, which is defined in the same manner as B_{ul} but is for the radiation-induced absorption process, is given by

$$g_u B_{ul} = g_l B_{lu} \tag{A2-3}$$

Therefore, the relation between A_{ul} and B_{lu} is

$$A_{ul} = \frac{8\pi hcg_l}{g_u\lambda_0{}^3} B_{lu} \tag{A2-4}$$

322

The relations between B_{ul} (and B_{lu}) and f_{lu} (and f_{ul}) can be found by use of equations II-C-15, II-C-16, II-C-18, and II-C-20.

Probably the most useful expression is the one for the Einstein coefficient of spontaneous emission and the absorption oscillator strength, f_{lu}:

$$A_{ul} = \left(\frac{8\pi^2 e^2 g_l}{mcg_u}\right)\frac{f_{lu}}{\lambda_o{}^2} = 0.67\left(\frac{g_l}{g_u}\right)\frac{f_{lu}}{\lambda_o{}^2} \tag{A2-5}$$

(units: A_{ul}, $\sec^{-1}$; f_{lu}, dimensionless; g_u and g_l dimensionless, λ_o, Å)

and since the absorption oscillator strength f_{lu} is related to the emission oscillator strength f_{ul} by

$$f_{lu}g_l = f_{ul}g_u \tag{A2-6}$$

then

$$A_{ul} = 0.67\frac{f_{ul}}{\lambda_o{}^2} \tag{A2-7}$$

The other terms in equation A2-5 are m and e the mass and charge of the electron, c the speed of light, and g_l and g_u the statistical weights of the lower and upper states involved in the absorption (or emission) transition.

APPENDIX

III

GENERAL EXPRESSIONS FOR THE MEASURED LUMINESCENCE SIGNAL OF A SPECTROMETRIC SYSTEM*

A. THE OPTICAL SYSTEM UNDER CONSIDERATION

The arrangement of the optical system schematically shown in Figure A3-1 is used to represent any spectrometric system. Even if more than one lens or mirror is used in the entrance optics or in the spectrometer, the following general treatment will still apply, except that certain factors must be changed to account for the additional optical components (e.g., the transmission of the optical components).

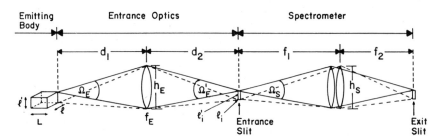

Fig. A-3-1 Arrangement of optical system for measurement of luminescence (or any source of radiation).

Assumptions and conditions necessary for luminescence signal expressions to be valid are:

1. The solid angle of the entrance beam Ω'_E must exceed or at least be equal to the solid angle collected by the spectrometer Ω_s; that is,

$$\Omega'_E \geqslant \Omega_s \tag{A3-1}$$

In other words, the spectrometer optics (lens, mirror, or dispersing device) *determines* the linear aperture (the size of the light beam transmitted through the spectrometer).

* The contents of this section are a result of informal discussion with V. Svoboda during his stay at the University of Florida. (V. Svoboda is now at Dvorakova 10, Praha 4-Modrany, Czechoslovakia.)

324

2. The height of the source image l_i on the entrance slit of the spectrometer must be equal to or greater than the slit height H_s; that is,

$$l_i' \geqslant H_s \qquad (A3\text{-}2)$$

3. The width of the image l_i on the entrance slit of the spectrometer must be equal to or greater than the entrance slit width W_s; that is

$$l_i \geqslant W_s \qquad (A3\text{-}3)$$

4. The trivial assumption that the transmittance of the entrance optics t_E and the transmittance of the spectrometer t_s (t_E and t_s are less than unity at all wavelengths owing to absorption and reflection losses) must be constant over the spectral range of interest; that is, over the wavelength range from the low wavelength cutoff λ_l to the upper wavelength cutoff λ_u. These wavelengths are determined by either the spectrometer, the light source, or a combination of the two; that is

$$t_s]_{\lambda_l}^{\lambda_u} = \text{constant} \qquad (A3\text{-}4)$$

$$t_E]_{\lambda_l}^{\lambda_u} = \text{constant} \qquad (A3\text{-}5)$$

5. The luminescing sample* must be completely within the solid angle collected by the entrance optics.

In addition to the conditions and assumptions just described, certain basic relationships are valid:

$$\frac{d_2}{d_1} = \frac{l_i'}{l'} \qquad (A3\text{-}6)$$

$$\frac{d_2}{d_1} = \frac{l_i}{l} \qquad (A3\text{-}7)$$

$$\Omega_E = \frac{h_E{}^2}{d_1{}^2} \qquad (A3\text{-}8)$$

$$\Omega_E' = \frac{h_E{}^2}{d_2{}^2} \qquad (A3\text{-}9)$$

$$\Omega_s = \frac{h_s{}^2}{f_1{}^2} \qquad (A3\text{-}10)$$

where h_E is the diameter (minimum linear aperture) of the lens (or mirror) used in the entrance optics, h_s is the diameter (minimum linear aperture) of the collimating lens (or mirror) used in the spectrometer, d_2 and d_1 are the image and object distances from the entrance lens (see Figure A3-1), and f_1 is the focal length of the collimator in the spectrometer.

---

* The luminescing sample is in a cell described by the dimensions in Figure II-C-1a.

Using the foregoing conditions, assumptions, and relationships, the following equations can be derived.* From condition 1, it can be shown that

$$h_E \gg \frac{h_s d_1 f_E}{f_1} \sqrt{1/(d_1{}^2 - 2d_1 f_E + f_E{}^2)} \qquad \text{(A3-11)}$$

and from conditions 2 and 3 it can be shown that

$$l' \gg H_s \frac{d_1}{d^2} = H_s \left(\frac{f_E}{d_1 - f_E}\right) \qquad \text{(A3-12)}$$

$$l \gg W_s \frac{d_1}{d_2} = W_s \left(\frac{f_E}{d_1 - f_E}\right) \qquad \text{(A3-13)}$$

As long *as these expressions are valid, then the general treatment to follow is valid.* It should be stressed, however, that the experimental conditions needed for validity of the expressions used *are almost always experimentally* attainable for analytical spectrometric systems (luminescence as well as absorption spectrometers). Finally, it should be pointed out that the expressions to be derived are valid for the measurement of radiation from any (line or continuum) source of radiation (source of excitation such as a hollow cathode lamp or a xenon arc lamp or an emitting or luminescing media).

B. GENERAL EXPRESSION FOR RADIANT FLUX EMERGING FROM EXIT SLIT OF SPECTROMETER SET AT A FIXED WAVELENGTH

The radiant flux Φ_D (watt) emerging from the exit slit of a spectrometer and striking a detector is given, for a luminescent sample and a spectrometric system as shown in Figure A3-1, by

$$\Phi_D = t_E t_s \Omega_s H_s \int_{\lambda_l}^{\lambda_u} B_{L_\lambda} F_{\lambda_s}(\lambda)\, d\lambda \qquad \text{(A3-14)}$$

where t_E is the transmittance of the entrance optics and t_s is the transmittance of the spectrometric system between the limiting wavelengths of λ_l and λ_u; H_s is the spectrometer* entrance slit height (cm); $F_{\lambda_s}(\lambda)$ is the spectrometric or apparatus function,† in (cm), which depends on the wavelength setting of the monochromator λ_s and the wavelength range (essentially the spectral bandpass s of the monochromator) when the spectrometer is *not* being wavelength scanned; and $B_{L\lambda}$ is the spectral radiance‡

---

* The thin lens approximation is used: $1/d_1 + 1/d_2 = 1/f_E$, where f_E is the focal length of the entrance optics lens.

* It is also assumed in this treatment that the entrance and exit slit widths are identical, for this is the case in most analytical spectrometers.

† Also called the slit function of the spectrometer. This function should have a triangular distribution if it is true that the extrance and exit slits are identical (except for spectrometer slit widths near the diffraction limited slit width).

‡ In this section λ rather than λ' as in the text is used to denote emission wavelength.

of luminescence being emitted by the sample (see Section II-C-1). The product of the terms $B_{L\lambda}$ and $F_{\lambda_s}(\lambda)$ is said to be folded (convoluted). The wavelength range λ_l to λ_u designates the range over which $F_{\lambda_s}(\lambda)$ is greater than zero; because $F_{\lambda_s}(\lambda) \rightarrow 0$ for wavelengths outside this range, the integration limits can be changed to zero to infinity (more convenient to handle). Nevertheless, the integral in equation A-314 can only be readily solved by numerical methods.

It is important to point out that the value of Φ_D depends on the product of $B_{L\lambda}$ and $F_{\lambda_s}(\lambda)$ integrated from λ_l to λ_u. For the limiting cases in which the spectral line from the emitter is much narrower than the spectral bandpass of the spectrometer or for the limiting case in which spectral radiance of the emitter is constant over the spectral bandwidth of the spectrometer, the simplified expressions given in Chapter III can be used with good accuracy. However, such expressions as in Chapter III *are* limiting cases and give only an approximation of the measured signal. Fortunately, the expressions in Chapter III are analytically useful for most real experimental cases and can be used in most real cases for estimating the influence of variables on the luminescence signal. The general expressions provided here are useful for all cases and give useful results for intermediate cases, but the expressions in this appendix are more unwieldy and of less interest to the analyst.

To convert the radiant flux Φ_D (watt) passing through the spectrometer exit slit and striking the photocathode of a multiplier phototube into an electrical signal S_L, in the following expression is used:

$$S_L = \gamma_\lambda G \Phi_D R_L \qquad (A3\text{-}15)$$

where R_L is the load resistance, in ohm, γ_λ is the photoanodic sensitivity of the multiplier phototube [A (at anode) watt^{-1} (radiant flux at cathode)] and G is the overall gain of the measurement circuit [A (at readout) to A (at anode of phototube)].

C. TOTAL RADIANT ENERGY EMERGING FROM EXIT SLIT OF SPECTROMETER WHEN WAVELENGTH SCANNING THE SPECTRAL LINE OR BAND

In the scanning mode, the total amount of radiant energy Q_D (watt sec) corresponding to some luminescence line or band that emerges from the spectrometer exit slit and reaches the photocathode of the multiplier phototube is given by

$$Q_D = \int_{t_1}^{t_2} \Phi_D \, dt = t_E \, t_s \Omega_s H_s \int_{t_1}^{t_2} \int_{\lambda_l}^{\lambda_u} B_{L\lambda} F_{\lambda_s}(\lambda) \, dt \, d\lambda \qquad (A3\text{-}16)$$

where $F_{\lambda_s}(\lambda)$ is a function of the spectral bandwidth of the monochromator s, the wavelength limits λ_u and λ_l, and the wavelength setting of the spectrometer λ_s, which is a function of time. All terms before the integral are assumed to be constant with time between the initiation time t_1 and the final time t_2 of the wavelength scan and constant with wavelength.

If it is now assumed that the scanning speed is constant, then

$$\lambda_s = a + bt \qquad (A3\text{-}17)$$

where a and b are constants for the scanning system. Therefore

$$\frac{d\lambda_s}{dt} = b \tag{A3-18}$$

Assuming a triangular slit (apparatus) function,* $F_{\lambda_s}(\lambda)$ is given† by

$$F_{\lambda_s}(\lambda) = F_{\lambda_s}(\lambda_s) \left[1 - \frac{|\Delta\lambda|}{s} \right] \tag{A3-19}$$

where $\Delta\lambda = |\lambda - \lambda_s|$, where λ is any wavelength between the limits of λ_l and λ_u which are imposed by the scanning limits and s is the spectral bandpass of the spectrometer.

To convert Q_D into a measured readout signal, it is necessary to use an integrating system and to measure stored charge (coulomb). If the photoanodic sensitivity of the multiplier phototube is γ_{λ_s} A watt^{-1}, then the charge q_D (coulomb), stored in a capacitor connected to the photoanode is

$$q_D = Q_D \gamma_{\lambda_s} \tag{A3-20}$$

It was assumed that γ_{λ_s} was a constant* over the wavelength limits (λ_u to λ_l) of the

---

* We assume that the spectrometer has equal entrance and exit slit widths (and heights) and that the slit widths are greater than the diffraction-limited slit width; that is,

$$W_s \quad \text{or} \quad W_e \gg \frac{2\lambda_s f_1}{H_s}$$

where λ_s is the spectrometer wavelength setting (cm), f_1 is the collimator focal length (cm), and H_s is the slit height (cm). If this condition is *not* valid, see K. D. Mielenz, *J. Opt. Soc. Amer.*, **57**, 66 (1967).

† If $W_s \neq W_e$, then the apparatus function will be trapezoidal rather than triangular, and the wavelength width of the top $\Delta\lambda_{\text{top}}$ and base $\Delta\lambda_{\text{bcse}}$ of the trapezoid will be

$$\Delta\lambda_{\text{top}} = \left[W_s \left(\frac{f_2}{f_1} \right) - W_e \right] R_d$$

$$\Delta\lambda_{\text{base}} = \left[W_s \left(\frac{f_2}{f_1} \right) + W_e \right] R_d$$

where R_d is the reciprocal linear dispersion of the spectrometer (nm cm^{-1}). For this case, the peak value of F_{λ_s} at λ_s, which is designated $F_{\lambda_s}(\lambda_s)$, is a constant for the wavelength range over $\Delta\lambda_{\text{top}}$. For $\lambda - \lambda_s > \Delta\lambda_{\text{top}}$ or $\lambda - \lambda_s < \Delta\lambda_{\text{top}}$, $F_{\lambda_s}(\lambda)$ drops off in the same manner as for the triangular slit function. Also, for the limiting case of a triangular slit function—that is, $W_s = W_e$ and $f_2 = f_1$; $\Delta\lambda_{\text{top}} = 0$ and $\Delta\lambda_{\text{base}} = 2R_d W_s = 2s$ (twice the spectral bandpass of the spectrometer).

* The value of γ_{λ_s} is approximately a constant over the wavelength range of $\lambda_p + 1000$ Å to $\lambda_p - 1000$ Å, where λ_p is the peak response wavelength for the specific multiplier phototube being used. For some phototubes, the range is even greater. Outside of this range, especially at the low wavelength end, γ_{λ_s} varies rapidly with λ_s, and so the wavelength dependence of γ_{λ_s} must be accounted for.

scan; if γ_{λ_s} varies over the wavelength range, then it would be necessary to account for the wavelength sensitivity by including γ_{λ_s} within the integral in equations A3-20 and A3-16 (i.e., convolution of the photoanodic sensitivity factor with the luminescence spectral radiance and the apparatus function). The luminescence spectral radiance function $B_{L\lambda}$ is known or can be estimated for most atomic lines—it can be approximated by the expressions in Section II-C—but is known much less accurately for molecular bands. The expressions in Section II-C are useful for obtaining the integrated spectral luminescence radiance; that is, B_L (erg sec^{-1} cm^{-2} sr^{-1}) but the spectral luminescence radiance $B_{L\lambda}$ (erg sec^{-1} cm^{-2} sr^{-1} nm^{-1}) can only be obtained if the spectral distribution of molecular luminescence is known in detail.

IV

PRINCIPLES OF OPERATION OF
NONPHOTOEMISSIVE DETECTORS

A. PHOTOVOLTAIC OR BARRIER-LAYER CELLS

One of the simplest types of photodetectors is the photovoltaic cell; it is self-generating and requires no auxiliary power supply. However, it is one of the least satisfactory photodetectors for luminescence work because of its limited sensitivity and relatively narrow linear dynamic range. A description is included here mainly for completeness.

The usual type of photovoltaic cell consists of a thin layer of crystalline selenium deposited on a metal base plate. The selenium layer is covered with a very thin, semi-transparent layer of metal (usually silver) which forms the other electrode. Photons passing through the semitransparent metal layer are absorbed at the " barrier layer " between the selenium and the thin metal layer and produce electron-hole pairs. The migration and resulting separation of these holes and electrons cause a small difference in potential to develop between the base plate and the thin metal layer.

Either the open-circuit voltage or the short-circuit current from the photovoltaic cell may be measured and used as a gauge of light intensity. Only the short-circuit current, however, increases linearly with increasing illumination; the open-circuit voltage increases approximately as the logarithm of the incident intensity.

The short-circuit current can be measured accurately only with a low resistance current meter (milliammeter or microammeter); that is, the input resistance of the current meter must be very small compared with the internal resistance of the photo-voltaic cell. Unfortunately, the internal resistance is quite low (of the order of 100 ohm) and this requires that the meter input resistance be of the order of an ohm. Such a low value is difficult to obtain in a sensitive meter. A sensitive meter is required because the sensitivity of the photocell is low—of the order of 0.1 mA watt^{-1} of incident radiation in the visible and near-ultraviolet regions. The useful spectral range extends from 250 to about 750 nm.

The logarithmic characteristic of the open-circuit voltage is useful for some applications: certainly the voltage would be easier to measure, because high resistance volt-meters are commonly available. The output voltage is typically of the order of several millivolts. Unfortunately, the logarithmic characteristic is limited in dynamic range, and the large capacitance of the cell increases the response time seriously when loaded by a high resistance.

Photovoltaic cells are occasionally found on simple colorimeters. Their dynamic range is too restricted for use in most fluorescence instruments. Spectral ranges of several cells are given in Figure A4-1.

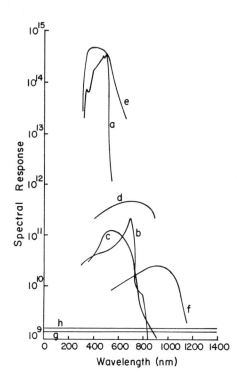

Fig. A-4-1 Spectral ranges of photo-detectors; plots are of spectral detectivity (called response here) D_λ^* versus wavelength λ. Spectral detectivity is a figure of merit of the phototube and is defined as the reciprocal of the noise equivalent power, P_N, (watts Hz^{-1}) times the square root of the detector area A_D (cm^2) onto which radiation falls. Thus spectral detectivity has units of cm Hz$^{\frac{1}{2}}$watt^{-1}. It is related to spectral responsivity R_λ (volt watt^{-1}) by $D_\lambda^* = R_\lambda A_D \, \Delta f / V_N$, where Δf is the noise bandwith of the measurement system and V_N is the rms noise voltage. Plots are: (a) CdS photoconductivity cell modulated at 90 Hz, (b) CdSe photoconductivity cell modulated at 90 Hz, (c) Se-SeO photovoltaic cell modulated at 90 Hz, (d) GaAs photovoltaic cell modulated at 90 Hz, (e) 1P21 multiplier tube (for reference), (f) 1N217 silicon photo-duodiode photovoltaic cell modulated at 400 Hz, (g) radiation thermocouple modulated at 5 Hz, (h) Golay cell modulated at 10 Hz. Curves from P. W. Kruse, L. N. McGlauchlin, and R. B. Quistan, *Elements of Infrared Technology*, Wiley, New York, 1962, with permission of the authors.

B. PHOTOCONDUCTIVE CELLS

Photoconductive cells are essentially light-variable resistors. They are not self-generating and so they require an auxiliary voltage supply. Nevertheless, they are much more sensitive than barrier-layer cells and have a wider dynamic range.

Photoconductive cells may be constructed of almost any crystalline semiconductive material. Selenium, silicon, germanium, and some metal oxides, sulfides, and halides are commonly used. Electrical connections are made to each end of a bar or plate of semiconductor material, and a small potential is applied across these electrodes by means of an external battery or power supply. When light strikes the semiconductor material, its resistance decreases and a current flows that is proportional to the incident light intensity. In the dark, the resistance of the cell is very high, and only a small "dark" current flows.

Photoconductivity is caused by the excitation of valence-band electrons to the conduction band by the absorption of incident photons. The energy of the photon $h\nu$ must be greater than the energy gap ΔE between the valence and conduction bands. This sets a lower limit on the energy (and an upper limit on the wavelength) of an incident photon. Semiconductors, whose values of ΔE are comparatively small, are used as photoconductors in order to obtain good response to reasonably long wavelengths.

The transfer function of a photoconductive detector (i.e., the characteristic of photocurrent i versus incident light flux Φ) is sometimes linear, as in the case of lead sulfide, silicon, and germanium photoconductors. But the current i often varies with Φ^n, where n may be either less than or greater than unity. In grey selenium, for example, n is about 0.5. In cadmium sulfide and selenide detectors, n decreases with increasing light flux. An additional source of nonlinearity may arise because as the current i increases, the voltage V across the element is reduced by the drop across the resistance of the current meter. This effect can be reduced by using a very low resistance meter, as for the photovoltaic cells.

Photoconductive detectors may be used over a spectral range extending from the infrared into the ultraviolet. Some types even respond to X-rays, gamma rays, and to α and β particles. The sensitivity usually drops off sharply at long wavelengths, when the energy of the photons drops below the gap energy ΔE. The long wavelength response can often be improved by cooling the detector. The spectral ranges of a variety of photoconductors are summarized in Figure A4-1.

Radiant sensitivities of commercial photoconductive detectors are generally of the order of a few hundred amperes per watt at the wavelength of maximum sensitivity. The sensitivity can be increased by increasing the voltage across the cell; however, a limitation is imposed by the maximum power which the element can dissipate without damage, which is typically less than 1 watt cm^{-2}.

If an appreciable amount of power is dissipated, the resulting temperature increase will increase the dark current. In general, the photocurrent increases linearly with the applied voltage unless polarization or barrier-layer formation occurs.

The dark resistance of photoconductors varies widely with the type of material. Typical values range from about 10^5 ohms for selenium and lead sulfide cells to greater than 10^{10} ohms for cadmium sulfide cells. In general, materials with small gap energies ΔE have relatively low dark resistances, since room temperature thermal excitation can promote more electrons to the conduction band across a narrow energy gap. Cooling the detector with dry ice or liquid nitrogen will reduce the thermal excitation and thus increase the dark resistance.

The electrical properties of photoconductors are affected greatly by extremely small amounts of impurities, whether accidentally or intentionally added. Certain types of impurities provide current carriers at an energy level only very slightly below the conduction bands, thereby increasing the infrared response and the dark current. Pure germanium, for instance, which has a gap energy of 0.7 eV, responds out to about 2 μ; whereas in gold-doped germanium the energy gap is reduced to about 0.01 eV and the infrared response extended to about 40 μ.

The response time of a photoconductive detector may range from microseconds to many seconds, depending on the particular material.

Noise currents in photoconductors consist primarily of Johnson (thermal agitation) noise, photocurrent noise,* and dark current noise. The last-named noise increases with decreasing gap energy and increasing temperature and can be reduced by cooling the detector. An additional noise source may arise from the formation of barrier surfaces at the point of contact between the photoconductive material and the two electrodes. This noise,† which exhibits a $1/f$ frequency characteristic, may be reduced by the use of special electrode materials.

Although photoconductive cells cannot compare for sensitivity, linearity, and response time with photomultiplier detectors, they do find some service as source intensity monitors in ratio fluorometers. They have the advantage of small size, low cost, and low voltage requirements.

C. THERMAL DETECTORS‡

A disadvantage of many types of detectors (the so-called photon detectors) is that their sensitivity varies strongly with the wavelength of the incident light. This is not the case for the various types of thermal detectors. Thermal detectors operate by first converting the light energy into an equivalent amount of heat and then measuring the resulting temperature rise by means of a voltage change, resistance change, or expansion of a gas. Thus thermal detectors respond to the total energy absorbed by the detector. Fortunately, it is possible to construct very effective and wavelength-insensitive black body absorbers simply by coating the absorbing surfaces with special coatings of high blackness. Commonly used coatings include platinum black and evaporated gold, bismuth, antimony, and aluminum.

Thermocouple detectors convert heat produced by incident light into a small voltage. They are made by attaching a small blackened piece of metal foil, usually gold, to a small thermocouple junction. All the parts are made quite small in order to reduce the heat capacity and thermal inertia of the system. As a consequence, the incident radiation must be carefully condensed and focused onto the tiny element.

The output voltage of a single thermojunction is of the order of 0.1 volt watt^{-1} of incident flux over a wide range of incident powers and wavelengths. In a device called a thermopile several thermojunctions can be connected in series in order to increase the sensitivity. However, since this arrangement increases the size, heat capacity, and radiation losses, and increases the response time compared with a single junction, the overall performance is not necessarily improved.

Heat losses by the thermocouple detector and by conduction, convection, or reradiation, limit the maximum temperature to which a given detector will rise under irradiation by light of a given intensity. These losses can be reduced, and the sensitivity consequently increased, in several ways. Conduction losses are lessened by making the external electrical connections to the junction by means of very thin wires. Convection losses are reduced by enclosing the thermojunction in an evacuated housing at a pressure of the order of 10^{-4} torr. A disadvantage, however, is that absorption in the window may compromise the wide spectral response unless the window is made of a very thin piece of some relatively transparent material (e.g., silica). Radiation losses are proportional to the temperature of the junction and are therefore reduced by cooling the detector (in liquid nitrogen, e.g.). This is rarely done in practice, however.

Thermocouple detectors are most effectively used in AC detection systems; the

* Also called generation-recombination noise.
† Also called contact noise.
‡ Spectral response curves of several thermal detectors appear in Figure A4-1.

incident radiation is chopped, and the AC component of the output voltage is measured. In this way, the effects of slow changes in the ambient temperature are reduced. The response time of most thermocouple detectors is poor, however, so the chopping frequency is limited to a few hertz at most. Because the resulting signal is AC, it can be coupled to the amplifier electronics by means of a step-up input transformer which increases the signal voltage and matches the low junction resistance (a few tens of ohms) to the relatively higher amplifier input impedance (a few thousand ohms). The transformer must be carefully shielded and must be capable of responding to the low chopping frequency.

Bolometers are essentially temperature-sensitive resistors. They may be made of any material whose temperature coefficient of resistivity is sufficiently high. Examples are platinum and various semiconductor materials ("thermistors"). As for the thermocouple detector, the bolometer resistance element must be small and light, and it must be blackened bo absorb efficiently. The bolometer usually forms one arm of a resistance (Wheatstone) bridge. The bridge imbalance voltage is used as a measure of light flux. The relation between resistance and light flux is essentially linear for metal bolometers, which simplifies calibration. Bolometers typically have response times of the order of milliseconds—much faster than thermocouples. The sensitivity of a bolometer is of the same order as that of a thermocouple detector.

Neither the thermocouple detector nor the bolometer is used to measure luminescence intensities in quantitative analytical work. The photomultiplier is preferred because of its much greater sensitivity. However, the wavelength-independent response of thermal detectors is very valuable, and the instruments find much use in the calibration of light sources and more sensitive photodetectors and in monitoring excitation source radiances.

Thermal detectors are also an important component of corrected-spectra *luminescence spectrometers*, where they are used to measure the absolute radiant flux of the excitation beam emerging from the exit slit of the excitation monochromator. In some cases, the thermal detector is used in a servo loop controlling the monochromator slit width in order to maintain a constant excitation radiant flux at all wavelengths.

Thermal detectors themselves may be calibrated by comparison with a calibrated light source (such as a black body radiator) or to another calibrated detector. Alternately, they may be calibrated more or less directly by attaching the sensitive element to an electrically heated resistance and calculating the sensitivity from the known Joule heat (voltage times current) dissipated in the resistor and the output of the detector.

D. CHEMICAL DETECTORS: THE FERRIOXALATE ACTINOMETER

A chemical actinometer is a light detector system consisting of a solution of a photoreactive substance (photolyte) capable of undergoing an irreversible photoreaction of known and reproducible quantum efficiency (the yield of products for a given photon flux) upon exposure to light. After exposure, the products of the reaction are determined by an appropriate quantitative analytical method and the photon flux of the light is calculated from the known quantum efficiency and stoichiometry of the reaction. A satisfactory photolyte should have the following characteristics:

1. The photolyte solution should absorb strongly over the wavelength region of interest; but the photolysis products should absorb weakly or not at all, to prevent interference with the linearity of the system.

2. Both photolyte and photolysis products should be stable in the dark and the photoreaction must be irreversible.

3. The quantum efficiency should be high and reasonably constant over the wavelength range of interest.

4. The photolysis products should be easily measurable with speed and precision by a convenient method.

5. The photolyte should be readily available in a pure form.

The advantages which an actinometer would be expected to have over a conventional electrical detector include: improved interlaboratory reproducibility, the fact that the integrating system acts as a dosimeter, and flexibility—the physical size and shape of the solution vessel may be chosen to suit the dimensions of the light beam.

Parker* has recommended a 0.006 to 0.15M solution of potassium ferrioxalate $K_3Fe(C_2O_4)_3 \cdot 3H_2O$ in 0.05M sulfuric acid. Upon exposure to light of the appropriate wavelength region (approximately 200–580 nm), the following overall reaction occurs:

$$2Fe(C_2O_4)^+ \xrightarrow{\;h\nu\;} 2Fe^{+2} + C_2O_4{}^{2-} + 2CO_2$$

After irradiation, the ferrous ion is determined spectrophotometrically following conversion to the o-phenanthroline complex. The quantum efficiency of the foregoing photoreaction remains constant over a very wide range of incident quantum fluxes† (5×10^{-11} to 2×10^{-4} einstein cm^{-2} sec^{-1}) and total radiation doses (up to 5×10^{-6} einstein ml^{-1}). The minimum detectable dose is about 2×10^{-10} einstein ml^{-1}. The quantum efficiency varies smoothly with wavelength and remains within 1.0 ± 0.25 from 250 to about 500 nm. Several workers have independently determined the quantum efficiency of this system at various wavelengths by reference to calibrated thermocouples or other standards; the results have indicated a remarkable degree of interlaboratory consistency. Thus the ferrioxalate actinometer would seem to be a valuable and highly versatile tool for absolute light-level measurements.

E. QUANTUM COUNTERS

A quantum counter is a detector whose response is directly proportional to the *quantum irradiance* of a light beam (e.g., photon sec^{-1} cm^{-2} or einstein sec^{-1} cm^{-2}) independent of the wavelength of the light. We can contrast this to a thermal detector, whose response is proportional to irradiance (e.g., watt cm^{-2}). A chemical quantum counter may be made by utilizing the following data: that the spectral distribution of the fluorescence emission of all known pure fluorescent compounds (in the condensed phase) is independent of the wavelength of excitation, and that the fluorescence

---

* C. A. Parker, *Photoluminescence of Solutions*, Elsevier, Amsterdam, 1968.

† An einstein is equal to 6×10^{23} $h\nu$ erg or 1 einstein is equivalent to $(1.196 \times 10^8)/\lambda$ watt, where λ is the wavelength (nm).

quantum yield of some fluorescent species is also independent of the wavelength of excitation. A sufficiently concentrated solution so such a substance will absorb all the radiant flux of an incident light beam and convert it into a proportional number of quanta of fluorescence radiation whose spectral distribution is independent of the wavelength of the incident beam (at least over the spectral region in which the substance absorbs appreciably). Any convenient photodetector (e.g., a photomultiplier tube) may then be used to measure the radiance of the resulting fluorescence emission; the wavelength dependence of the sensitivity of this detector is unimportant, since the spectral distribution of the light it views does not change.

In luminescence spectrometry, quantum counters are sometimes used to monitor the relative quantum irradiance of the excitation beam, thereby allowing the direct measurement of corrected excitation spectra and the compensation for uncontrolled fluctuations in the source intensity. A portion of the excitation beam is sampled by means of a quartz plate beam splitter and monitored by the quantum counter. The quantum counter phototube is best positioned slightly off-axis from the direction of the sampled beam in order to avoid obtaining an erroneous response due to any stray light in the beam that is not absorbed by the quantum counter solution.

Some of the solutions that have been used as quantum counters include a 1 g liter^{-1} solution of esculin in water, a 3 g liter^{-1} solution of rhodamine B in ethylene glycol, and a $0.004M$ solution of fluorescein in water. The concentration of these solutions must be sufficiently high that all the incident radiation is absorbed in a relatively thin layer at the front part of the cell. In this way, the fluorescence radiation passes through a relatively constant depth of remaining solution before reaching the phototube. Thus the filtering effect of this solution is relatively independent of the depth of penetration of the incident beam into the solution, which varies with the incident wavelength and with the molar extinction of the solution. Very pure materials must be used in quantum counter solutions; even small amounts of absorbing impurities may have significant effect in spectral regions in which the molar extinction of the quantum counter substance is low.

F. PHOTODIODES AND PHOTOTRANSISTORS

An ordinary semiconductor "*pn*" junction, of the type found in transistors and diodes, can also be used as a photosensitive element. The reverse leakage current (i.e., the current flowing through a *pn* junction when it is reverse biased) is generally quite small in the dark but increases when the junction is illuminated. The principle of operation is similar to that of the photoconductive or photovoltaic detector, except that *doped* semiconductor materials are used in a photodiode, the junction being between an *n*-doped region and a *p*-doped region of the same semiconductor (e.g., germanium or silicon). An external voltage supply is required.

Photodiodes have several important advantages over photoconductive and photovoltaic cells. They are more sensitive and have better long wavelength response than photovoltaic cells. The response time is also much shorter, since the junction area and capacitance are small. Photodiodes are physically very small.

A phototransistor is essentially a photodiode with built-in current amplification. It is typically several times more sensitive than, but otherwise similar to, a photodiode.

APPENDIX

V

ELECTRONIC SIGNAL PROCESSING INSTRUMENTS

A. SINGLE-SECTION LOW-PASS FILTER

A single-section RC low-pass filter has a signal bandwidth of $(2\pi RC)^{-1}$ and a noise bandwidth of $(4RC)^{-1}$. It is the simplest and cheapest noise-reduction device, but it is also quite limited in some respects. The 1% response time of a low-pass filter is relatively long, about $4.6RC$, and thus no signal components occurring in a time faster than about $4.6RC$ will be recorded with fidelity (see the discussion of the effects of response time in Section III-D-1-d).

B. DC INTEGRATOR

A DC integrator is a circuit whose output voltage e_o is proportional to the time integral of the input voltage e_i

$$e_o = k \int_0^{t_i} e_i \, dt \qquad (A5-1)$$

where k is a circuit constant and $e_o = 0$ at $t = 0$. If t_i is made equal to k^{-1}, then

$$e_o = \frac{1}{t_i} \int_0^{t_i} e_i \, dt = \overline{e_i} \qquad (A5-2)$$

that is, e_o is the mean value of e_i during the interval $t = 0$ to $t = t_i$. Such a circuit is easily constricted from an operational amplifier with a capacitor in the feedback loop. The response time of a DC integrator is simply the integration time t_i. The effective noise bandwidth is $(2t_i)^{-1}$. The main advantage of a DC integrator over a low-pass filter is that the *response time of the integrator is less than half that of the filter for the same noise bandwidth*. For example, compare a low-pass filter with a time constant of 2.5 sec to a DC integrator with an integration time of 5.0 sec. Both devices have noise bandwidths of 0.1 Hz and thus would result in equal signal-to-noise ratios; but the filter has a response time of 11.5 sec, whereas the integrator may be read after only 5 sec. A disadvantage of the integrator, however, is that its output signal is inherently discontinuous and is thus unsuitable for continuous recording of spectra and other time-variant signals.

C. HIGHER ORDER FILTERS

Although the DC integrator is more efficient than a simple single-section (first-order) low-pass filter, higher order filters are often more efficient than integrators. High-order filters of several types are available commercially, but they are relatively expensive and their bandwidths may not be changed easily.

D. AC SYSTEMS

We discussed AC detection and modulation methods in Section III-A-5-b. In terms of noise reduction capability, AC systems have the advantage that the signal frequency is transformed by modulation from DC to some convenient AC frequency high enough to avoid low frequency noise sources such as drift and " $1/f$ noise." However, it should be realized that this is effective only for drift and noise components that *add* to but do not *modulate* (multiply by) the AC signal waveform. For example, variations (drift and noise) in phototube dark current, amplifier offset, and background radiation reaching the photodetector (e.g., flame background emission in atomic fluorescence) are all additive type noises; that is, they affect only the DC level of the AC signal waveform and not its amplitude. Thus any drift and $1/f$ characteristics of these noise sources will be reduced by an AC detection system. However, variations in phototube supply voltage, amplifier gain, excitation-source radiance, sample introduction rate into the cell in atomic fluorescence spectrometry, or scattering signals (in all types of luminescence spectrometry) are multiplicative in nature when source intensity modulation (chopping) is used; that is, these factors influence the amplitude of the AC signal waveform. Drift and $1/f$ noise in these factors are reduced no more by AC detection systems than by DC systems of the same noise bandwidth. Other modulation methods may have specific advantages in this respect. For example, wavelength modulation used in continuum-source atomic fluorescence should reduce drift and $1/f$ noise caused by scattering of incident radiation by droplets in the flame.

E. BOXCAR INTEGRATOR

A boxcar integrator is essentially a gated amplifier followed by a low-pass filter or integrator. It is useful when only a small interval of a repetitive signal waveform is of interest. The boxcar is gated *on* only during the interval of interest and thus will ignore all the noise components outside that interval. Normally the gate width and delay time are adjustable. By slowly varying the delay time between the initiation of the signal waveform and the generation of the gate pulse, a complex repetitive waveform can be extracted from a great deal of random noise. The effective signal bandwidth, which determines the ability to resolve small time elements in the signal waveform, is inversely proportional to the gate width, whereas the noise bandwidth is simply that of the low-pass filter or integrator following the gated amplifier. Since these two factors are unrelated in a boxcar circuit, it is possible to have a large signal bandwidth and a small noise bandwidth simultaneously, thus preserving the signal waveform as well as filtering out noise. In nongated DC or AC systems, on the other hand, low noise bandwidth and poor time resolution (long response time) are inseparable. A boxcar integrator is useful in luminescence decay studies as well as in stroboscobic flash spectroscopy.

A disadvantage of using the boxcar circuit in the sweep-delay mode to extract a waveform is that all the signal information falling outside the gate interval in any one instant is wasted. The efficiency of information collection is essentially the gate width divided by the total delay sweep time; thus this efficiency becomes very small if an attempt is made to achieve good time resolution (necessitating short gate widths) and low noise bandwidth (resulting in long output response time and long sweep times).

F. MULTICHANNEL SIGNAL AVERAGERS

The last-mentioned disadvantage of boxcar circuits is eliminated in the multi-channel signal averager, which is in effect a collection of boxcar integrators gated sequentially in synchronization with the signal waveform. Each boxcar is "assigned" a particular time segment of the signal waveform and is gated *on* only during that particular segment. In this way, the signal is "sliced up" into many time segments, and each segment is averaged over many repetitions of the signal waveform. Thus none of the signal information is wasted. Random noise in the signal waveform will be averaged out toward zero as the number of repetitions is increased; only the net signal waveform accumulates point by point in each channel. Multichannel signal averagers have the same kinds of applications as boxcar integrators, but their superior efficiency permits shorter measuring times and, consequently, much less trouble from long term drifts in the instrumental system.

G. SUMMARY

The relation between the various types of sample signal processing systems may be seen in the following summary, based on the frequency components of the signal of interest.

1. If the sample signal of interest occurs within a frequency region centered about DC (zero frequency), a low-pass filter or DC integrating system may be used. If the dependence of the sample signal on time or some other time-related variable is of interest (e.g., spectrum scanning, phosphorescence decays), then the response time of the system must be faster than the shortest time interval to be resolved in the sample signal.

2. If the sample signal of interest occurs or can be made to occur within a frequency region centered about some nonzero frequency and if only the amplitude, not the waveform, or the sample signal is of interest (e.g., luminescence spectrometry with chopped source, derivative spectroscopy, sample introduction modulation in flame methods), then an AC detection system may be used to convert the AC signal to DC, whereupon it may be processed by a DC filter or integrator system as previously. If a reference signal of high signal-to-noise ratio and definite frequency and phase relation to the sample signal can be obtained, then a lock-in system may be used. The response time of an AC system, which is determined by *both* the pre- *and* the post detector bandwidths, must satisfy the same requirements as the DC systems discussed earlier.

3. If the sample signal is a repetitive transient occurring over a relatively wide band of frequencies and if the detailed waveform of the sample signal is of interest, then a gated signal averager, either single-channel (boxcar) or multichannel, can be used. A reference (trigger) signal related in frequency and phase to the sample signal is required. In contrast to linear AC and DC systems, the time resolution of gated signal averagers is independent of and can be made much shorter than the total averaging time (which determines the effective noise bandwidth).

APPENDIX

VI

PRINCIPLES OF
SINGLE PHOTON COUNTING (SPC)

If the average anode current of a multiplier phototube is i_a, then the average count ráte of anode pulses $\bar{n}$ is given by

$$\bar{n} = \frac{i_a}{Me} = \frac{\gamma_a \Phi}{Me} = \frac{\gamma_c \Phi}{e} \qquad (A6\text{-}1)$$

where Φ is the light flux on the cathode (watt), γ_a and γ_c are the anodic and cathodic sensitivities of the phototube, respectively (A watt^{-1}), M is the multiplication factor of the phototube, and e is the charge on the electron. Thus we see that $\bar{n}$ is directly proportional to Φ but is not a function of M. Therefore, $\bar{n}$ is not a function of the phototube supply voltage, which is an advantage over the analog methods.

As an illustration of the range of count rates to be expected, consider the following example. A typical photomultiplier with a multiplication factor of 10^6 is used as a detector in a luminescence spectrometer at a wavelength at which its anodic radiant sensitivity is 10^4 A watt^{-1}. In luminescence spectrometry, it is reasonable to expect values of incident light flux from about 10^{-14} watt near the limit of detection to about 10^{-9} watt near the upper end of the analytical curve. This would correspond to average count rates of 100 to 10^6 sec^{-1}. Because these count rates are within the range of modern electronic counters, the SPC technique is well suited to applications in luminescence spectrometry.

It is important to realize that equation A6-1 gives only the *average* count rate of photoelectron pulses. Because photoemission is a statistical process, the instantaneous count rate over a short period of time varies statistically above and below the average value $\bar{n}$. This, of course, is ultimately the cause of shot noise as observed in the analog methods. The effect in photon counting is quite analogous; the standard deviation of the number of photoelectron counts accumulated in a series of equal time intervals t_c from a light source of constant intensity is

$$\sigma_c = \sqrt{C_t} = \sqrt{\bar{n} t_c} \qquad (A6\text{-}2)$$

where C_t is the total counts accumulated per time interval t_c. The signal-to-noise ratio S/N is therefore

$$\frac{S}{N} = \frac{C_t}{\sqrt{C_t}} = \sqrt{C_t} = \sqrt{\bar{n} t_c} \qquad (A6\text{-}3)$$

This puts a lower limit on the total number of photoelectrons to be counted and, therefore, on the minimum count rate measurable in order to obtain sufficient accuracy in a given observation time. For example, if a relative standard deviation of

340

10% is tolerable, at least 100 pulses must be totaled; and if the observation time is not to exceed 100 sec, the lowest count rate measurable is 1 sec^{-1}. The signal-to-noise ratio increases with $\bar{n}$ for a given counting time, but the maximum allowable count rate is limited by the speed of the electronics. One other result of the statistical variation of instantaneous count rate is that the *maximum count rate of the counter must exceed $\bar{n}$ by at least 20-fold* in order to avoid losing an appreciable number of counts during those instants when the instantaneous count rate exceeds the counter capability. The bandwidth of the phototube amplifier should also exceed the maximum expected count rate by at least 20-fold.

Furthermore, the effect of the input time constant $R_L C_S$ must be considered. The input time constant $R_L C_S$ (formed by the phototube load resistance R_L and the shunt capacitance C_S) influences both the *height* and the *width* of the *photoelectron pulses*. This in turn affects the amount of required amplification, the setting of the pulse height discriminator, and the maximum count rate. Usually C_S is determined by the available equipment, and R_L is the only convenient variable. If R_L is too large, the pulse width will be too large, increasing the danger of pulse overlap and loss of counts. If R_L is too small, the pulse height may be too low, requiring excessive amplification before counting. Usually $R_L C_S$ is much greater than the transit time spread of the phototube, and in this case, the pulse height V_p (volt) and the pulse width t_p (sec) are given by

$$V_p \cong \frac{Me}{C_S} \qquad \text{(A6-4)}$$

and

$$t_p \cong 0.7 R_L C_S \qquad \text{(A6-5)}$$

For this case, the maximum count rate R_M (Hz) must never exceed

$$R_M \cong \frac{1}{20 R_L C_S} \qquad \text{(A6-6)}$$

For a typical luminescence spectrometer, the phototube might have a multiplication factor M of 10^6, a transit time spread of 2×10^{-9} sec, and the stray capacitance of the system might be 100 pF (10^{-10} farad). For this case, an R_L of 10^3 ohms would result in pulses of height of $1.6 \times 10^{-19} \times 10^6/10^{-10}$ or 1.6 mV and of width of $0.7 \times 10^3 \times 10^{-10}$ or 0.07 μsec. The maximum counting rate would be $1/20 \times 10^3 \times 10^{-10} = 5 \times 10^5$ Hz.

It is desirable that only photoelectron pulses be counted in a photon counting experiment. In order to reject pulses with heights other than that expected for photoelectron pulses (i.e., approximately Me/C_S volt), a *pulse-height discriminator* is often inserted between the amplifier and the counter. This is an electronic device that passes only those pulses whose heights fall within a certain voltage range (the "window"). The window is adjustable in both position and width to suit the particular phototube; in practice, it is usually centered on the single photoelectron pulse height. Unfortunately, not every photoelectron pulse has exactly the same pulse height, as calculated previously, because of the statistical variations in M from pulse to pulse. Thus the discriminator window must be wide enough to avoid losing an appreciable number of photoelectron pulses, yet not so wide that stray pulses and noise are picked up. The optimum conditions must usually be determined experimentally.

APPENDIX

VII

SPECIAL SPECTRAL TECHNIQUES

A. FOURIER TRANSFORM SPECTROMETRY

Fourier transform spectrometry (FTS) is a technique of measuring visible and infrared spectra that is fundamentally different in concept and instrumentation from conventional dispersive spectrometry. The principal advantages of FTS are greatly increased signal-to-noise ratio, optical speed, resolution, and analysis time.

A Fourier transform spectrometer is based on the familiar Michelson interferometer. The entering light beam is collimated and falls on a 45° beam splitter. The two resulting perpendicular beams strike two plane mirrors at normal incidence and are reflected back to the beam splitter, where the two beams are recombined into one beam which leaves the instrument at 90° to the incident beam. The emerging beam passes through the sample cell (in absorption studies) and on to the detector. One of the plane mirrors is fixed and the other is made movable in order to change the path difference between the two beams. So far this is just a conventional Michelson interferometer. In an FTS, the movable mirror is caused to move smoothly in one direction with constant velocity by means of an electromechanical drive mechanism.

The operation of the FTS system is best understood by considering first the special case in which the incident radiation is a *monochromatic* beam of wavelength λ. In this case, destructive interference (zero detector signal) will occur for all those positions of the movable mirror for which the beam path difference is an odd multiple of $\lambda/2$. The mirror positions for which this condition is satisfied are separated evenly by a distance of $\lambda/2$. Thus, if the mirror is moved at a constant velocity V, destructive interference will occur $2V/\lambda$ times per second. Similarly, constructive interference (maximum signal) will occur at path differences that are even multiples of $\lambda/2$, and partial interference will occur for all intermediate positions. In fact, it is not difficult to see that the detector ouput will be a sine (or cosine) wave of frequency $2V/\lambda$ sec^{-1}. In this way, the wavelength or frequency of the incoming light could be continuously measured by simply measuring the detector output frequency by means of a frequency meter or similar instrument. If the incident light is polychromatic (i.e., a mixture of monochromatic lines of various frequencies and intensities), then the detector output will be a more complex waveform consisting of the sum of sinusoidal components, each one having a frequency and intensity corresponding to one of the spectral components of the incident light. Such a waveform could be analyzed with a spectrum analyzer in order to separate the various frequency components.

If a *polychromatic* (white) light source is used, the detector output will contain a single burst at the instant the beam path difference is zero. In general, the detector output is a very complex waveform; it does not appear at all like a conventional spec-

342

trum but does nevertheless contain the same frequency and intensity information. It is, in fact, the Fourier transform of the intensity-path difference information which can be converted to the more recognizable form by means of a well-known but unfortunately very tedious mathematical transformation. In recent years, however, it has become feasible to perform the required computations by means of a small, relatively low-cost digital computer. The computer is connected on-line and is, in fact, considered an integral part of the instrument. In addition, multichannel signal averaging can be done in the computer memory in order to increase the signal-to-noise ratio. Furthermore, the computer is capable of correcting, ratioing, and subtracting spectra, computing absorbances, and performing other desired operations.

In order for an FTS system to be useful, the mirror position must be known to a fraction of a wavelength of the light to be studied. This can cause considerable experimental problems, particularly at short wavelengths. The most successful commercial systems use the "fringe-reference" system. In these systems, another mirror is attached to the moving mirror drive so that the two mirrors move exactly together. This additional mirror forms part of a separate "reference" interferometer which accepts monochromatic light from a laser source. The sinusoidal interference "fringe" signals from the reference interferometer detector are counted and serve as a very precise measure of the mirror velocity. A beam of white light from a small incandescent lamp is also passed through the reference interferometer and produces a burst at a precisely reproducible point on the sample interferogram. This burst serves as a trigger signal for the signal averaging.

The resolution of an FTS instrument is determined by the quality of the optics, by the degree of collimation of the incident light, and, most important, by the length of the mirror sweep. In fact, resolution is inversely proportional to the mirror excursion. Fortunately, the fringe-reference system allows long mirror excursions to be measured with precision, and so resolutions of 0.1 cm^{-1} can be obtained.

The advantages of FTS over conventional spectrometry are several:

1. *Higher throughput.* An interferometer has an aperture of at least several square centimeters, much larger than the area of even the largest slits used in dispersive spectrometry. Thus much more light can be put into the FTS system, which is a considerable advantage at low light levels.

2. *Fellgett's advantage.* The so-called Fellgett's advantage arises from the fact that the entire spectrum is measured continuously, whereas in a dispersive instrument only that portion of the spectrum falling on the exit slit is measured at any one time. If there are M resolution elements in the spectrum (i.e., M equals the ratio of spectral range covered in a given spectrum with a dispersive spectrometer to spectral bandwidth), then the TFS system can achieve a signal-to-noise ratio $\sqrt{M}$ higher than a dispersive instrument under the same conditions of resolution and measurement time, simply because a dispersive instrument wastes most of the light that enters its exit slit. This is quite significant; $\sqrt{M}$ is often of the order of 10 to 100.

3. *Rapid response.* The signal-to-noise advantage of an FTS system may be traded off for greatly increased speed. That is, an FTS instrument can obtain a spectrum approximately M times faster than an equivalent dispersive spectrometer. It is routinely possible to obtain a good spectrum in only 1 sec. This opens up tremendous

possibilities; for example, in gas or liquid column chromatography systems, the complete infrared spectrum of each component can be measured "on the fly" as it comes off the column.

At the present time, FTS systems are limited to the visible through far-infrared regions (40,000–10 cm^{-1}) and a resolution of 0.1 cm^{-1}. The possibilities of application of FTS to luminescence spectrometry are numerous but yet largely unexplored.

B. HADAMARD SPECTROMETRY

Hadamard spectrometry involves the use of a multiplexed dispersive spectrometer. The system consists of a dispersive spectrometer (generally a grating instrument) with a mask containing numerous open and closed slots (slits) in place of the normal exit slit. The size of the slots is equal to or a multiple of the spectral bandpass of the dispersive spectrometer. Actually the mask is cyclic with $2n - 1$ total spectral elements. (A spectral element consists of either an opaque or an open slot equal to the spectral bandpass of the instrument.) The mask is then moved (by a stepping motor) across the exit slit plane in front of the photomultiplier tube. At each position of the mask, the detector "sees" a linear combination of radiant fluxes corresponding to each spectral component allowed to reach the detector surface. By making a total of n measurements* (n different positions of the mask) and performing a Hadamard transformation with a digital computer, the spectral information (radiant flux versus wavelength) can be obtained.

Hadamard spectrometry (HS) has less precise mechanical and alignment problems than Fourier transform spectrometry (FTS). Nor does HS exhibit the intense zero-order spike characteristic of FTS. Finally, HS involves a simpler mathematical computation than FTS. On the other hand, both the optical throughput (speed) and the signal-to-noise ratio of FTS are greater than HS with one Hadamard mask. However, if in addition to the exit plane Hadamard mask an additional mask is placed at the entrance focal plane in place of the entrance slit and if these two masks are independently stepped, then the doubly multiplexed Hadamard spectrometer approaches the FTS system in optical throughput and in signal-to-noise ratio. The doubly multiplexed Hadamard system is rather mechanically complex. The greatest advantage of FTS over HS is certainly the much greater wavelength range in the former; for example, FTS can cover 1 to 1000 μ in one measurement period, whereas HS covers a much smaller range as 1 to 3 μ.

At the time of writing this book, neither FTS nor HS had been utilized for luminescence spectrometry. However, because of the mechanical difficulties of maintaining plate parallelism and of measuring path differences at small optical path difference, measurements below about 0.8 μ have not been made with FTS. No such limitations seem to exist for HS, and so HS may find considerable use in luminescence spectrometry—particularly for atomic fluorescence spectrometry.

---

* The grating angle is constant during any given series of measurements. Thus the Hadamard spectrometer covers a range determined by the width of the detector (sensitive area) and the reciprocal linear dispersion of the grating monochromator.

C. OTHER SYSTEMS

Fabry-Perot spectrometers will probably find little use in luminescence spectrometry because of their small wavelength ranges—typically a few nanometers. Such spectrometers, however, because of their great resolving powers, will certainly be used for studies related to luminescence spectrometry, such as measurement of the structure of spectral lines from various sources (electrodeless discharge lamps, hollow cathode discharge lamps, flames, lasers, etc.).

Rapid-scan spectrometers involve dispersive spectrometers in which one of the spectral elements (a mirror, a dispersive element, or a slit) is moved rapidly (rotated, oscillated, vibrated, etc.). Rapid-scan spectrometers are capable of covering a fairly wide wavelength range—for example, 2000 to 6000 Å in time periods less than one second (e.g., milliseconds). Such devices are useful for studying plasmas, flash photolysis, explosions, gas chromatographic effluents, and so on. Because of the vast store of spectral information in the infrared region, most rapid-scan spectrometers have been utilized in the infrared wavelength region. Nevertheless, it would seem that rapid-scan dispersive luminescence spectrometers could serve for investigating some rapid kinetic processes. Of course, if either FTS or HS instrumentation becomes available for luminescence spectrometry, rapid-scan spectrometers with their inferior resolution, optical speed, and mechanical complications would probably find little use even for kinetic studies.

INDEX

A–antenna, 192, 193
Absorbed radiant flux, 83
Absorption coefficient, 84, 87, 89–91, 95, 100, 109
 effective or average, 97, 98, 106
Absorption factor, 106
Absorption law, 83
Absorption line half-width, 107
Absorption line profile, 107
Acid-base reactions, 50
AC systems, 338, 339
Actinometer, ferrioxalate, 334, 335
Alpha factor for phosphoroscope, 205, 206, 221
Amplification factor, multiplier phototubes, 174
Amplifier, ac, 180, 181, 338, 339
 dc, 179, 180, 337–339
 drift, 212
 electrometer, *see* Electrometer amplifier
 gain, 220, 223
 readout systems, 178–185
Amplifier time constant, 179; *see also* Response time
Analysis time, 212
Analytical curves, 130, 142, 220, 229, 232, 240, 245, 264–266, 293–295
Angle of diffraction, *see* Diffraction angle
Angular dispersion, 162
Angular momentum, 5
Anharmonic oscillator, 42
Anti-Stokes sensitized delayed fluorescence, 312, 313
a-parameter, 87, 88, 100
Aperiodic signal, 209
Aperture, effective, 155
Apex angle, prism, 163, 164
Apparatus function, *see* Spectrometer function
Applications (general), atomic fluorescence spectrometry, 296
 inorganic fluorimetry, 296

organic fluorimetry, 297–299
organic phosphorimetry, 297–299
Arc area, 149
Arc current, 149
Arc lamps, ac, 146, 147
 compact, 148
 dc, 146, 147, 150
 high pressure, 145, 146
 low pressure, 145, 146
 short, 148
Arc resistance, effective, 151
Arc wander, 146, 147, 213
Aspirator yield, 123, 130, 195, 220
Association equilibrium, 136
Astigmatism, 167
Atomic absorption coefficient, *see* Absorption coefficient
Atomic absorption instrument, 188
Atomic orbitals, 7–10
Atomic partition function, *see* Partition function
Atomic term, 14
Atomization, 124–130, 218
Atoms, atomic fluorescence of, *see* Elements, atomic fluoresence
Auxiliary electrodes, 191
Averaging time, 172, 210
Azimuthal quantum number, 7

Background luminescence, 283
Bandwidth, *see* Electrical bandwidth; Noise bandwidth
Barrier layer cell, *see* Photovoltaic cell
Beer-Lambert Law, 49
Bias error, 209
Bimolecular processes, 22, 24, 25, 66–72
Black body, 106, 144, 145, 149
Blank, errors, 231
 internal, 241
 luminescence, 269, 283, 285, 286, 288
 reagent, 231
 solvent, 231